Satellite Systems Engineering in an IPv6 Environment

OTHER TELECOMMUNICATIONS BOOKS FROM AUERBACH

AUERBACH PUBLICATIONS
www.auerbach-publications.com
To Order Call: 1-800-272-7737 • Fax: 1-800-374-3401
E-mail: orders@crcpress.com

Satellite Systems Engineering in an IPv6 Environment

DANIEL MINOLI

CRC Press
Taylor & Francis Group
Boca Raton London New York

CRC Press is an imprint of the
Taylor & Francis Group, an **informa** business

AN AUERBACH BOOK

CRC Press
Taylor & Francis Group
6000 Broken Sound Parkway NW, Suite 300
Boca Raton, FL 33487-2742

First issued in paperback 2019

© 2009 by Taylor & Francis Group, LLC
CRC Press is an imprint of Taylor & Francis Group, an Informa business

No claim to original U.S. Government works

ISBN-13: 978-1-4200-7868-8 (hbk)
ISBN-13: 978-0-367-38599-6 (pbk)

Library of Congress Cataloging-in-Publication Data

Minoli, Daniel, 1952-
    Satellite systems engineering in an IPv6 environment / Daniel Minoli.
       p. cm.
    Includes bibliographical references and index.
    ISBN 978-1-4200-7868-8 (alk. paper)
    1. Artificial satellites in telecommunication. 2. TCP/IP (Computer network protocol) I. Title.

TK5104.M54 2009
621.382'5--dc22

2008048556

Visit the Taylor & Francis Web site at
http://www.taylorandfrancis.com

and the CRC Press Web site at
http://www.crcpress.com

Dedication

For Anna and the kids.

And for my parents, Gino and Angela

Contents

Preface

This book provides a practical guide to satellite transmission engineering. Although there are a number of textbooks in the field, this book has two distinguishing features: (1) the focus is more on practical results and less on the actual derivation of the mathematical equations, and (2) usage of satellite transmission in an IPv6 environment is highlighted. This is the first book on the market to address the issue of satellite communications in IPv6 networks. Key aspects to consider include transmission theory, impairments, antennas' geometry/size, and reception techniques. Modulation is fundamental to any transmission system and high symbol-rate digital modulation in satellite transmission is now the norm. Multiplexing is also an important capability in any modern communication system. Multiplexing can take place at the physical layer, at the data-link layer, and at the packet layer. Many variables control the quality, bandwidth, and reliability of the received signal, such as transmit power, antenna/Low Noise Amplifier gain, antenna size, fade phenomena, and Forward Error Correction techniques, among others. A Link Budget Analysis determines the kind of tradeoffs that can be made to achieve engineering objectives. IPv6 is increasingly being deployed around the world. Because of intrinsic latency in satellite transmission, special considerations have to be taken into account for TCP traffic, in order to optimize throughput. As a corollary of multiplexing techniques, Very Small Aperture Terminals make use of statistical in-channel multiplexing to support a relatively large base of medium-throughput users, especially for data applications. All of these topics are discussed at a pragmatic level in this text.

There is now a global interest by (all) the telcos in Europe, Asia, and North America to enter the Internet Protocol TV (IPTV) distribution, and Digital Video Broadcast–Handheld (DVB-H), or OMA BCAST mobile video markets in order to replace revenues that have eroded to cable TV companies and wireless providers. Nearly all the traditional telcos worldwide are looking into these technologies at this juncture. Telcos need to compete with cable companies and IPTV, and DVB-H is the way to do it. In fact, even the cable TV companies themselves are looking into upgrading their ATM technology to IP. While these services are now starting out by using IPv4, IPv6 is just around the corner. Finally, government agencies looking to deploy IPv6 and also use satellite communication can benefit from this text.

After the Introduction, Chapter 2 covers electromagnetic propagation. Chapter 3 discusses basic antenna theory. Modulation and multiplexing techniques are discussed in Chapter 4. Chapter 5 covers Forward Error Correction. The critical topic of Link Budget Analysis is discussed in Chapter 6. IPv6 is discussed in Chapter 7. TCP/IPv6 issues are covered in Chapter 8. Considerations related to IPv6 support in satellite environments are surveyed in Chapter 9 and in Appendix A.

Telephone carriers (telcos), equipment manufacturers, content providers, content aggregators, satellite companies, venture capitalists, and colleges and technical schools can make use of this text. The text can be used for a college course on satellite applications to video distribution, specifically IPv6, IPTV, DVB-H, and datacasting.

It is not the goal of this book to present an exhaustive view of the satellite field. There is a very extensive literature on the topic of satellite communications; however, this text looks at the issues from a forward-looking and pragmatic perspective.

Acknowledgments

The author would like to thank William B. McDonald, WBMSAT P.S., Inc., for his review of Chapters 1 through 4 of the manuscript.

Also thanking Mike Noon, SES Engineering.

About the Author

Daniel Minoli has many years of technical hands-on and managerial experience (including budget or PL responsibility) in networking, telecom, wireless, video, Enterprise Architecture, and security for global Best-In-Class carriers, service providers, and financial companies. He has worked at financial firms such as AIG, Prudential Securities, Capital One Financial and service provider firms such as Bell Telephone Laboratories, ITT, Bell Communications Research (now Telcordia), AT&T, Leading Edge Networks, Inc., and SES Engineering, where he is director of Terrestrial Systems Engineering. SES is the largest satellite communications company in the world. He also played a founding role in the launching of two companies through the high-tech incubator Leading Edge Networks, Inc., which he ran in the early 2000s: Global Wireless Services, a provider of secure broadband hotspot mobile Internet and hotspot VoIP services; and, InfoPort Communications Group, an optical and Gigabit Ethernet metropolitan carrier supporting Data Center/SAN/channel extension and Grid Computing network access services. For several years he has been session, tutorial, and now overall technical program chair for the IEEE ENTNET (Enterprise Networking) conference. ENTNET, part of IEEE Globecom, focuses on enterprise networking requirements for large financial firms and other corporate institutions.

At SES Engineering Mr. Minoli has been responsible for engineering satellite-based video, Internet, IPTV, and DVB-H systems. This includes overall engineering design, deployment, and operation of SD/HD encoding, inner/outer AES encryption, Conditional Access Systems, video middleware, Set Top boxes, Headends, and related terrestrial connectivity. At Bellcore/Telcordia he did extensive work on broadband, on video-on-demand for the RBOCs (then known as Video Dialtone); on multimedia over ISDN/ATM, and on distance learning (satellite) networks. At DVI he deployed a (satellite-based) distance learning system for William Patterson College. At Stevens Institute of Technology (where he was an adjunct professor) he taught approximately a dozen graduate courses on digital video. At AT&T he deployed large broadband networks to also support video applications, for example, video over ATM. At Capital One he was involved with the deployment of corporate video-on-demand over the IP-based intranet. As a consultant he handled the technology-assessment function of several high-tech companies seeking funding, developing multimedia, digital video, physical layer switching, VSATs, telemedicine, Java-based CTI, VoFR and VPNs, HDTV, optical chips, H.323 gateways, nanofabrication/(Quantum Cascade Lasers), wireless, and TMN mediation.

He has also written columns for *ComputerWorld, NetworkWorld*, and *Network Computing* (1985–2006). He has taught at New York University (Information Technology Institute), Rutgers University, and Stevens Institute of Technology (1984–2006). Also, he was a Technology Analyst at Large for Gartner/DataPro (1985–2001); based on extensive hands-on work at financial firms and carriers, he tracked technologies and wrote CTO/CIO-level technical scans in the area of telephony and data systems, including topics on security, disaster recovery, network management, LANs, WANs (ATM and MPLS), wireless (LAN and public hotspot), VoIP, network design/economics, carrier networks (such as metro Ethernet and CWDM/DWDM), and E-Commerce. Over the years he has advised venture capitalists for investments of $150M in a dozen high-tech companies. He has acted as expert witness in a (won) $11B lawsuit regarding a VoIP-based wireless air-to-ground communication system, and has been involved as a technical expert in a number of patent infringement proceedings.

Chapter 1

Introduction to Satellite Communications

Satellite communication plays, and will continue to play, a key role in commercial, TV/media, government, and military communications because of its intrinsic multicast/broadcast capabilities, mobility aspects, global reach, reliability, and ability to quickly support connectivity in open-space and/or hostile environments. At a different level, Internet Protocol version 6 (IPv6) is a technology now being deployed in various parts of the world that allows true explicit end-to-end device addressability. As the number of intelligent systems that need direct access expands to the multiple billions (e.g., cell phones, personal digital assistants (PDAs), appliances, sensors/actuators/Smart dust, and even body-worn biometric devices), IPv6 becomes an institutional imperative in the final analysis. The integration of satellite communication and IPv6 capabilities promises to provide a powerful networking infrastructure that can serve the evolving needs of government, military, IP-based television (IPTV), and mobile Digital Video Broadcast Handhelds (DVB-H) stakeholders, to name just a few.

This text provides a pragmatic assessment of satellite communication and engineering in an IPv6 environment and in light of newly evolving applications. Because the U.S. government is a major user of satellite systems and a proponent of IPv6, this text may be of interest to this community of users, among others. The satellites of the future will not only be signal regenerators in space but will contain onboard IP and IPv6 routers to facilitate intelligent traffic distribution; hence, it is important to understand the interplay and overlaying of IPv6 routing over a satellite-based transmission channel. The first part of the text (Chapters 1 through 6) focuses on traditional engineering issues, and the second part (Chapters 7 through 9) focuses on IPv6.

This chapter provides an introductory overview of the field, whereas chapters that follow provide more details on some key aspects of the technology, particularly those that have relevance to the IPv6 and related, or evolving, services. After this introduction, Chapter 2 covers electromagnetic propagation. Chapter 3 discusses basic antenna

1

theory. Modulation and multiplexing techniques are discussed in Chapter 4. Chapter 5 covers Forward Error Correction (FEC). The critical topic of Link Budget Analysis is discussed in Chapter 6. IPv6 is discussed in Chapter 7. Transmission Control Protocol (TCP)/IPv6 issues are covered in Chapter 8. Initiatives and considerations related to IPv6 support in satellite environments are surveyed in Chapter 9 and Appendix A. There is an extensive body of literature on the topic of satellite communications (including such minor contributions as [MIN197901], [MIN197801], [MIN198601], and [MIN199101]); however, this chapter looks at the issues from a forward-looking but pragmatic perspective.

1.1 Satellite Orbits

Satellite communication is a line-of-sight (LOS) one-way or two-way radio frequency (RF) transmission system that comprises a transmitting station (uplink), a satellite system that acts as a signal regeneration node, and one or more receiving stations (downlink). (See Figure 1.1.) Satellites can reside in a number of orbits. A geosynchronous (GEO) satellite* circles the earth at the earth's rotational speed and in the same direction of rotation, therefore appearing at the same position in the sky at a particular time each day. When the satellite is in the equatorial plane, it appears to be permanently stationary when observed from the earth's surface, so that an antenna pointed to it will not require tracking or (major) positional adjustments at periodic intervals of time (this satellite arrangement is also known as *geostationary*[†,‡]). The geostationary orbit is at an altitude of 35,786 km (22,236 mi.) from the earth's surface (42,164 km from the earth's center, the earth's radius being 6,378 km). See Figure 1.2.

The major consequence of the geostationary orbital position is that signals experience a propagation delay of no less than 119 ms on an uplink (longer for earth stations at northern latitudes or for earth stations looking at satellites that are significantly offset longitudinally compared with the earth station itself[§]), and no less than 238 ms for an uplink and a downlink or a one-way end-to-end transmission path. A two-way interactive session with a typical communications protocol, such as TCP, will experience this roundabout delay twice (no less than 476 ms) because the information is making two round trips to the satellite and back. One-way or broadcast (video or data) applications easily deal with this issue, as the delay is not noticeable to the video viewer or the receive data user. However, interactive data applications and voice backhaul applications typically have to accept (and adjust to) this predicament imposed by the limitations

---

* In this book, whenever we use the term *satellite*, we mean a geostationary communications satellite, unless noted otherwise by the context.

† In practice, the terms *geosynchronous* and *geostationary* are used interchangeably.

‡ A geostationary orbit is a circular prograde orbit (prograde is an orbital motion in the same direction as the primary's rotation) in the equatorial plane, with an orbital period equal to that of the earth; hence, a satellite in a geostationary orbit appears to be fixed above the surface of the earth, that is, it is at a fixed latitude and longitude.

§ Depending on the location of the earth station and the target satellite (which determines the look angle), the path length (and so the propagation delay) can vary by several thousand kilometers. (e.g., for a satellite at 101°W and an antenna in Denver, Co., the "slant" range is 37,571.99 km; for an antenna in Van Buren, ME, the range is 38,959.54 km.)

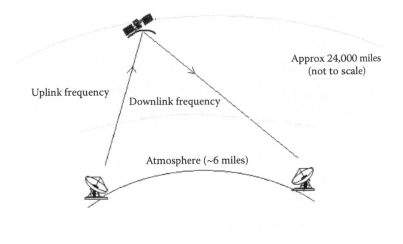

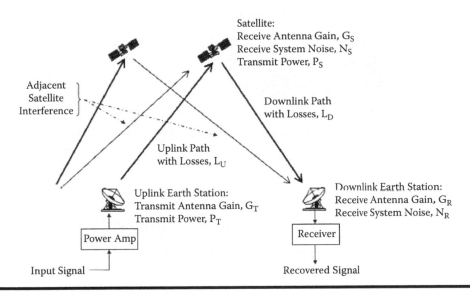

Figure 1.1   A typical satellite link.

of the speed of light, which is the speed that radio waves travel. Satellite delay compensation units and "spoofing" technology have successfully been used to compensate for these delays in data circuits. Voice transmission via satellite presently accounts for only a tiny fraction of overall transponder capacity, and users are left to deal with the satellite delay individually; only a few find it to be objectionable.

At the practical level, the orbit has a small nonzero inclination and eccentricity, which causes the satellite to trace out a small but manageable "figure eight" in the sky. Orbital positions are defined by international regulation as longitude values on the "geosynchronous circle," for example, 101°W, 129°W, and so on. Satellites are spaced at 2° or 3° to allow sufficient separation to support frequency reuse (see Figure 1.3). In actuality, an orbital position is a box of about 150 × 150 km,

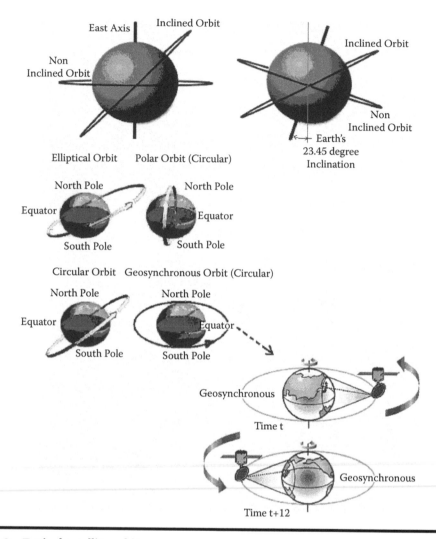

Figure 1.2   Typical satellite orbits.

within which the satellite is maintained by ground control. Table 1.1 lists some key concepts related to orbits [SAT200501]. Table 1.2 identifies some satellites that cover the United States. (A similar worldwide tabulation can also easily be compiled.)

1.2 Satellite Transmission Bands

The transmission channel of a satellite system is a radio channel using a direct-wave approach, operating at specific RF bands within the overall electromagnetic spectrum (see Figure 1.4 [MIN199101]). The frequency of operation is in the super high frequency (SHF) range (3–30 GHz), as defined in Table 1.3. Regulation and practice dictates the frequency of operation, the channel

Top: Global Array
Bottom: U.S. Orbital Slots (Longitudes)

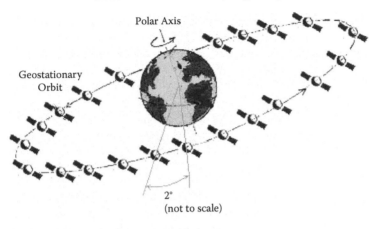

Polar Axis

Geostationary
Orbit

2°
(not to scale)

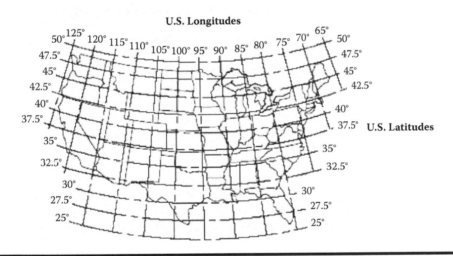

U.S. Longitudes

U.S. Latitudes

Figure 1.3   Worldwide population of geostationary satellites (illustrative).

bandwidth, and the bandwidth of the subchannels within the larger channel. Different frequencies are used for the uplink and the downlink. A satellite link is a radio link between a transmitting earth station and a receiving earth station through a communications satellite. A satellite link consists of one uplink and one downlink; the satellite electronics (i.e., the transponder) will remap the uplink frequency to the downlink frequency.

Note that $c = \lambda f$, where c is the speed of light (3×10^8 m/s), λ is the wavelength, and f is the frequency.

Frequencies above about 30 MHz can pass through the ionosphere and therefore can be utilized for communicating with satellites. (Frequencies below 30 MHz are reflected by the ionosphere at certain stages of the sunspot cycle.) However, commercial satellite services use

Table 1.1   Key concepts related to orbits

Circular orbit	A satellite orbit where the distance between the center of mass of the satellite and the center of mass of the earth is constant.
Clarke belt	The circular orbit (geostationary orbit) at approximately 35,786 km above the equator, where the satellites travel at the same speed as the earth's rotation and thus appear to be stationary to an observer on earth (named after Arthur C. Clarke, who was the first to describe the concept of geostationary communication satellites).
	Note: Sir Arthur C. Clarke died on March 19, 2008, at the age of 90.
Collocated satellites	Two or more satellites occupying approximately the same geostationary orbital position such that the angular separation between them is effectively zero when viewed from the ground. To a small receiving antenna, the satellites appear to be exactly collocated; in reality, the satellites are kept several kilometers apart in space to avoid collisions. Different operating frequencies and/or polarizations are used.
Geostationary orbit/ satellite	The orbit of a geosynchronous satellite, which lies in the plane of the earth's equator. A satellite orbiting the earth at such speed that it permanently appears to remain stationary with respect to the earth's surface.
Geosynchronous object	An object orbiting the earth at the earth's rotational speed and in the same direction of rotation. The object appears at the same position in the sky at a particular time each day but will not appear stationary if it is not orbiting in the equatorial plane.
Inclination	The angle between the plane of the orbit of a satellite and the earth's equatorial plane. An orbit of a perfectly geostationary satellite has an inclination of 0.
Inclined orbit	An orbit that approximates the geostationary orbit but whose plane is tilted slightly with respect to the equatorial plane. The satellite appears to move about its nominal position in a daily "figure-of-eight" motion when viewed from the ground. Spacecrafts (satellites) are often allowed to drift into an inclined orbit near the end of their nominal lifetime to conserve on-board fuel, which would otherwise be used to correct this natural drift caused by the gravitational pull of the sun and moon. North–south maneuvers are not conducted, allowing the orbit to become highly inclined.
Orbit	The path described by the center of mass of a satellite in space, subjected to natural forces, principally gravitational attraction, but occasional low-energy corrective forces exerted by a propulsive device to achieve and maintain the desired path.
Orbital plane	The plane containing the center of mass of the earth and the velocity vector (direction of motion) of a satellite.

much higher frequencies. The range 3–30 GHz represents a useful set of frequencies for geostationary satellite communication; these frequencies are also called *microwave frequencies*.* Above about 30 GHz, the attenuation in the atmosphere due to clouds, rain, hydrometeors, sand, and

---

* From 30 to 300 GHz, the frequencies are referred to as *millimeter wave*; above 300 GHz, optical techniques take over; these frequencies are known as *far infrared* or *quasi optical*.

Table 1.2 Partial list of geostationary satellites that cover the United States/ North America

Satellite name	Location	Notes
SES-Americom 6	72 W	
SES-Americom 9	83 W	
SES-Americom 3	87 W	
Intelsat Americas 8	89 W	
Galaxy 11	91 W	
Intelsat Americas 6	93 W	
Galaxy 3C	95 W	
Galaxy 16	99 W	
SES-Americom 4	101 W	
DirecTV Television	101 W	Primary and additional programming: 110 and 119
SatMex5	117 W	
Dish Network Television	119 W	Primary and additional programming: 61.5, 110, and 148
Galaxy 10R	123 W	
Horizon 1	127 W	
Intelsat Americas 7	129 W	

Note: W stands for West, which refers to the longitude West of Greenwich, England. For example, 101°W L = 259°EL.

dust makes a ground-to-satellite link unreliable. (Such frequencies may still be used for satellite-to-satellite links in space, although these applications have not yet developed commercially [JEF200401]).

The actual frequencies of operation of commercial (U.S.) satellites are*

- C band: 3.7–4.2 GHz for downlink frequencies, and 5.925–6.425 GHz for uplink frequencies (Extended C band operates at frequencies of 5.850–6.425 GHz and 3.625–4.200 GHz, respectively.)
- Ku band: 11.7–12.2 GHz for downlink frequencies, and 14–14.5 GHz for uplink frequencies
- Broadcast satellite service: 12.2–12.7 GHz for downlink frequencies
- Ka band: 18.3–18.8 GHz and 19.7–20.2 GHz for downlink frequencies, and 27.5–31 GHz for the uplink frequencies

See Table 1.4 for details about frequency bands. These bands are further subdivided into smaller channels that can be independently used for a variety of applications. Table 1.5 depicts a typical subdivision of the C band into these channels, which are also called colloquially "transponders." (*Transponder* as a proper term is defined later in the chapter.) The nominal subchannel bandwidth is (typically) 40 MHz, with a (typical) usable bandwidth of 36 MHz. (Also see Figure 1.5.) Similar frequency allocations have been established for the Ku and Ka bands. Many satellites simultaneously support a C-band and a Ku-band infrastructure. (They have dedicated feeds and transponders for each band.) Most communications systems fall into one of three

---

* The international set of microwave bands is as follows: L band (0.39–1.55 GHz); S band (1.55–5.20 GHz); C band (3.70–6.20 GHz); X band (5.20–10.9 GHz); and K band (10.99–36 GHz).

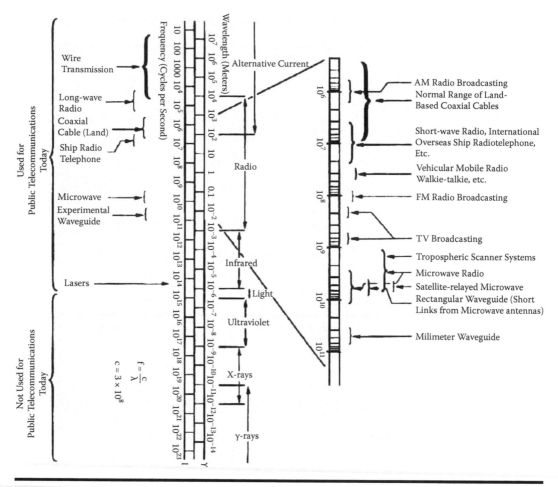

Figure 1.4   Electromagnetic spectrum.

categories: bandwidth efficient, power efficient, or cost efficient. Bandwidth efficiency describes the ability of a modulation scheme to accommodate data within a limited bandwidth. Power efficiency describes the ability of the system to reliably send information at the lowest practical power level. In satellite communications, both bandwidth efficiency and power efficiency are important [AGI200101].

Figure 1.6 depicts a two-way satellite link. The end-to-end (remote to central point) link makes use of a radio channel, as described previously, for the transmitting station uplink to the satellite; additionally, it uses a downlink radio channel to the receiving station (this is also generally called the inbound link). The outbound link from the central point to a remote point also makes use of a radio channel comprised of an uplink and a downlink.

From an application's perspective, the link may be point-to-point (effectively, where both ends of the link are peers), or it may be point-to-aggregation-point, for example, for handoff to a corporate network or to the Internet. Some applications are simplex, typically making use of an outbound link; other applications are duplex, using both an inbound and outbound link.

Table 1.3   Traditional classification of radio frequencies

Frequency band	Frequency range	Propagation modes	Systems/uses/characteristics
ELF (Extremely low frequency)	Less than 3 KHz	Surface wave	Worldwide, military, and submarine communication
VLF (Very low frequency)	3–30 kHz	Earth-ionosphere guided	Worldwide, military, and navigation
LF (Low frequency)	30–300 kHz	Surface wave	Stable signal, distances up to 1500 km
MF (Medium frequency)	300 kHz–3 MHz	Surface/sky wave for short/long distances, respectively	Radio broadcasting. Long-distance sky-wave signals are subjected to fading
HF (High frequency)	3–30 MHz	Sky wave, but very limited, short-distance ground wave also	3–6 MHz: Continental; 6–30 MHz: Intercontinental. Land and ship-to-shore communications
VHF (Very high frequency)	30–300 MHz	Space wave	Close to line-of-sight over short distances; broadcasting and land mobile
	30–60 MHz	Scatter wave	Ionospheric scatter over 900–2000 km distances
UHF (Ultrahigh frequency)	300 MHz–3 GHz	Space wave	Essentially line-of-sight over short distances; broadcasting and land mobile
	Above 300 MHz	Scatter wave	Tropospheric scatter over 150–800 km distances
SHF (Super-high frequency)	3–30 GHz	Space wave	The "workhorse" microwave band; Line-of-sight; terrestrial and satellite relay links
EHF (Extremely high frequency)	30–300 GHz	Space wave	Line-of-sight millimeter waves. Space-to-space links, military uses, and possible future use

Increasingly, satellite communications make use of digital modulation. Modulation is the process of overlaying intelligence (say, a bit stream) over an underlying carrier so that the information can be relayed at a distance. Demodulation is the recovery from a modulated carrier of a signal having the same characteristics as the original modulating signal. The underlying analog carrier is superimposed with a digital signal, typically using 4- or 8-point phase shift keying (PSK) techniques, or 16-point quadrature amplitude modulation (QAM). In addition, the original signal is fairly routinely encrypted and invariably protected with forward error correction (FEC) techniques. These topics are discussed in Chapters 4 and 5.

As noted, different frequencies are used for the uplink and downlink to avoid self-interference, following the terrestrial microwave transmission architecture developed by the Bell System in the

Table 1.4 Satellite band details

Band	Characteristics	Considerations
C band (6 GHz uplink and 4 GHz downlink)	• Relatively immune to atmospheric effects • Popular band, but on occasion it is congested on the ground (see note at right) • Bandwidth (~500 MHz/36 MHz transponders) allows video and high data rates • Provides good performance for video transmission • Proven technology with long heritage and good track record • Common in heavy rain zones	• Requires large antennas (3.8–4.5 m or larger, especially on the transmit side) • Large footprints • Best-performing band in the context of rain attenuation • Potential interference due to terrestrial microwave systems
Ku band (14–14.5 GHz uplink and 11.7–12.2 GHz downlink)	• Moderate to low cost hardware • Highly suited to VSAT networks • Spot beam footprint permits use of smaller earth terminals, 1–3 m wide, in moderate rain zones	• Attenuated by rain and other atmospheric moisture • Spot beams generally focused on land masses • Not ideal in heavy rain zones
DBS band (12.2–12.7 GHz downlink)	• Simplex • Multiple feeds for access to satellite neighborhoods • Small Receive Only antennas	• Attenuated by rain and other atmospheric moisture
Ka band (18.3–18.8 GHz and 19.7–20.2 GHz downlink)	• Microspot footprint • Very small terminals, much less than 1 m • High data rates are possible: 500–1000 Mbps	• Rain attenuation • Obstruction interference due to heavy rainfall (black out)

1940s and 1950s. In systems using the C band, the basic parameters are 4 GHz in the downlink, 6 GHz in the uplink, and 500 MHz bandwidth over 24 transponders using vertical and horizontal polarization (a form of frequency reuse discussed later on), resulting in a transponder capacity of 36 MHz, or 45–75 Mbps, depending on the modulation and FEC scheme. Table 1.6 depicts some key physical parameters of relevance to satellite communication [SAT200501]. C-band has been used for several decades and has good transmission characteristics, particularly in the presence of rain, which typically affects high-frequency transmission. Generally, C-band links are used for TV and video distribution to headends and for military applications, among others. A number of antenna types are utilized in satellite communication, but the most commonly used narrow beam antenna type is the dish reflector antenna. C-band receive dishes for broadcast-quality video reception are typically 3.8–4.5 m in diameter. The size is selected to optimize reception under normal (clear sky) or medium-to-severe rain conditions; however, smaller antennas of 1.5–2.4 m

Table 1.5 Typical subchannel ("transponder") allocation for C-band satellites

xpdr	UP Center frequency (MHz)	UP Lower-end frequency (MHz)	UP Higher-end frequency (MHz)	DOWN Center frequency (MHz)	DOWN Lower-end frequency (MHz)	DOWN Higher-end frequency (MHz)	POL
			By xpdr				
1	5945	5925	5965	3720	3700	3740	V
2	5965	5945	5985	3740	3720	3760	H
3	5985	5965	6005	3760	3740	3780	V
4	6005	5985	6025	3780	3760	3800	H
5	6025	6005	6045	3800	3780	3820	V
6	6045	6025	6065	3820	3800	3840	H
7	6065	6045	6085	3840	3820	3860	V
8	6085	6065	6105	3860	3840	3880	H
9	6105	6085	6125	3880	3860	3900	V
10	6125	6105	6145	3900	3880	3920	H
11	6145	6125	6165	3920	3900	3940	V
12	6165	6145	6185	3940	3920	3960	H
13	6185	6165	6205	3960	3940	3980	V
14	6205	6185	6225	3980	3960	4000	H
15	6225	6205	6245	4000	3980	4020	V
16	6245	6225	6265	4020	4000	4040	H
17	6265	6245	6285	4040	4020	4060	V
18	6285	6265	6305	4060	4040	4080	H
19	6305	6285	6325	4080	4060	4100	V
20	6325	6305	6345	4100	4080	4120	H
21	6345	6325	6365	4120	4100	4140	V
22	6365	6345	6385	4140	4120	4160	H
23	6385	6365	6405	4160	4140	4180	V
24	6405	6385	6425	4180	4160	4200	H
			By frequency				
2	5965	5945	5985	3740	3720	3760	H
4	6005	5985	6025	3780	3760	3800	H
6	6045	6025	6065	3820	3800	3840	H
8	6085	6065	6105	3860	3840	3880	H
10	6125	6105	6145	3900	3880	3920	H
12	6165	6145	6185	3940	3920	3960	H
14	6205	6185	6225	3980	3960	4000	H
16	6245	6225	6265	4020	4000	4040	H
18	6285	6265	6305	4060	4040	4080	H
20	6325	6305	6345	4100	4080	4120	H
22	6365	6345	6385	4140	4120	4160	H
24	6405	6385	6425	4180	4160	4200	H
1	5945	5925	5965	3720	3700	3740	V
3	5985	5965	6005	3760	3740	3780	V
5	6025	6005	6045	3800	3780	3820	V

(Continued)

Table 1.5   Typical subchannel ("transponder") allocation for C-band satellites

xpdr	UP Center frequency (MHz)	UP Lower-end frequency (MHz)	UP Higher-end frequency (MHz)	DOWN Center frequency (MHz)	DOWN Lower-end frequency (MHz)	DOWN Higher-end frequency (MHz)	POL
7	6065	6045	6085	3840	3820	3860	V
9	6105	6085	6125	3880	3860	3900	V
11	6145	6125	6165	3920	3900	3940	V
13	6185	6165	6205	3960	3940	3980	V
15	6225	6205	6245	4000	3980	4020	V
17	6265	6245	6285	4040	4020	4060	V
19	6305	6285	6325	4080	4060	4100	V
21	6345	6325	6365	4120	4100	4140	V
23	6385	6365	6405	4160	4140	4180	V

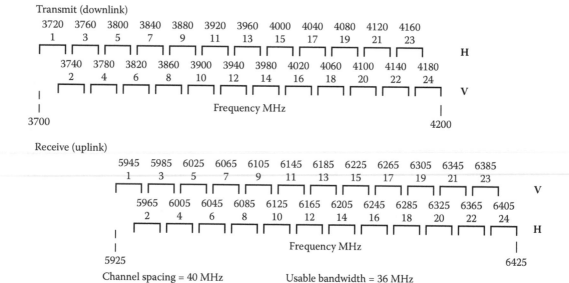

Figure 1.5   Subchannel ("transponder") allocation for C-band satellites.

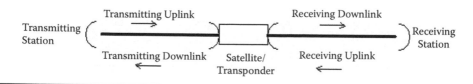

Figure 1.6   A satellite (radio) (microwave) link.

Table 1.6   Some key physical parameters of relevance to satellite communication

Frequency	The number of times that an electrical or electromagnetic signal repeats itself in a specified time. It is usually expressed in cycles per second (hertz [Hz]). Satellite transmission frequencies are in the gigahertz (GHz) range.
Frequency band	A range of frequencies used for transmission or reception of radio waves (e.g., 3.7–4.2 GHz).
Frequency spectrum	A continuous range of frequencies.
Hertz (Hz)	SI unit of frequency, equivalent to one cycle per second. The frequency of a periodic phenomenon that has a periodic time of 1 s.
Kelvin (K)	SI unit of thermodynamic temperature.
Msymbol/s	Unit of data transmission rate for a radio link, equal to 1,000,000 symbol/s. Actual channel throughput is related to the modulation scheme employed.
Symbol	A unique signal state of a modulation scheme used on a transmission link that encodes one or more information bits to the receiver.
Watt (W)	SI unit of power, equal to 1 J/s.

can also be used, depending on the intended application, service availability goals, and satellite footprint. For two-way transmission, the same size and considerations apply (although larger antennas can also be used in some applications, especially at major earth stations); availability, acceptable bit error rate, satellite-radiated power, and rain mitigation goals drive the design/size of the antenna and ground transmission power.

Enterprise applications tend to make use of the Ku band because smaller antennas can be employed, typically in the 0.6–2.4 m range (depending on application, desired availability, rain zone, and throughput, among other factors). Newer applications, typically for direct-to-home (DTH) video distribution, look to make use of the Ka band, where antenna size can range from 0.3–1.2 m. (See Table 1.7.) Spread-spectrum techniques and other digital signal processing are

Table 1.7   Frequency and wavelength of satellite bands

Frequency (GHz)	Wavelength (m)	Typical antenna size (m)
3.7	0.081081081	1.2–4.8
4.2	0.071428571	
5.925	0.050632911	
6.425	0.046692607	
11.7	0.025641026	0.6–2.4
12.2	0.024590164	
12.7	0.023622047	
18.3	0.016393443	0.3–1.2
18.8	0.015957447	
19.7	0.015228426	
20.2	0.014851485	
27.5	0.010909091	

being used in some applications to reduce the antenna size by reducing unwanted signals (e.g., either in the uplink with spread spectrum, or in the downlink with adjacent satellite signal cancellation using digital signal processing).

Related to the issue of orbits, note that, as stated, there are multiple satellites in the geostationary orbit, typically every 2° on the arc, and even collocated at the "same" location, when different operating frequencies are used (as depicted in Figure 1.3). Effectively, satellite systems may employ cross-satellite frequency reuse via space-division multiplexing; this implies that a large number of satellites (even neighbors) make use of the same frequency operating bands as long as the antennas are highly directional. Some applications (e.g., direct broadcast to homes) or jurisdictions (non-U.S.) allow spacing at 3°; higher separation reduces the technical requirements on the antenna system but results in fewer satellites in space.

Unfortunately, unless the system is properly "tuned" by following all applicable regulations and technical guidelines, adjacent satellite interference (ASI) can occur. A transmit earth station can inadvertently direct a proportion of its radiated power toward satellites that are operating at orbital positions adjacent to that of the wanted satellite. This can occur because the transmit antenna is incorrectly pointed toward the wanted satellite, or because the earth station antenna beam is not sufficiently concentrated in the direction of the satellite of interest (e.g., the antenna being too small). This unintended radiation can interfere with services that use the same frequency on the adjacent satellites. Interference into adjacent satellite systems is controlled to an acceptable level by ensuring that the transmit earth station antenna is accurately pointed toward the satellite and that its performance (radiation pattern) is sufficient to suppress radiation toward the adjacent satellites. In general, a larger uplink antenna will have less potential for causing adjacent satellite interference but will generally be more expensive and may require a satellite tracking system. Similarly, a receive earth station can inadvertently receive transmissions from adjacent satellite systems, which then interfere with the wanted signal. This happens because the receive antenna, while being very sensitive to signals coming from the direction of the wanted satellite, is also sensitive to transmissions coming from other directions. In general, this sensitivity reduces as the antenna size increases. As for a transmit earth station, it is also very important to accurately point the antenna toward the satellite to minimize ASI effects [FOC200701]. As noted, spread-spectrum techniques and other digital signal processing are being used in some advanced (but not typical) applications to reduce unwanted signals (e.g., with spread spectrum, or with ASI cancellation using digital signal processing).

The sharing of a channel (colloquially, a "transponder") is achieved, at this juncture, using Time Division Multiple Access (TDMA), random access techniques, Demand Access Multiple Access (DAMA), or Code Division Multiple Access (CDMA) (spread spectrum). Increasingly, the information being carried, whether voice, video, or data, is IP based. Multiplexing techniques are covered in Chapter 4.

1.3 Satellite Signal Regeneration

In general, the information transfer function entails bit transmission across a channel (medium). Because there is a variety of media in use in communication, many of the transmission techniques are specific to the medium at hand. Functions include, but are not limited to, modulation, timing, noise/impairments management, and signal level management. In the context of

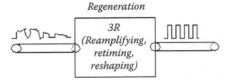

Figure 1.7   Regeneration as a general concept.

this book, the transmission channel is a radio channel. Typical transmission problems include the following:

- Signal attenuation (e.g., free space loss)
- Signal dispersion
- Signal nonlinearities (due to, e.g., amplification or propagation phenomena)
- Internal or external noise
- Cross talk (e.g., spectral regrowth), intersymbol interference, and intermodulation
- External interference and adjacent satellite interference

In general, some of these impairments, but not all, can be dealt with by using a regenerator. Regeneration is the function of restoring the signal (and/or bit stream) to its original shape and power level. These techniques are specific to the medium (e.g., radio channel, fiber channel, twisted-pair copper channel, and so on). Regeneration correctively addresses signal attenuation, signal dispersion, and cross talk; this is done via signal reamplification, retiming, and reshaping. Regeneration is generally considered a layer 1 function in the Open Systems Interconnection Reference Model (OSIRM). Figure 1.7 depicts a signal regeneration function pictorially. Regeneration and amplification are critical functions in satellite systems because the attenuation through space and the atmosphere is in the order of 200 dB (i.e., the power is reduced by 20 orders of magnitude).

Figure 1.8 depicts the basic building blocks of various regenerators. A "low-end" regenerator includes only the reamplification function. These are known as 1R regenerators. A "high-end" regenerator includes the reamplification, retiming, and reshaping functionalities. These are known as 3R regenerators. The functions of a 3R regenerator are (see Figure 1.9)

- Reamplification—increases power levels above the system sensitivity
- Retiming—suppresses timing jitter by optical clock recovery
- Reshaping—suppresses noise and amplitude fluctuations by decision stage

Regenerators are invariably technology specific. Hence, one has LAN repeaters (even if rarely used), Wi-Fi repeaters, copper-line (T1 channel) repeaters, cable TV repeaters, and optical regenerators (of the 1R, 2R, or 3R kind). Figure 1.10 depicts a regenerator in the satellite environment; this regenerator is the satellite transponder. The term *satellite transponder* refers properly to a transmitter–receiver subsystem on board the satellite that uses a single high-power amplification chain and processes a particular range of frequencies (the *transponder bandwidth*). There are many transponders on a typical satellite, each being capable

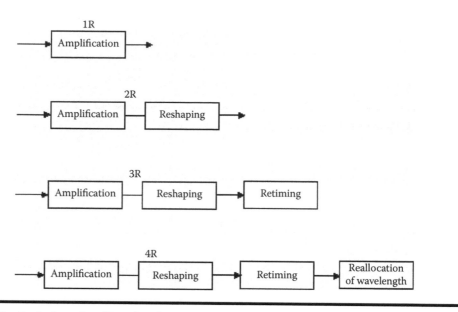

Figure 1.8   Basic functionality of various regenerators.

of supporting one or more communication channels [SAT200501]. In fact, a typical satellite will have 24 transponders: 12 to regenerate the consecutive 12–36 MHz frequency blocks that comprise the segments assigned for operation at the C band, Ku band, or Ka band (500 MHz total) for use in the *vertical signal polarization* mode, and 12 for the frequency blocks that comprise the segments assigned to the *horizontal signal polarization* mode. By utilizing

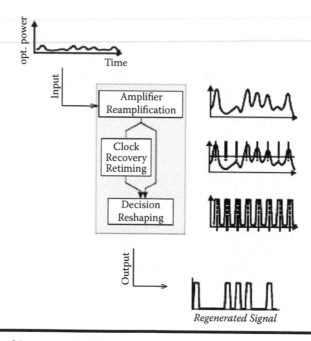

Figure 1.9   Basic architecture of a 3R regenerator.

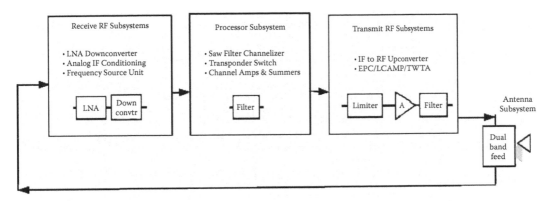

**C-Band
Transponder Block Diagram**

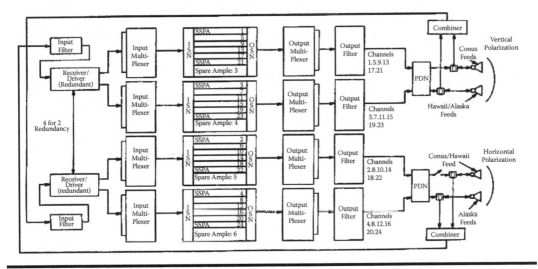

Figure 1.10 Architecture satellite regenerator—a transponder. ISN = input switching network; OSN = output switching network; PDN = power divider network.

transponders, commercial communication satellites perform the following functions:

- Receive signals from the ground station (uplink beam)
- Separate, amplify, and recombine the signals (the regeneration function)
- Transmit the signals back to (another) earth station (downlink beam)

Some advanced functions include digital signal processing and are called regenerative and nonregenerative onboard processors.

1.4 Satellite Transmission Chain

The ground segment of a satellite transmission chain consists of the earth stations (also known as ground stations) that are operating within a particular satellite system or network. Earth stations are typically connected to the end user's equipment directly with local cabling or via a

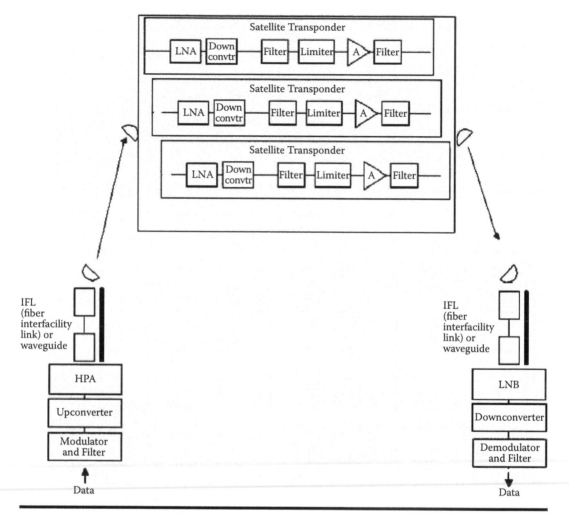

Figure 1.11 Satellite link (some details). LNA = low noise amplifier; HPA = high power amplifier; LNB = low noise block downconverter.

terrestrial network. The ground segment supports either one or both of the following: (i) an uplink, (ii) a downlink. Figure 1.11 depicts an end-to-end link. A satellite link is comprised of the following elements:

> Link = modulation equipment + upconversion equipment + amplification equipment
> + uplink transmission channel + frequency shift (conversion) equipment
> + downlink transmission channel + signal reception/antenna
> + amplification equipment + downconversion equipment
> + demodulation equipment.

The uplink is that portion of a satellite communications link that involves signal transmission from the ground and reception on board the satellite. The downlink is that portion of a satellite communications link that involves signal (re-)transmission from the satellite and reception on the ground. Downconversion is the process of converting the frequency of a signal to a lower frequency; it is performed at the reception point to permit the recovery of the original signal. The opposite, upconversion, is the process of converting the frequency of a signal into a higher frequency; it is done at the point of transmission. Signal management on the ground is better handled at lower frequencies, and hence, the purpose/need for frequency conversion (to a higher frequency level for transmission and down to a lower frequency level at reception). The transponder is a repeater that takes in the signal from the uplink at a frequency f_1, amplifies it, and sends it back on a second frequency f_2.

An uplink system consists of the following subsystems (often in redundant mode):

- Network interface devices such as routers, encryptors, conditional access systems, and encapsulators.
- Modulators (devices that superimpose the amplitude, frequency, or phase of a wave or signal onto another wave—the carrier—which is then used to convey the original signal over the satellite link. QPSK and 8-PSK are typical, but other methods are also used. FEC is typically handled at this point in the chain.)
- Upconverters (devices for converting the frequency of a signal into a higher frequency; transceivers that take a 70/140/900 MHz signal and frequency and convert it to either C-, Ku-, or Ka-band final frequency)
- PA (power amplifiers), specifically high-power amplifiers (HPAs); for example, solid-state power amplifier (SSPA or a klyston). Transmit power amplifiers provide amplification of signals to be transmitted to the satellite (typically 750–3000 W for high end applications)

A downlink link includes the following components:

- Downconverters (devices for converting the frequency of a signal into a lower frequency; transceivers that take a C-, Ku-, or Ka-band signal and frequency and convert it to either 70-, 140-, or 900-MHz final frequency)
- LNA (low noise amplifier), LNB (low-noise block downconverter), or LNC (low noise converter)
- Modem (customer site)

The LNA amplifies the RF signal from the antenna and feeds it into a frequency converter, the output of which is typically the intermediate frequency (IF) of 70/140/900 MHz. It provides 50–60 dB of amplification. An LNA is more precise and stable, but more expensive than an LNB. The LNB amplifies the RF signal from the antenna and converts it to an L-band signal. An LNB provides 50–60 dB of amplification and also converts from one block of frequencies to another (what goes in comes out amplified and at a different frequency). The LNC is similar to an LNB, but it has a variable shift. It provides 50–60 dB of amplification and also converts a block of frequencies to a specific portion of that block of frequencies (what goes into the LNC comes out amplified and at a different frequency within a given range; often it uses a local oscillator to help in conversion). LNA/LNB/LNCs are typically used for one-way field antennas. See Figure 1.12 [MUC200201].

Antennas in a satellite environment are reflective systems, typically parabolic in shape. A highly directional antenna concentrates most of the radiated power along the antenna "boresight." It follows that a high-gain antenna is very directional and needs to be pointed with reasonably high precision. As an example, at 12 GHz the pointing accuracy needed for a 1-m diameter dish is of the order of a degree or two of arc. The antenna uses a three-axis

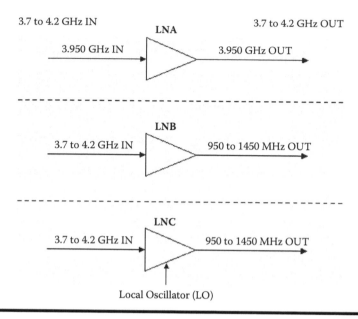

Figure 1.12   LNA/LNB/LNC operation (C-band example).

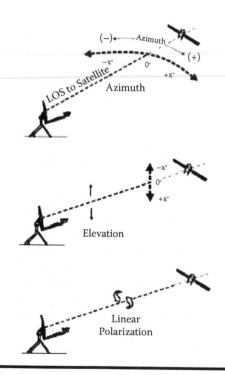

Figure 1.13   Antenna pointing.

pointing system that one needs to adjust when pointing it to a satellite (see Figure 1.13 [LAU200701]):

- Azimuth—This is the magnetic compass direction (angle of sighting) at which one points the dish. It is a side-to-side adjustment. Azimuth is the angular distance from true north along the horizon to a satellite, measured in degrees.
- Elevation—This is the angle above the horizon at which one points the dish. This is an up-and-down adjustment.
- Polarization—The (linear) polarization or *skew* represents the feed's alignment needed to capture the maximum signal from the satellite consistent with the satellite signal's transmit signal orientation (polarization*). This is a rotational adjustment. Polarization prevents interference with signals on the same satellite at the same frequency but opposite polarization.

Antennas are addressed in detail in Chapter 3.

A Block Upconverter (BUC) takes the L-band signal and converts it to either the C- or Ku-band final frequency. This is typically used for two-way field antennas, and operates in the 5–25 W range for commercial applications (but can also operate at other power levels in special circumstances).

1.5 Satellite Services

Traditionally, satellite services have been classified in the following categories:

- Fixed Satellite Service (FSS): This is a satellite service between satellite terminals at specific fixed points using one or more satellites. Typically, FSS is used for the transmission of video, voice, and IP data over long distances from fixed sites. It makes use of geostationary satellites with fixed ground stations.[†] Signals are transmitted from one point on the globe either to a single point (point-to-point) or from one transmitter to multiple receivers (point-to-multipoint). FSS may include satellite-to-satellite links (not commercially common) or feeder links for other satellite services such as the Mobile Satellite Service or the Broadcasting Satellite Service.
- Broadcast Satellite Service (BSS): This is a satellite service that supports the transmission and reception via satellite of signals that are intended for direct reception by the general public. The best example is Direct Broadcast Service (DBS), which supports the direct broadcast of TV and audio channels to homes or businesses directly from satellites. BSS/DBS make use of geostationary satellites. Unlike FSS, which has both point-to-point and point-to-multipoint communications, BSS is only a point-to-multipoint service. Therefore, a smaller number of satellites are enough to service the market.
- Mobile Satellite Service (MSS): This is a satellite service intended to provide wireless communication to any point on the globe. With the broad penetration of the cellular

---

* Incorrect alignment results in picking up the undesired cross-pol (XP) signal, which will severely impact performance/quality of the intended signal.
† Some mobile applications are also possible with FSS, but the antenna system is generally very complex and expensive.

telephone, users have started to take for granted the ability to use the telephone anywhere in the world, including rural areas in developed countries. MSS is a satellite service that enhances this capability. It uses low earth orbit and medium earth orbit satellite systems.
■ Maritime Mobile Satellite Service (MMSS): This is a satellite service between mobile-satellite earth stations and one or more satellites.
■ Although not formally a service in the regulatory sense, one can add global positioning service/system (GPS) to this list; this service uses an array of satellites to provide global positioning information to properly equipped terminals.

1.6 Satellite Applications with IPv6 Implications

There are many traditional and emerging satellite applications. Major commercial applications include, but are not limited to, the ones listed here; each of these may need to support IPv6 modes in the not-too-distant future.

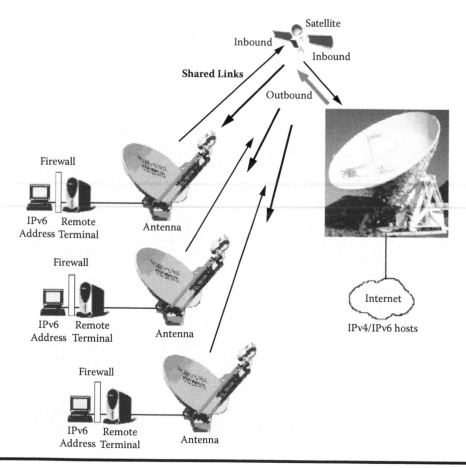

Figure 1.14 Typical (enterprise) (two-way) (very small aperture terminal) satellite communications.

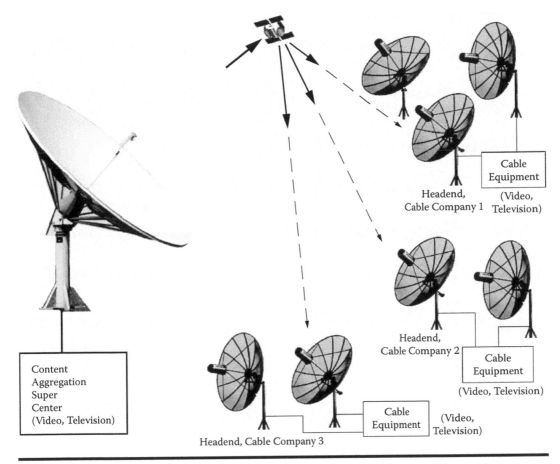

Figure 1.15 Typical satellite video distribution to cable headends.

- Two-way enterprise (very small aperture terminal) satellite communications for intranet/ Internet access connectivity (see Figure 1.14). Enterprise customers and/or government agencies may want to use IPv6 in the future (IPv4 is common today).
- Video distribution to cable headends (see Figure 1.15). Cable TV companies may be interested in IPv6 environments in the future.
- IPTV video distribution to telco headends (see Figure 1.16). Telephone companies may want to use IPv6 in the future, especially to support "Triple Play" or "Quadruple Play" (IPv4 is common today).
- DTH video reception by consumers. Some ancillary applications may be able to make effective use of an IPv6 infrastructure.

1.7 Non-Geostationary Satellites

Up to now we have covered satellite systems in the geostationary orbits (aka as GEO); some other orbits/satellites of interest include low earth orbits (LEOs), medium earth orbits (MEO)

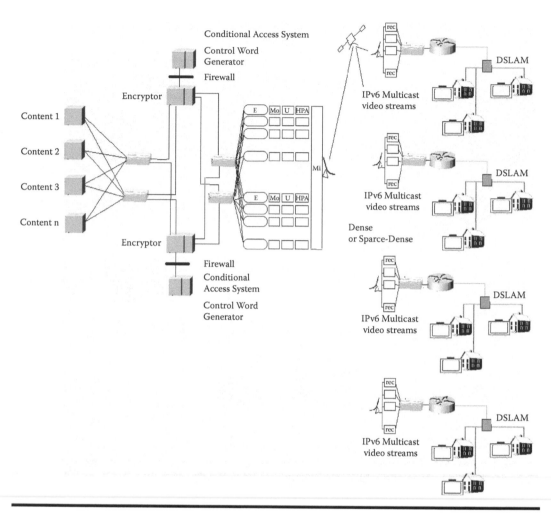

Figure 1.16   IPTV example (IPv6).

(aka intermediate circular orbits [ICO]), polar orbits, and highly elliptical orbits (HEOs). (See Figure 1.17.) This section has a brief overview of these systems, but they are not the focus of our coverage in this text.

As we saw earlier in the chapter, geostationary orbits are circular orbits that are oriented in the plane of the earth's equator. A geostationary satellite completes one orbit revolution around the earth every 24 hr; hence, given that the satellite spacecraft is rotating at the same angular velocity as the earth, it overflies the same point on the globe on a permanent basis (unless the satellite is repositioned by the operator). In a geostationary orbit, the satellite appears stationary, that is, in a fixed position, to an observer on the earth. The maximum footprint (service area) of a geostationary satellite covers almost one-third of the earth's surface; in practice, however, except for the oceanic satellites, most have a footprint optimized for a continent and/or portion of a continent (e.g., North America, or even continental United States). Non-GEO orbits are discussed next.

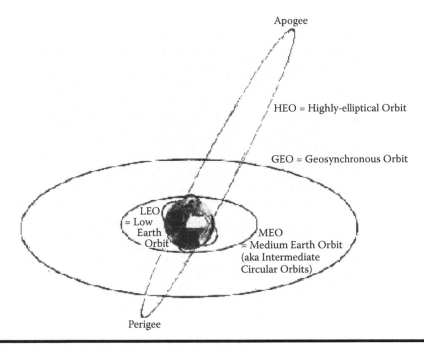

Figure 1.17   Satellite classes (based on orbit).

1.7.1   *Low Earth Orbits (LEOs)*

These are either elliptical or, more commonly, circular orbits that are at a height of 2000 km or less above the surface of the earth. The orbital periods at these altitudes vary between 90 min and 2 hr, and the maximum time during which a satellite in LEO is above the local horizon for an observer on the earth is up to 20 min. With LEOs there are long periods during which a given satellite is out of view of a particular ground station; this may be acceptable for some applications, for example, for earth monitoring. Coverage can be extended by deploying more than one satellite and using multiple orbital planes. A complete global coverage system using LEOs requires a large number of satellites (40–80) in multiple orbital planes and in various inclined orbits. Most small LEO systems employ polar or near-polar orbits. Due to the relatively large movement of a satellite in LEO with respect to an observer on the earth, satellite systems using this type of orbit need to be able to cope with Doppler shifts. Satellites in LEOs are also affected by atmospheric drag that causes the orbit to deteriorate (the typical life of an LEO satellite is 5–8 years, whereas the typical life of a GEO satellite is 14–18 years). However, launches into LEOs are less costly than to GEOs, due to their much lighter weight, and multiple LEO satellites can be launched at one time [GEO200101].

1.7.2   *Polar Satellites*

Polar orbits are LEOs that are in a plane of the two poles. Their applications include the ability to view only the poles (e.g., to fill in gaps of GEO coverage), or to view the same place on earth at the same time each 24-hr day. By placing a satellite at an altitude of about 850 km, a polar orbit period of about 100 min can be achieved (for more continuous coverage, more than one polar orbiting satellite is employed.)

A special polar orbit that crosses the equator and every latitude at the same time each day is called a sun-synchronous (SS) orbit; this orbit can make data collection a convenient task. Satellites in polar orbits are mostly used for earth-sensing applications. Typically, such a satellite moves at an altitude of 1000 km. In an SS orbit, the angle between the orbital plane and the sun remains constant. This orbit can be achieved by an appropriate selection of orbital height, eccentricity, and inclination that produces a precession of the orbit (node rotation) of approximately 1° eastward each day, equal to the apparent motion of the sun; this condition can only be achieved for a satellite in a retrograde orbit. As noted, the SS low-altitude polar orbit is widely used for monitoring the earth because, each day, as the earth rotates below it, the entire surface is covered, and the satellite views the same earth location at the same time each 24-hr period. All SS orbits are polar orbits, but not all polar orbits are SS orbits. All polar orbits are LEOs [GEO200101].

A special SS orbit, called a dawn-to-dusk orbit, is where the satellite trails the earth's shadow. Because the satellite never moves into this shadow, the sun's light is always on it. These satellites can, therefore, rely mostly on solar power and not on batteries; they are useful for agriculture, oceanography, forestry, hydrology, geology, cartography, and meteorology.

1.7.3 Medium Earth Orbits (MEO)/Intermediate Circular Orbits

These are circular orbits at an altitude of around 10,000 km. Their orbit period is in the range of 6 hr. The maximum period a satellite in MEO is above the local horizon for an observer on the earth is in the order of a couple of hours. A global communications system using this type of orbit requires a small number satellites in two or three orbital planes to achieve global coverage. The U.S. GPS is example of a MEO system.

1.7.4 Highly Elliptical Orbits (HEOs)

These typically have a perigee (point in each orbit which is closest to the earth) at about 500 km above the surface of the earth, and an apogee (the point in its orbit which is farthest from the earth) as high as 50,000 km. The orbits are inclined at 63.4° to provide communications services to locations at high northern latitudes. The orbit period varies from 8 to 24 hr. Owing to the high eccentricity of the orbit, a satellite spends about two-thirds of the orbital period near the apogee, and during that time it appears to be almost stationary to an observer on the earth (this is referred to as *apogee dwell*). A well-designed HEO system places each apogee to correspond to a service area of interest. After the apogee period of orbit, a switchover needs to occur to another satellite in the same orbit to avoid loss of communications. Due to the relatively large movement of a satellite in HEO with respect to an observer on the earth, satellite systems using this type of orbit need to be able to cope with Doppler shifts. An example of an HEO system is the Russian Molnya system; it employs three satellites in three 12-hr orbits separated by 120° around the earth, with apogee distance at 39,354 km, and perigee at 1,000 km [GEO200101].

1.8 Glossary of Key Satellite Concepts and Terms

Table 1.8 provides a basic satellite glossary for the concepts covered in this chapter from a variety of industry sources (including [SAT200501], [ANS20001], [EUT200701], and [GEO200101] among others).

Table 1.8 Basic satellite glossary

Amplifier	A highly linear device designed to increase the amplitude (power) of a signal. In an earth station, the main amplifier is often called the high-power amplifier (HPA); for a very small aperture terminal (VSAT) antenna, the amplifier may be included in the Block Upconverter (BUC). Although a variety of power values are available, HPAs typically range from 500 to 3000 W; BUCs range from 5 to 25 W for commercial applications. Other amplifiers are used throughout the earth station to amplify various stages of the signal.
Antenna	A passive device for transmitting or receiving radio waves. In commercial satellite communication, the antenna almost invariably consists of a parabolic reflector and a feed horn. On the receiving link, the reflector focuses radio waves onto the feed horn; the feed horn detects the signal and converts it into an electrical signal. On the transmitting side, the reflector concentrates the radio waves emitted by the feed horn into a narrow beam that is aimed at the satellite.
Antenna aperture	The effective area of an antenna capable of radiating or receiving RF energy. In a typical parabolic antenna, this dimension is equivalent to the diameter of the main reflector.
Antenna efficiency	Also known as radiation efficiency, the ratio of the power applied to an antenna to the power actually radiated by the antenna, stated as a percentage.
Apogee	Point in a satellite orbit (especially for highly elliptical ones) that is farthest from the earth.
Attenuator	In electrical systems, a resistive network that reduces the amplitude of a signal without appreciably distorting its waveform. Electrical attenuators are usually passive devices. The degree of attenuation may be fixed, continuously adjustable, or incrementally adjustable. Fixed attenuators are often called pads, especially in telephony.
	Note: The input and output impedances of an attenuator are usually matched to the impedances of the signal source and load, respectively. The amount by which the signal power is reduced is usually expressed in dB [ANS200001].
Availability	The ratio of the total time a functional unit is capable of being used during a given interval to the length of the interval.
	Note: An example of availability is 100/168 if the unit is capable of being used for 100 hr in a week.
	Note: Typical availability objectives are specified in decimal fractions, such as 0.9998 [ANS200001].
Band pass filter	A filter that ideally passes all frequencies between two nonzero finite limits and blocks all frequencies not within the limits.
	Note: The cutoff frequencies are usually taken to be the 3-dB points. The band pass filter allows only a specified range of frequencies to pass from input to output, rejecting all signals at lower or higher frequencies.
Block Upconverter (BUC)	Transmitter device that combines signal upconversion and power amplification in a single unit. The BUC is typically located directly at the antenna output, or relatively close to it.

(Continued)

Table 1.8   Basic satellite glossary (*Continued*)

Broadcast Satellite Service (BSS)	A satellite service that supports the transmission and reception via satellite of signals that are intended for direct reception by consumers. BSS systems employ satellites capable of transmitting high power, enabling reception by very small aperture terminals (VSAT).
C band	Original operating frequency band for communications satellites. Lowest frequency range that occupies roughly from 4 GHz to 6 GHz. This band is most resistant to rain fade.
Coax(ial) cable	A cable used principally as a transmission line for RF signals that has low loss at mid-to-high frequencies (3–20 GHz). It supports mid-range power levels (waveguide is used at high power high frequency.) A coax cable consists of an inner conductor that may be solid or stranded wire, and a dialectic filler covered by a flexible wire braid. (Metal foil may also be used.) All of these are covered by an insulating material.
Dawn-to-dusk orbit	Also called a heliosynchronous orbit, in this orbit, the satellite is in perpetual sunlight, allowing it to rely totally on solar panels to generate power. This orbit is used for earth observation, imaging, and research satellites.
Demodulation	The process of retrieving information impressed upon a carrier wave during the process of modulation. Demodulation along with FEC aims at recovery of the original signal or data being transmitted with a high level of fidelity (extremely low BER).
Demodulator	A device that creates a facsimile of the original signal or data transmitted by a modulator.
Descrambler	A device that recovers the original signal from one that has been scrambled (encrypted).
Digital-to-analog converter (DAC)	A device that converts a digital signal (a series of numbers or characters) into its equivalent analog waveform (a continuously varying signal voltage).
Dish	A parabolic antenna used for transmitting and/or receiving satellite signals. The term typically describes the entire antenna system, including the feed horn reflectors and all associated antenna structures.
Downconverter	A device for converting the frequency of a signal to a lower frequency; transceivers that take a C-, Ku-, or Ka-band signal and frequency and converts it to either 70-, 140-, or L-band final frequency.
Dual feed	An antenna system capable of simultaneously receiving from two different satellites at different orbital positions (the angular separation between satellite positions is typically 4–6°). This system consists of a reflector, a support structure, and two LNBs, each equipped with a separate feed horn or sharing an integrated feed assembly.
Dual-band feed	A feed horn that can simultaneously receive signals in two different frequency bands, typically the C band and the Ku band, often employed at earth stations working with hybrid satellites.
Earth station antenna	A specialized, highly directional, high-gain antenna designed to transmit and receive radio signals to and from an orbiting spacecraft. Typically, earth station antennas are parabolic; however, new-type antennas are also being deployed including flat-panel (phased-array antennas) and spherical antennas (Luneberg lens).

Table 1.8   Basic satellite glossary (*Continued*)

Effective isotropic radiated power (EIRP)	The effective power radiated by a satellite or earth station. Unit of measure is dBW or dBm.
End-of-life (EOL)	Point in the life of a communications satellite when it is no longer capable of carrying out its mission and must be retired.
F/D ratio	The ratio of an antenna's focal length F to its diameter D. This ratio describes the geometric architecture of the antenna, which impacts its physical size, design, and electrical performance.
Feed horn (feed)	A device that (i) in a transmitting system directs the signals onto the reflector surface for focusing into a narrow beam aimed at the satellite, and/or (ii) in a receiving system collects signals reflected from the surface of the antenna. The feed is designed to match a particular antenna geometry (F/D ratio) and is mounted at the focus of the parabolic reflector.
Filter	A passive electrical device that blocks signals or radiation of certain frequencies while allowing others to pass unaltered.
Fixed Satellite Service (FSS)	A satellite service between satellite terminals at specific fixed points using one or more satellites. Typically, FSS is used for the transmission of video, voice, and IP data over long distances. FSS makes use of geostationary satellites with fixed ground stations.
Focal length	The distance F from the reflective surface of an antenna to its focal point, usually measured in the horizontal plane.
Frequency reuse	A technique for utilizing a specified range of frequencies more than once within the same satellite system so that the total capacity of the system is increased without increasing its allocated bandwidth. Frequency reuse schemes require sufficient isolation between the signals that use the same frequencies so that mutual Interference between them is controlled to an acceptable level.
G/T	The figure of merit for an earth station, where G = gain (of the antenna system in dB) and T = system noise temperature (in Kelvins).
Geostationary orbit (GEO)	(Also known as the Clarke Orbit after Arthur C. Clarke, its inventor.) A circular orbit orientated in the plane of the earth's equator, allowing the satellite to appear stationary in relation to an observer on the earth's surface. This allows a satellite to provide communications services 24 hr a day to an entire hemisphere.
Global positioning (service/) system (GPS)	A satellite service that uses a constellation of LEO satellites to provide global positioning information to earth stations small enough to be hand held.
High power amplifier (HPA)	High-reliability high-power (750–3000 W) amplifier used in large earth station environments. Can be an SSPA, a TWT, or klystron.
Highly elliptical orbits (HEOs)	Orbits that have a perigee (point in each orbit that is closest to the earth) at about 500 km above the surface of the earth, and an apogee (the point in its orbit which is farthest from the earth) as high as 50,000 km [GEO200101].
Hybrid satellite	A satellite capable of operating in two or more frequency bands simultaneously, for example, C band and Ku band.

(*Continued*)

Table 1.8   Basic satellite glossary (*Continued*)

Intermediate frequency (IF)	A frequency that a carrier frequency is translated to in the process of reception or transmission of the signal. Intermediate frequencies are found between a modulator and an upconverter, or between a downconverter and a demodulator. Typical intermediate frequencies are 70 MHz, 140 MHz, and L band 950–1450 MHz.
Isotropic antenna	A theoretical point source of radiation used as a reference antenna when calculating antenna gain.
Ka band	The third and most recent band of frequencies authorized for satellite communications. It occupies roughly from 20 to 30 GHz in the radio spectrum. This band is the most susceptible to rain fade of all the three satellite bands.
Kelvin (Kelvin scale)	The Kelvin scale is a measure of temperature where the lowest value is absolute zero, the point where all molecular motion stops. Because all electrical devices contribute noise to a system based upon their temperature, noise contribution in satellite systems is expressed in Kelvins.
Klystron	A power amplifier tube used to amplify microwave energy (provided by an RF exciter) to a high power level. A klystron is characterized by high power, large size, high stability, high gain, relatively narrow bandwidth, and high operating voltages. Electrons are formed into a beam that is velocity modulated by the input waveform to produce microwave energy. A klystron is sometimes referred to as a linear beam tube because the direction of the electric field that accelerates the electron beam coincides with the axis of the magnetic field [AMS200001].
Ku band	The second band of frequencies authorized for satellite communications. It occupies roughly from 10 to 14.5 GHz in the radio spectrum. This band is more susceptible to rain fade than C band.
L band	An intermediate frequency (IF), typically employed at an earth station to route traffic between various points over coaxial/waveguide facilities. The frequency range coves the 950–1450 MHz spectrum. Note that over-the-air L-band ranges are (slightly) different and are defined by various regulatory agencies. Satellite signals (at C-band and Ku-band frequencies) are downconverted to L band in the focal point of many dish antennas by the LNB for further distribution within the electronics subsystem or the earth station. At C band, the downconversion is typically as follows: 4200–950 MHz; 4180–970 MHz; 4160–990 MHz, and so on up to 3700–1450 MHz. At the Ku band, the downconversion is typically as follows: 11700–950 MHz; 11720–970 MHz; 11740–990 MHz, and so on, up to 12200–1450 MHz. The upconverter handles the opposite function.
Line amplifier	An amplifier in a transmission line that boosts the strength of a signal level.
Line splitter	An active or passive device that divides a signal into two or more signals containing all the original information. A passive splitter feeds an attenuated version of the input signal to the output ports; an active splitter amplifies the input signal to overcome the splitter's loss.

Table 1.8 Basic satellite glossary (*Continued*)

Link budget	A collection of the various system parameters of a satellite link that is used to determine either the link performance from a fixed set of system parameters or some aspect of the system parameters given particular link performance criteria.
Local oscillator (LO)	A single-frequency reference signal that is used by a mixer to convert a communications signal to a higher or lower frequency band.
Low earth orbits (LEO)	Either elliptical or, more commonly, circular orbits that are at a height of 2000 km or less above the surface of the earth. The orbit period at these altitudes varies between 90 min and 2 hr, and the maximum time during which a satellite in LEO orbit is above the local horizon for an observer on the earth is up to 20 min [GEO200101].
Low-noise amplifier (LNA)	A low-noise device that receives and amplifies satellite signals at the output of the antenna feed horn; an LNA does not change the frequency of the received signal. LNAs are designed to contribute a minimum amount of noise to the signal received from the satellite to minimize the overall system noise temperature.
Low-noise block downconverter (LNB)	A low-noise device that receives and amplifies satellite signals at the output of a feed horn while also performing other functions such as signal detection, high-gain low-noise amplification, and frequency conversion. The frequency conversion downconverts a block of frequencies to a lower intermediate frequency range (typically in the L band). The feed horn is often integrated with LNB in a single mechanical unit.
Maritime Mobile Satellite Service (MMSS)	A satellite service between mobile satellite earth stations and one or more satellites.
Medium earth orbits/ intermediate circular orbits (MEO/ICOs)	Circular orbits at an altitude of around 10,000 km. Their orbital period is around 6 hr.
Mixer	A nonlinear device in which two or more input signals are combined to generate a single output signal.
Mobile Satellite Service (MSS)	A satellite service intended to provide wireless communication to any point on the globe.
Modulator	A device that superimposes a signal (intelligence) onto a wave or signal (called a carrier), which is then used to convey the original signal via a transmission medium. Modulation techniques include amplitude modulation, frequency modulation, or phase modulation. For satellite applications, phase shift keying (PSK) is fairly common.
Offset antenna/ feed	A parabolic antenna that has its feed horn offset from the center of the reflector in an effort to improve the performance of the antenna and reduce unwanted signals from adjacent satellites. Offset antennas can be easily modified to accept dual or multiple feeds, allowing them to receive signals from more than one satellite.
Perigee	Point in a satellite orbit (especially for highly elliptical ones) that is closest to the earth.

(Continued)

Table 1.8 Basic satellite glossary (*Continued*)

Polar satellites	LEOs orbits that are in a plane of the two poles. Their applications include the ability to view only the poles (e.g., to fill in gaps of GEO coverage), or to view the same place on earth at the same time each 24-hr day. Polar orbits are typically used by LEO communications satellites as well as research, weather, and spy satellites.
Polarization	Transmission approach where radio waves are restricted to certain directions of electrical and magnetic field variations, where these directions are perpendicular to the direction of wave travel. By convention, the polarization of a radio wave is defined by the direction of the electric field vector. Four senses of polarization are used in satellite transmissions: horizontal linear polarization, vertical linear polarization, right-hand circular polarization, and left-hand circular polarization.
Rain zone (aka precipitation zone)	Rain is precipitation that falls to earth in drops more than 0.5 mm in diameter. Rainfall is the amount of precipitation of any type, primarily liquid. It is usually the amount that is measured by a rain gauge. To aid in calculating the effect of precipitation loss, the world is divided into precipitation zones or rain climatic zones, each of which has a numerical value defined by the International Telecommunication Union (ITU), used in the calculation of a link budget.
Satellite receiver	A receiver designed for satellite reception. It receives modulated signals from an LNA or LNB and converts them into their original form.
Scrambler	A device that renders a signal unintelligible; of interest for modern communication is the process of encryption that has high cryptographic strength.
Solid-state power amplifier (SSPA)	A high-power amplifier using solid-state technology (i.e., transistors). Originally used for low- and medium-power applications; however, reliable medium- to high-power SSPA technology has emerged and is used routinely at earth stations and on spacecraft.
Splitter	A device that takes an input signal and splits it into two or more identical output signals, each a replica of the input signal (typically, with reduced amplitude, for example, –3 dB, but active devices can also operate at 0 dB loss).
Spread spectrum	(1) Telecommunications techniques in which a signal is transmitted in a bandwidth considerably greater than the frequency content of the original information. *Note:* Frequency hopping, direct-sequence spreading, time scrambling, and combinations of these techniques are forms of spread spectrum. (2) A signal-structuring technique that employs direct sequence, frequency hopping, or a hybrid of these, which can be used for multiple access and/or multiple functions. This technique decreases the potential interference to other receivers while achieving privacy and increasing the immunity of spread-spectrum receivers to noise and interference. Spread spectrum generally makes use of a sequential noise-like signal structure to spread the normally narrowband information signal over a relatively wide band of frequencies. The receiver correlates the signals to retrieve the original information signal [ANS200001].

Table 1.8   Basic satellite glossary (*Continued*)

Sun-synchronous (SS) orbit	A special polar orbit that crosses the equator and each latitude at the same time each day; this orbit can make data collection a convenient task. Satellites in polar orbits are typically used for earth-sensing applications. Typically, such a satellite moves at an altitude of 1000 km [GEO200101].
System noise temperature	A value, expressed in Kelvins, that accounts for the noise contribution of all components in the earth station's receive chain. Often depicted as T_s or T_{sys}.
Traveling wave tube amplifier (TWTA)	A high-power amplifier based on vacuum-tube technology. Normally employed when high output power levels and wide bandwidths are required. Typically used on board satellites and often in earth stations.
Upconverter	A device for converting the frequency of a signal into a higher frequency.
Very small aperture terminal (VSAT)	A complete terminal (typically with a small 4–5 ft antenna) that is designed to interact with other terminals in a satellite-delivered data network, commonly in a "star" configuration through a hub. The term *small aperture* here actually refers to the occupied bandwidth of the VSATs' transmitted carrier, which is typically only a few kHz wide. The VSAT terminal uses a special and often proprietary modulation, scrambling and coding algorithms; this allows the hub or network operator to control the system and present billing based on a data throughput or other form of usage basis. VSATs are utilized in a variety of applications and are designed as low-cost units. Commonly, several VSAT networks are operated through the same hub (shared services), which reduces the initial installation/setup costs [BAR200101].
Waveguide	A material medium that confines and guides a propagating electromagnetic wave, a transmission line. At microwave frequencies, a waveguide normally consists of a hollow metallic conductor, usually rectangular, elliptical, or circular in cross section, usually a pipe about 1×2 in. This type of waveguide may, under certain conditions, contain a solid or gaseous dielectric material [ANS200001]. It is typically used in earth stations to connect HPAs to the antenna.

References

[AGI200101] Agilent Technologies, "Digital Modulation in Communications Systems—An Introduction," Application Note 1298, March 14, 2001, Doc. 5965-7160E, Santa Clara, CA.

[AMS200001] American Meteorological Society, *Glossary of Meteorology*, Todd S. Glickman Editor, AMS, Cambridge, MA, June 2000.

[ANS200001] ANS T1.523-2001, *Telecom Glossary 2000*, American National Standard (ANS), an outgrowth of the Federal Standard 1037 series, *Glossary of Telecommunication Terms*, 1996.

[BAR200101] M. Bartlett, Satellite FAQ, S+AS Limited, 6 The Walled Garden, Wallhouse, Torphichen, West Lothian, Scotland, July 2001.

[EUT200701] Eutelsat, Glossary, Paris, France.

[FOC200701] Staff, "An Introduction to Earth stations," Focalpoint 2007, http://focalpoint-consulting.com. St Gaudens, France.

[GEO200101] Staff, "Geostationary, LEO, MEO, HEO Orbits Including Polar and Sun-Synchronous Orbits with Example Systems and a Brief Section on Satellite History," 2001, http://www.geo-orbit.org/

[JEF200401] D. Jefferies, "Microwaves: Satcoms applications," MSc in Satcoms Notes, University Of Surrey, Department of Electronic Engineering, School of Electronics and Physical Sciences, Guildford, Surrey, UK, 18th March, 2004.

[LAU200701] J. E. Laube, HughesNet, "Introduction to the Satellite Mobility Support Network, HughesNet User Guide," 2005–2007.

[MIN197801] D. Minoli, K. Schneider, "An Optimal Receiver For Code Division Multiplexed Signals," *Alta Frequenza*, July 1978, Vol. XLVII, No. 7, pp. 587–591.

[MIN197901] D. Minoli, "Satellite On-Board Processing of Packetized Voice," International Communication Conference, 1979, pp. 58.4.1–58.4.5.

[MIN198601] D. Minoli, "Aloha Channels Throughput Degradation," Computer Networking Symposium, Washington, D.C., 1986. {Degradation in satellite/VSAT channels}.

[MIN199101] D. Minoli, "Satellite Transmission Systems," 70-page chapter in *Telecommunication Technologies Handbook*, 1st Ed. (Artech House, 1991).

[MUC200201] K. Muchorowski, CEENET Workshop 2002, Satellite Communications, NetSat Express, muchor@ceenet.org

[SAT200501] Satellite Internet Inc., Satellite Physical Units and Definitions, http://www.satellite-internet. ro/satellite-internet-terminology-definitions.htm

Chapter 2

Electromagnetic Propagation and Reception

This chapter covers some of the basic electromagnetic and transmission concepts that are required for the design, engineering, and implementation of satellite transmission systems. Other topics are covered in the chapters that follow. This chapter only presents a basic introduction to these topics and focuses pragmatically only on commercial satellite applications. Many of the topics addressed here would require an entire textbook to cover the discipline in a more complete manner.

2.1 Basic RF Terms and Concepts

The concept of decibel (dB) is fundamental to a discussion of transmission. One has

$$\text{Bel} = \log(P2/P1)$$

$$\text{Decibel (dB)} = 10 \log(P2/P1) = 0.1 \text{ Bel}$$

where $P2/P1$ represents the ratio of power values.
(Bel is a relatively large unit,* hence the smaller unit, the decibel is used; 10 decibels = 1 bel.)
   Typically,

$$\text{Decibels (dB)} = 10 \log(P_{out}/P_{in})$$

where P_{out} is the power after an amplification stage or at the far end of a transmission link, and P_{in} represents the input or initial power. $P1$ (P_{in}) is usually the reference power. As it can be plainly

---

* Think of the analogy that *liters* is too large a unit when talking about a medical injection of some medication, and hence, the unit *milliliter* (mL) is used, such as 1 mL, 5 mL, or 10 mL.

Table 2.1   Some dB values

P_{in} (W)	P_{out} (W)	dB
1	10000	40.00
1	1000	30.00
1	100	20.00
1	10	10.00
1	1	0.00
1	0.5	−3.00
1	0.1	−10.00
1	0.01	−20.00
1	0.001	−30.00
1	0.0001	−40.00
1	0.00001	−50.00
1	0.000001	−60.00
1	0.0000001	−70.00
1	0.00000001	−80.00
1	0.000000001	−90.00
1	1E–10	−100.00

seen from the definition, the decibel does not in itself indicate power but is a ratio or comparison between two power values. When the ratio is positive, one uses the term *gain*. If $P2$ (P_{out}) is less than $P1$ (P_{in}), the ratio is less than 1.0, and the resultant value in decibels is negative. For example, if $P2$ is one-tenth of $P1$, then dB = 10 log (0.1/1) = −10 dB. The terms *loss* and *attenuation* are used in the context of a negative sign. For example, an amplifier with an output of 100 W when the input is 0.1 W (100 mW), has a gain of

$$10 \log(P2/P1) = 10 \log(100/0.1) = 30 \text{ dB.}$$

As another example, if 100 W of power is applied to a cable but only 85 W is measured at the output because of resistive attenuation and other types of attenuations, the signal has been decreased by a factor of

$$10 \log (0.85) = -0.7 \text{ dB.}$$

An attenuator that reduces its input power by a factor of 0.001 has an attenuation of −30 dB (see Table 2.1).

It is often desirable to express power levels in decibels by using a fixed power as a reference for $P1$ (P_{in}). A common reference is the milliwatt (mW). That is, $P1$ = 1.0 mW. The nomenclature dBm indicates dB referenced to 1.0 mW. One milliwatt for $P2$ (P_{out}) results in zero dBm.

The abbreviation dBW indicates dB referenced to 1.0 W; $P1$ = 1.0 W. Then, if $P2$ = 1.0 W, it results in 0 dBW (or 30 dBm). An increase of 3 dBW represents a doubling of power; 10 dBW represents a tenfold increase, and 20 W dBW, a 100-fold increase.

For antenna gain, the reference is the linearly polarized isotropic radiator, dBLI. Usually, the "L" or "I", or both are understood and left out. The ability of an antenna to intercept or transmit a signal is expressed in dB referenced to an isotropic antenna rather than as a ratio. Hence, if the antenna has an effective gain ratio of 7.5, one states that it has a gain of 10 log 7.5 = 8.8 dB.

Table 2.2 Key concepts related to decibels

dBi	The relative gain of an antenna with respect to an equivalent isotropic antenna, expressed on the decibel logarithmic scale.
dBW	The power of a signal in watts (W), expressed on the decibel logarithmic scale.
dBW/m²	The power of a radio wave incident on a surface area of 1m², measured in watts and expressed on the decibel logarithmic scale.
dBK	The equivalent noise temperature of a device in Kelvins, expressed on the decibel logarithmic scale.
dB/K	Units used to express the figure of merit or G/T of an earth station, with dimensions of 1/Kelvin, expressed on the decibel logarithmic scale.
decibel (dB)	A unit for comparing two currents, voltages, or power levels based on a logarithmic scale. It is used particularly for expressing the difference between very large and very small values (Expression: $R = 10*\log_{10}(r)$, where r is the linear ratio, and R is the ratio in dB.)

The logarithmic definition of the dB has key intrinsic advantages; instead of having to multiply gain or loss factors as ratios, one can add them as positive or negative dB. Consider the example where one has a microwave system with a 13 W transmitter, and a cable with a 0.8 dB loss connected to a 12 dB gain transmit antenna; the signal loss through the atmosphere (here, absorption) is 137 dB to a receive antenna with a 10 dB gain connected by a cable and 1.7 dB loss to a receiver. To undertake the calculation of how much power is available at the receiver, one must convert the 13 W to dBm, which is by assumption the reference terms used in all of these specs:

$$10\text{ W} = 10{,}000\text{ mW; so, one has }10\log(13{,}000/1) = 41.1\text{ dBm.}$$

Then, applying the additive principle,

$$41.1\text{ dBm} - 0.8\text{ dB} + 12\text{ dB} - 137\text{ dB} + 10\text{ dB} - 1.7\text{ dB} = -76.3\text{ dBm.}$$

To know how much power is received at the far end, one needs to convert this figure back to milliwatts:

$$mW = 10^{(dBm/10)} = 10^{(-76.3/10)} = 10^{(-7.63)} = 0.000000023\text{ mW}$$

Table 2.2 summarizes some of the concepts based loosely on [SAT200501].

2.2 Basic Transmission Theory Concepts

This section provides a basic pragmatic overview of transmission theory.

2.2.1 Signal Propagation

Satellite communication occurs via the propagation of electromagnetic signals. Propagation is the motion of waves through or along a medium; for electromagnetic waves, propagation may occur in a vacuum as well as in material media; radiant power is the rate of flow of electromagnetic energy.

Put differently, propagation of a wave is the evolution in time and space of an electromagnetic wave. Transmission occurs via changes in time in the electromagnetic field (electric and magnetic field) that are projected over space; these changes can be intercepted remotely by an appropriately configured antenna. Figure 2.1 depicts this process pictorially.

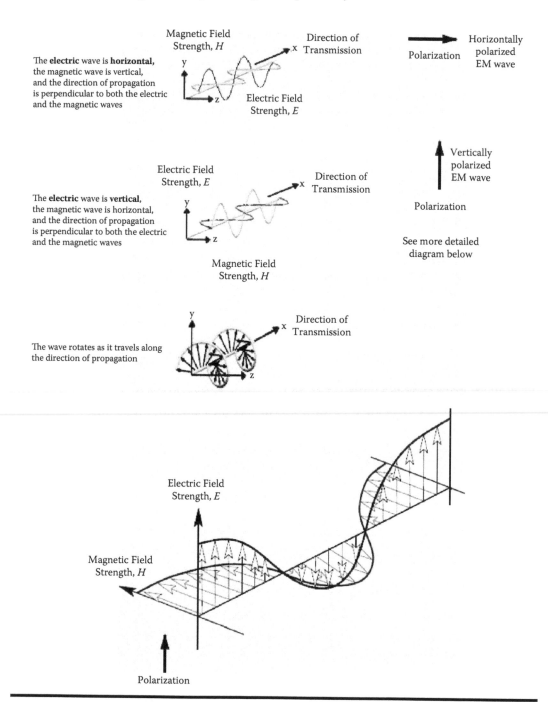

Figure 2.1 Electromagnetic propagation.

Table 2.3 Partial list of electrical quantities and their SI units and symbols

Electromagnetic quantities	*Symbol*
Electric charge (coulombs)	q or Q
Volume charge density (coulombs per cubic meter)	ρ
Surface charge density (coulombs per square meter)	σ
Linear charge density (coulombs per meter)	λ [not to be confused with the wavelength, also λ]
Electrostatic potential (volts)	ϕ
Electric field (volts per meter)	E
Electric induction	$D = \varepsilon \times E$ (coulombs per square meter) [not to be confused with the antenna diameter, also D]
Electric current (amps or coulombs per second)	I
Electric current density (amps per square meter)	J
Magnetic field (amps per meter or amp turns per meter)	H
Magnetic field	$B = \mu \times H$ (tesla)
Magnetic vector potential (tesla meters)	A [not to be confused with the antenna area, also A]
Capacitance (coulombs per volt, or Farads)	C
Inductance (volts-seconds per amp, or Henries)	L [not to be confused with the free space loss, also L]
Permittivity (Farads per meter)	ε
Permeability (Henries per meter)	μ
Velocity of light in vacuum (meters per second)	c

Electromagnetic signals propagate in (free) space according to Maxwell equations. Table 2.3 identifies key quantities involved in electromagnetic propagation; the most critical quantities are the following:

- Electric field **E**
- Electric flux density **D**
- Magnetic field **H**
- Magnetic flux density **B**

Electric fields form lines that are trajectories along which a very small free charge would travel in the absence of a magnetic field in the same place. Magnetic fields form lines along which little bar magnet "dipoles" or iron filings would align. The flow of charge constitutes a current. Charge is neither created or destroyed, so, if in a region there is a changing charge (with time), then there must be current flow in or out of that region. Currents often travel along electric field line directions and generate magnetic fields. In regions of constant charge density, all currents flow in closed loops. If the current loop does not close, then there must be accumulation of charge that varies with time [JEF199901]. Appendix 2.A provides a short overview of Maxwell's equations.

Note that, in considering the electromagnetic fields generated by a transmitting antenna, it is useful to distinguish between what is called the *near field*, which is the field within a distance of about $\lambda/2\pi$ of the antenna, and what is called the *far field*, which is the field at a distance greater than about $\lambda/2\pi$ from the antenna. The near field is basically an energy storage field; it is concerned with the storage of energy per unit volume in electromagnetic form in the region close to the antenna, and is not concerned, to a significant degree, with the propagation of energy away from the antenna. The near field can be regarded as being intimately associated with the charges on the antenna or the currents within it. Twice per cycle of oscillation, the energy stored in that field disappears from space for an instant and then reenters the antenna. In contrast, the far field is an energy propagation field: it is concerned with the propagation of energy per unit area per unit time in electromagnetic form away from the antenna. The far field is not intimately associated with the charges and currents in the antenna but can be thought of as being generated by the electromagnetic fields that stand between the antenna and the far field point. The far field is the field that tends to dominate at large distance [COL200201].

As the transmitted signal traverses a distance of space, its power level decreases at a rate inversely proportional to the distance traveled and proportional to the wavelength of the signal. This effect is due to the spreading of the radio waves as they propagate. As radio waves propagate in free space, the power falls off as the square of the range: for a doubling of range, the power reaching a receiver antenna is reduced by a factor of four. For line-of-sight free-space propagation, the loss L can be calculated by [ZYR199801]:

$$L = 20 \log_{10} (4\pi \, d/\lambda)$$

where
$d =$ the distance between receiver and transmitter
$\lambda =$ free space wavelength $= c/f$
(where $c =$ speed of light (3×10^8 m/s), $f =$ frequency (Hz))

Note 1: In this equation, when the distance (d) is equal to $1/4\pi$, the path loss is 0 dB, and at shorter distances the path loss is calculated as a negative number. Therefore, this equation has practical limits, and one should not use it for distances less than about a wavelength because, at that point, physical factors such as antenna dimensions dominate and force the use of electromagnetic field equations.

Note 2: Information in the transmitted signal is rarely concentrated at a single frequency, and hence, the path loss will actually be different for every frequency component in the signal. Fortunately, the ratio of the bandwidth to the center frequency is usually small enough so that the range of frequencies will not matter. Still, a signal that is transmitted with a constant power across a certain bandwidth will appear at the receiver with a power slope that decreases at the upper end of the band [BLA200701].

The Friis Equation gives a more complete accounting for all the factors from the transmitter to the receiver (the concept of antenna gain is defined in the sections that follow):

$$P_{Rx} = P_{Tx} \frac{G_{Tx} \cdot G_{Rx} \cdot \lambda^2}{16 \cdot \pi^2 \cdot d^2 \cdot L}$$

where

G_{Tx} = transmitter antenna gain
G_{Rx} = receiver antenna gain
λ = wavelength (same units as d)
d = distance separating Tx and Rx antennas
L = system loss factor (≥ 1)

In addition to free-space losses (attenuation), there are a multitude of other factors that have to be taken into account. These include, but are not limited to, attenuation, distortion, dispersion, intermodulation, fade, multipath, dropouts, and external and adjacent satellite interference, and cross-pol(arization) interference. Losses and interference degrade the reception of radio waves. Losses are typically due to absorption, for example rain fade, among other causes. Interference can be caused, among other possibilities, by ground-based sources (e.g., terrestrial microwave links operating at the same frequency), celestial sources (sun spots), elements within the satellite system (e.g., cross-pol interference), or by transmissions by other satellite systems that use the same frequency bands. Table 2.4 depicts a longer (but not exhaustive) list of impairments and issues that have to be taken into account when designing a radio frequency (RF) system in general and a satellite link in particular (many terms here are based on [ANS200101] and [SAT200501]).

Taking into account the main contributors to attenuation, the (pragmatic) total attenuation through unobstructed atmosphere is the sum of free space path loss, the attenuation caused by oxygen absorption, the attenuation caused by water vapor absorption, and the attenuation caused by rain, when present; thus,

$$\text{Atten}_{Total} = \text{Atten}_{FreeSpacePathLoss} + \text{Atten}_{Oxygen} + \text{Atten}_{WaterVapor} + \text{Atten}_{Rain}$$

Electromagnetic waves are absorbed in the atmosphere; the absorption is a function of the wavelength. Oxygen (O_2) and water vapor (H_2O) are responsible for the majority of signal absorptions. There are areas in the electromagnetic spectrum where there are local maxima; the first maximum occurs at 22 GHz due to water, and the second maximum occurs at 63 GHz due to oxygen. Note that the amount of water vapor and oxygen in the atmosphere decreases with an increase in altitude because of the decrease in pressure. See Figure 2.2 for a view of atmospheric absorption, and Figure 2.3 for a perspective on attenuation due to rain. Table 2.5 depicts the amount of rain (in mm/hr) in various rain zones (A through P) defined by the International Telecommunications Union (ITU), and the percentage of the time the specified precipitation rate is exceeded. One would first determine in which rain zones the antenna is located, then consult the table to determine the precipitation rate in millimeters per hour, and then use the graph to determine the attenuation. When one considers designing a distribution network consisting of one hub and multiple remote sites that are geographically distributed over the satellite downlink beam (e.g., for commercial digital video distribution), one needs to keep in mind that sites have typically a variation of G/T values in the downlink, as well as different rain-fade margin requirements.

The transmitted power path loss for the Ku-band satellites is higher than C-band because of the higher frequency at the Ku band. The path loss for a C-band signal is −196.5 dB, whereas the path loss for a Ku-band signal is −205.8 dB. Hence, under severe weather conditions such as heavy

Table 2.4   Partial list of RF-related impairments

Absorption	In the transmission of electrical, electromagnetic, or acoustic signals, absorption is the conversion of the transmitted energy into another form, usually thermal. Absorption is one cause of signal attenuation. The conversion takes place as a result of interaction between the incident energy and the material medium at the molecular or atomic level.
Adjacent channel interference (ACI)	Extraneous power from a signal in an adjacent channel. Adjacent channel interference may be caused by inadequate filtering, such as incomplete filtering of unwanted modulation products in frequency modulation (FM) systems, improper tuning, or poor frequency control, in either the reference channel or the interfering channel or both.
Adjacent satellite interference (ASI)	For a ground station, extraneous power from a signal in an adjacent satellite operating at the same band; may be because the antenna is too small and is not able to focus properly on the specific satellite of interest or because the antenna is mispointed. Another cause of ASI can be a carrier on an adjacent satellite that is exceeding its permitted power density.
Aperture-medium coupling loss	Coupling loss at the antenna-LNA/LNB interface.
Atmospheric absorption	The attenuation of signal due to absorption; conversion of the transmitted energy into thermal energy. The absorption coefficient is a measure of the attenuation caused by absorption of energy that results from its passage through a medium.
Atmospheric noise	Radio noise caused by natural atmospheric processes, primarily lightning discharges in thunderstorms.
Attenuation	The decrease in intensity of a signal, beam, or wave as a result of absorption of energy and of scattering out of the path to the detector, but not including the reduction due to geometric spreading. Attenuation is usually expressed in dB. The term *attenuation* is often used as a misnomer for *attenuation coefficient*, which is expressed in dB per kilometer.
Channel noise	All objects that have heat emit RF energy in the form of random (Gaussian) noise [ZYR199801]. The amount of radiation emitted is $N = kTB$, where N = noise power (W) k = Boltzmann's constant (1.38×10^{-23} J/K) T = system temperature, usually assumed to be 290K B = channel bandwidth (Hz)
Clear sky	A term describing the weather conditions encountered at the terrestrial end of an earth–space path of a satellite communication link. It is used to describe the condition where the attenuation of radio waves caused by precipitation (rain, snow, sleet, dew, etc.) is lowest (i.e., cloud-free sky and good visibility).
Clock error	The difference between the local clock time or value and a designated reference clock time or value. Subtracting the clock difference from the local clock brings the local clock into agreement with the reference clock.
Coupling loss	The loss that occurs when energy is transferred from one circuit, circuit element, or medium to another. Coupling loss is usually expressed in the same units—such as watts or dB—as in the originating circuit element or medium.

Table 2.4   Partial list of RF-related impairments (*Continued*)

Cross-pol interference	Interference caused by a received signal that has opposite polarization to the signal of interest.
Crosstalk (XT)	Undesired coupling of a signal from one circuit, part of a circuit, or channel to another. Any phenomenon by which a signal transmitted on one circuit or channel of a transmission system creates an undesired effect in another circuit or channel.
Cumulative transit delay	The total transit delay applicable for a data call obtained by summing the individual transit delays of all component portions of the data connection.
Diffraction	The deviation of an electromagnetic wave front from the path predicted by geometric optics when the wave front interacts with (i.e., it is restricted by) a physical object such as an opening (aperture) or an edge. *Note:* Diffraction is usually most noticeable for openings of the order of a wavelength; however, diffraction may still be important for apertures many orders of magnitude larger than the wavelength.
Dispersion	Any phenomenon in which the velocity of propagation of an electromagnetic wave is wavelength dependent. In communication technology, the tem *dispersion* is used to describe any process by which an electromagnetic signal propagating in a physical medium is degraded because the various wave components (i.e., frequencies) of the signal have different propagation velocities within the physical medium.
Distortion	In a system or device, any departure of the output signal waveform from that which should result from the input signal waveform being operated on by the systems specified (i.e., ideal) transfer function. Distortion may result from many mechanisms; examples include nonlinearities in the transfer function of an active device, such as a vacuum tube, transistor, or operational amplifier. It may also be caused by a passive component such as a coaxial cable or optical fiber, or by inhomogeneities, reflections, etc., in the propagation path.
Doppler effect	The change in the observed frequency (or wavelength) of a wave caused by a time rate of change in the effective path length between the source and the point of observation.
Dropouts	Momentary loss of signal. Dropouts are usually caused by noise, propagation anomalies, or system malfunctions, as well as rain fade.
External interference	Interference caused by other sources, such as terrestrial microwave links, cosmic rays, intentional jamming, etc.
Fade	In a received signal, the variation (with time) of the amplitude or relative phase, or both of one or more of the frequency components of the signal.
Fade margin	A design allowance that provides for sufficient system gain or sensitivity to accommodate expected fading for the purpose of ensuring that the required quality of service is maintained; the amount by which a received signal level may be reduced without causing system performance to fall below a specified threshold value.

(Continued)

Table 2.4   Partial list of RF-related impairments (*Continued*)

Free-space attenuation	See geometric spreading.
Free-space loss	The signal attenuation that would result if all absorbing, diffracting, obstructing, refracting, scattering, and reflecting influences were sufficiently removed so as to have no effect on propagation. Free-space loss is primarily caused by beam divergence, that is, signal energy spreading over larger areas at increased distances from the source.
Frequency deviation	The amount by which a frequency differs from a prescribed value, such as the amount an oscillator frequency drifts from its nominal frequency. In frequency modulation, this is the absolute difference between (a) the maximum or the minimum permissible instantaneous frequency of the modulated wave, and (b) the carrier frequency. In frequency modulation, this is the maximum absolute difference, during a specified period, between the instantaneous frequency of the modulated wave and the carrier frequency.
Frequency drift	An undesired progressive change in frequency with time. The causes of frequency drift include component aging and environmental changes. Frequency drift may be in either direction and is not necessarily linear.
Frequency shift	Any change in frequency. Any change in the frequency of a radio transmitter or oscillator. *Note:* In the radio environment, frequency shift is also called RF shift.
Fresnel diffraction pattern	See near-field diffraction pattern.
Fresnel zone	In radio communications, one of a (theoretically infinite) number of concentric ellipsoids of revolution that define volumes in the radiation pattern of a (usually) circular aperture. The cross section of the first Fresnel zone is circular. Subsequent Fresnel zones are annular in cross section and concentric with the first. Odd-numbered Fresnel zones have relatively intense field strengths, whereas even-numbered Fresnel zones are nulls. Fresnel zones result from diffraction by the circular aperture.
Geometric spreading (aka inverse-square law)	The physical law stating that irradiance, that is, the power per unit area in the direction of propagation, of a spherical wave front varies inversely as the square of the distance from the source, assuming there are no losses caused by absorption or scattering. For example, the power radiated from a point source (e.g., an omnidirectional isotropic antenna), or from any source at very large distances from the source compared to the size of the source, must spread itself over larger and larger spherical surfaces as the distance from the source increases. Diffuse and incoherent radiation are similarly affected.
Group delay	The rate of change of the total phase shift with respect to angular frequency $dQ/d\omega$ through a device or transmission medium, where Q is the total phase shift, ω is the angular frequency equal to $2\pi f$, and f is the frequency.

Table 2.4 Partial list of RF-related impairments (*Continued*)

Hit	A transient disturbance to, or momentary interruption of, a communication channel.
Interference	In general, extraneous energy from natural or man-made sources that impedes the reception of desired signals. Also, a coherent emission having a relatively narrow spectral content, for example, a radio emission from another transmitter at approximately the same frequency, or having a harmonic frequency approximately the same as another emission of interest to a given recipient, and which impedes reception of the desired signal by the intended recipient. In the context of this definition, interference is distinguished from noise in that the latter is an incoherent emission from a natural source (e.g., lightning) or a man-made source, of a character unlike that of the desired signal (e.g., commutator noise from rotating machinery) and which usually has a broad spectral content. Interference is the effect of unwanted energy due to any one or a combination of emissions, radiation, or inductions upon reception in a radiocommunication system, manifested by any performance degradation, misinterpretation, or loss of information that could be extracted in the absence of such unwanted energy. Interference is also the interaction of two or more coherent or partially coherent waves that produces a resultant wave that differs from the original waves in phase, amplitude, or both. *Note*: Interference may be constructive or destructive, that is, it may result in increased amplitude or decreased amplitude, respectively. Two waves equal in frequency and amplitude, and out of phase by 180°, will completely cancel one another; in phase, they create a resultant wave having twice the amplitude of either interfering beam.
Intermodulation	The production, in a nonlinear element of a system, of frequencies corresponding to the sum and difference frequencies of the fundamentals and harmonics thereof that are transmitted through the element.
Intermodulation distortion	Nonlinear distortion characterized by the appearance, in the output of a device, of frequencies that are linear combinations of the fundamental frequencies and all harmonics present in the input signals. Harmonic components themselves are not usually considered to characterize intermodulation distortion.
Inverse-square law	See geometric spreading.
Link budget analysis	Link budget is a generic term used to describe a series of mathematical calculations designed to model the performance of a communications link. In a typical simplex (one-way) satellite link, there are two link budget calculations: one link from the transmitting ground station to the satellite, and one link from the spacecraft to the receiving ground station.

(*Continued*)

Table 2.4   Partial list of RF-related impairments (*Continued*)

Link budget analysis tools include the following issues [USA199801]:

a. Link budget model
   A. Uplink and downlink carrier-to-receiver noise density (C/kT)
   B. Satellite receive power flux density
b. Positional data model
   A. Earth terminal to satellite slant range
   B. Earth terminal antenna elevation and azimuth
   C. Uplink and downlink Doppler frequency shift
c. Benign atmosphere attenuation
   • Clear air attenuation
   • Rainfall attenuation
   • Atmospheric signal scintillation
d. Modulation and channel encoding
   • Noise equivalent bandwidth
   • Modulation spectral efficiency
   • Demodulator implementation loss
   • Probability of detection of error
   • Convolution channel coding gain
e. Earth terminal model
   • Antenna model
   • Receive system model
   • Transmit system model
f. Satellite model
   • Uplink signal power
   • Downlink signal power
   • Transponder signal and noise power sharing
   • Transponder uplink power flux

The following input items are needed to produce a link budget calculation for a spacecraft-to-ground-station link

   • Earth station latitude
   • Earth station longitude
   • Spacecraft longitude
   • Downlink frequency
   • Antenna gain
   • Antenna noise temperature
   • Low noise amplifier
   • Ortho-mode transfer (OMT) loss
   • Effective isotropic radiated power (EIRP)
   • Intermediate frequency (IF) receive bandwidth
   • Transmit data rate
   • Link margin

Multipath | The propagation phenomenon that results in radio signals reaching the receiving antenna by two or more paths. Causes of multipath include atmospheric ducting, ionospheric reflection and refraction, and reflection from terrestrial objects, such as mountains and buildings. The effects of multipath include constructive and destructive interference, and phase shifting of the signal.

Table 2.4 Partial list of RF-related impairments (*Continued*)

Near-field diffraction pattern	The diffraction pattern of an electromagnetic wave that is observed close to a source or aperture, as distinguished from a far-field diffraction pattern. The pattern in the output plane is called the near-field radiation pattern.
Noise	Any undesired electrical disturbance in a circuit or communication channel. When combined with a received signal, it affects the receiver's ability to correctly reproduce the original signal. Also known as thermal noise.
Noise figure	A method for quantifying the electrical noise generated by a practical device. The noise figure is the ratio of the noise power at the output of a device to the noise power at the input to the device, where the input noise temperature is equal to the reference temperature (290 K). The noise figure is usually expressed in dB.
Noise temperature	A mathematical construct for predicting the influence of noise in a communications system. It is a measure of the noise power generated by a practical device, expressed as the equivalent temperature of a resistor that when placed at the input of a perfect noise-free device, generates the same amount of output noise. The noise temperature is usually expressed in kelvin or dBK.
Path loss	In a communication system, the attenuation undergone by an electromagnetic wave in transit between a transmitter and a receiver. Path loss may be due to many effects such as free-space loss, refraction, reflection, aperture-medium coupling loss, and absorption. Path loss is usually expressed in dB.
Phase shift	The change in phase of a periodic signal with respect to a reference.
Precipitation loss	Loss due to rain attenuating the signal, particularly in the Ku band, and further dispersing the signal as it passes through drops of water.
Radio frequency interference (RFI)	(*Synonym* of electromagnetic interference [EMI]). RFI is any electromagnetic disturbance that interrupts, obstructs, or otherwise degrades or limits the effective performance of electronics/electrical equipment. RFI can be induced intentionally, as in some forms of electronic warfare, or unintentionally, as a result of spurious emissions and responses, intermodulation products, etc.
Rain fade	Fade caused by heavy downpours of rain, related to absorption. Impacts Ku and Ka bands more so than the C-band operation.
Rain outage	Loss of signal due to absorption and increased sky-noise temperature caused by heavy rainfall.
Reflection	The abrupt change in direction of a wave front at an interface between two dissimilar media so that the wave front returns into the medium from which it originated. Reflection may be specular (i.e., mirror-like) or diffuse (i.e., not retaining the image, only the energy) according to the nature of the interface. Depending on the nature of the interface, that is, dielectric–conductor or dielectric–dielectric, the phase of the reflected wave may or may not be inverted.

(Continued)

Table 2.4 Partial list of RF-related impairments (*Continued*)

Refraction	Retardation and, in the general case, redirection of a wave front passing through (a) a boundary between two dissimilar media or (b) a medium having a refractive index that is a continuous function of position, for example, a graded-index optical fiber. For two media of different refractive indices, the angle of refraction is closely approximated by Snell's law.
Scattering	The scattering of a wave propagating in a material medium, a phenomenon in which the direction, frequency, or polarization of the wave is changed when the wave encounters discontinuities in the medium, or interacts with the material at the atomic or molecular level. *Note*: Scattering results in a disordered or random change in the incident energy distribution.
Sun outage	Sun outage is a natural phenomenon that occurs twice a year (in the spring and fall), when the sun appears to be passing directly behind the satellite, as seen from a receiving earth station. As the sun is a strong source of RF energy, the earth station's receivers may become overwhelmed by the sun's "noise" output, and reception becomes degraded and eventually impossible for a brief period, usually less than 15 min. An observer at the earth station will notice that the antenna feed's shadow will fall exactly in the center of the reflector during the peak of the sun outage period. This indicates that the antenna, the satellite, and the sun are in direct alignment. At this point in time, the sun's radio signals are being focused directly into the antenna's receive feed. This results in a temporary degradation in the signal-to-noise ratio (SNR) of the signal received from the satellite. The sun–satellite conjunction is the alignment of the sun with the satellite as seen from an earth station, which takes place twice a year for several minutes around local midday. This event can affect the performance of receiving earth stations.
System noise temperature	The equivalent noise temperature of a complete receiving system, taking into account contributions from the antenna, the receiver, and the transmission line that interconnects them, referred to the receiver input.
Thermal noise	The noise generated by thermal agitation of electrons in a conductor. The noise power P, in watts, is given by $P = kT\,\Delta f$, where k is Boltzmann's constant in joules per kelvin, T is the conductor temperature in kelvins, and f is the bandwidth in hertz. As can be seen, the magnitude of the noise generated by an object is dependent upon the object's physical temperature. Thermal noise power, per hertz, is equal throughout the frequency spectrum, depending only on k and T. It is an undesired electrical disturbance in a circuit or communication channel.

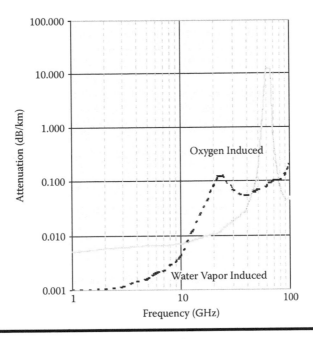

Figure 2.2 Atmospheric absorption (at 1 km altitude).

rain and snow, the link loss for the Ku band is even higher; Ku-band systems, therefore, need more conservative rain-fade margins designed into the link budget.

Uplink or downlink impairments such as rain fade can be addressed using automatic power control. Four types of power control are available: uplink power control (ULPC), end-to-end power control (EEPC), downlink power control (DLPC), and on-board beam shaping (OBBS).

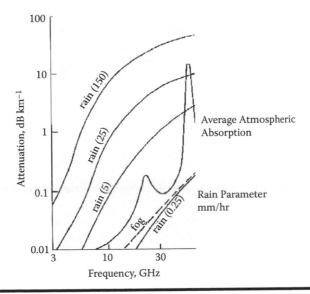

Figure 2.3 Attenuation due to rain.

Table 2.5   Rain rates in various zones and percentages of the time that the specified rate is exceeded

Percentage time the amount given is exceeded	Time (hr)	Rain rates in mm/hr													
		A	*B*	*C*	*D*	*E*	*F*	*G*	*H*	*J*	*K*	*L*	*M*	*N*	*P*
1.0%	87.60	0	1	2	3	1	2	3	2	8	2	2	4	5	12
0.3%	26.28	1	2	3	5	3	4	7	4	13	6	7	11	15	34
0.1%	8.76	2	3	5	8	6	8	12	10	20	12	15	22	35	65
0.03%	2.63	5	6	9	13	12	15	20	18	28	23	33	40	65	105
0.01%	0.86	8	12	15	19	22	28	30	32	35	42	60	63	95	145

With ULPC, the output power of a transmitting earth station is matched to uplink impairments. The transmitter power is increased to counteract fade, or decreased when more favorable propagation conditions are recovered, so as to limit interference in clear sky conditions and, therefore, optimize transmission performance. In the case of transparent payloads, ULPC can prevent reductions of satellite radiated power* caused by the decreased uplink power level that would occur in the absence of ULPC. EEPC can be used for transparent configurations only. Indeed, the output power of a transmitting earth station is matched to uplink or downlink impairments. In the case of regenerative repeaters, uplink and downlink budgets are independent; hence, the concept of EEPC is not applicable (EEPC is used to maintain a constant overall margin of the system). Here, transmitter power is increased to counteract fade or decreased when more favorable propagation conditions are recovered to limit interference and optimize satellite capacity. With DLPC, the on-board channel output power is adjusted to the magnitude of downlink attenuation. DLPC allocates a limited extra power on board to compensate a possible degradation in terms of downlink carrier power due to propagation conditions in a particular region. In this case, all earth stations in the same spot beam benefit from the improvement in radiated power. The OBBS technique is based on active antennas, which allows spot beam gains to be adapted to propagation conditions. Actually, the objective is to radiate extra power and to compensate rain attenuation only on spot beams where rain is likely to occur [CAS200701].

2.2.2 Polarization

Polarization is a transmission approach in which radio waves are restricted to certain directions of electrical and magnetic field variations; these directions are perpendicular to the direction of wave travel. Linear polarization of an electromagnetic wave is the confinement of the E-field vector or H-field vector to a given plane. By convention, the polarization of a radio wave is defined by the direction of the electric field vector. In this approach, radio waves travel

---

* Specifically, EIRP, defined in the next few sections.

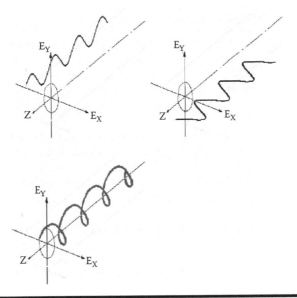

Figure 2.4   Polarization modes.

polarized (that is to say, oriented) in a vertical, horizontal, or circular fashion. (See Figure 2.4; also refer back to Figure 2.1.) The polarization of an antenna is the direction of the electric field it generates for propagation through space (or it captures from space). Polarization may be linear (vertical/horizontal), circular, or elliptical. (In the last two cases, the electric field vector is constantly changing clockwise—also known as right-hand circularly polarized—or counterclockwise—also known as left-hand circularly polarized). Some make the case that, for satellite communications, circular/elliptical polarization is better because of the impact of the upper atmosphere [DUR200501].

In cellular telephony and most commercial applications, signals are vertically polarized; as noted, in satellite applications, both linear and circular polarizations are used. For U.S. GEO satellites, linear polarization is typical; for international satellites, circular polarization is typical. Specifically, satellite systems make use of horizontal (X) linear polarization, vertical (Y) linear polarization, right-hand circular polarization (RHC-P), and left-hand circular polarization (LHC-P). Polarization must be taken into account by the (design, choice, and orientation of the) antenna. *Cross-polar* (XP) is a term used to refer to a signal that has the opposite (orthogonal) polarization to a given signal; typically, there is interest in optimizing the cross-polar discrimination (XPD) and XP isolation. Table 2.6 provides some key concepts related to polarization that the reader should master [SAT200501].

2.2.3 Basic Channel Operation

As noted, a channel is a band of radio frequencies within the overall electromagnetic spectrum that support band-limited signal propagation. The channel supports the communication link or a

Table 2.6   Key polarization concepts

Circular polarization	A circularly polarized wave in which the electric field vector, observed in any fixed plane normal to the direction of propagation, rotates with time and traces a circle in the plane of observation. Unlike linear polarization, circular polarization does not require alignment of earth station and satellite antennas with the polarization of the radio waves.
Co-polar(ized) (CP)	Of the same polarization. In contrast to XP (cross-polar).
Cross polar(ized) (XP)	Term to describe signals of the opposite polarity to another desired signal being transmitted and received. For example, cross-polarization discrimination refers to the ability of a feed to detect one polarity and reject signals having the opposite sense of polarity.
Cross-polar discrimination (XPD)	The ratio of the signal power received (or transmitted) by an antenna on one polarization (the polarization of the desired signal) to the signal power received (transmitted) on the opposite polarization. This ratio is usually expressed in decibels. XPD is a measure of the ability of the antenna to detect (emit) signals on one polarization and to reject signals at the same frequency having the opposite polarization.
Cross-polar isolation (XPI)	The ratio of the signal power received (or transmitted) by an earth station on one polarization (the desired signal) to the signal power received (transmitted) on the same polarization but originating from an XP signal. This ratio is usually expressed in decibels. XPI is a measure of interference from XP signals into the desired signal that occurs in all practical systems that exploit both orthogonal polarization. The terms *cross-polar isolation and cross-polar discrimination* have different meanings but are often used interchangeably.
Frequency reuse	A technique for utilizing a specified range of frequencies more than once within the same satellite system so that the total capacity of the system is increased without increasing its allocated bandwidth. Frequency reuse schemes require sufficient isolation between the signals that use the same frequencies so that mutual interference between them is controlled to an acceptable level. It is achieved by using orthogonal polarization states (X/Y for linear, or LHC/RHC for circular) for transmission and/or by using satellite antenna (spot) beams that serve separate, nonoverlapping geographic regions.
Horizontal polarization	Type of linear polarization where the electric field is approximately aligned with the local horizontal plane at an on-ground transmission or reception point.
Left-hand circularly polarized (LHC(P)) wave	An elliptically or circularly polarized wave in which the electric field vector, observed in any fixed plane normal to the direction of propagation, while looking in the direction of propagation, rotates with time in a left-hand or anticlockwise direction.
Linear polarization	Propagation of a wave in which the electric field vector, observed in any fixed plane normal to the direction of propagation, maintains a constant direction with time. With linear polarization, the earth station and satellite antennas of a particular earth–space link must be precisely aligned so that their reference polarization directions coincide to obtain maximum reception quality.

Table 2.6 Key polarization concepts (*Continued*)

Ortho Mode Transducer (OMT)	Physical element that is directly behind the feed horn and supports functions relating to reception and transmission of satellite signals. The main function of the OMT is to transfer RF to individual ports (e.g., transmit and receive) and provide an isolation between them; for example, to isolate 90° orthogonal (vertical and horizontal) signals.
Polarization	The approach in which radio waves are restricted to certain directions of electrical and magnetic field variations, where these directions are perpendicular to the direction of wave travel. By convention, the polarization of a radio wave is defined by the direction of the electric field vector. Four senses of polarization are used in satellite transmissions: horizontal (X or H) linear polarization, vertical (Y or V) linear polarization, right-hand circular polarization, and left-hand circular polarization.
Polarization alignment	The process of aligning the reference polarization plane of a linearly polarized antenna with a particular reference direction. For individual and collective systems receiving linearly polarized signals, this consists of rotating the LNB about the feed axis so that its radio wave detector is aligned with the electric field vector of the incoming signal (to achieve detected signal strength).
Polarization switching	The process of selecting one of two orthogonal polarizations (e.g., X or Y) for reception of satellite signals. Polarization switching is implemented in the LNB or, more rarely, in a separate device inserted between the feed horn and the LNA/LNB or integrated with the feed horn.
Right-hand circularly polarized (RHC(P)) wave	An elliptically or circularly polarized wave in which the electric field vector, observed in any fixed plane normal to the direction of propagation, while looking in the direction of propagation, rotates with time in a right-hand or clockwise direction.
Vertical polarization	Type of linear polarization where the electric field is approximately aligned with the local vertical plane at an on-ground transmission or reception point. See also frequency reuse.

path for an electrical signal. In general, the channel bandwidth is less than the transponder bandwidth (specifically, 36 MHz versus 40 MHz; the guard band of 4 MHz is used to assure that the transponders do not interact with each other and be plagued by intermodulation). See Figure 2.5 for the general concept and Figure 2.6 for an actual example.

As noted earlier, intermodulation is the cointerference between signals in adjoining frequency bands (channels) after nonlinear amplification by a common amplifier. Nonlinearities are not purposely built into an amplifier, but they are an intrinsic limitation of the amplifier as it is driven into saturation, namely, its maximum operating point. Saturation is the operation of a power amplifier, most often a satellite Traveling Wave Tube Amplifier (TWTA), at its maximum output power level ("saturated" power level). (See Figure 2.7.) In satellite communication systems, intermodulation usually occurs at the earth station's HPA or at a satellite transponder. It is controlled by means of input back-off/output back-off (IBO/OBO)

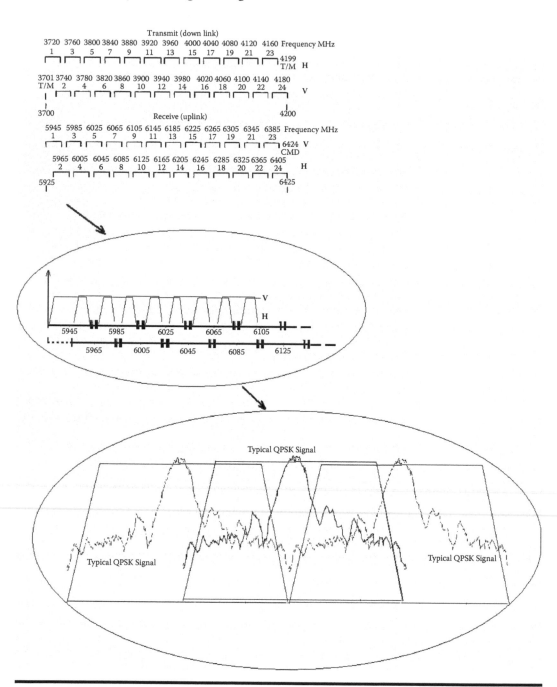

Figure 2.5   Example of RF channel.

of the amplifier. IBO is the ratio of the signal power measured at the input to an HPA to the input signal power that produces the maximum signal power at the amplifier's output. The input back-off is expressed in dB; IBO can be positive or negative. It can be applied to a single carrier at the input to the HPA (carrier IBO) or to the ensemble of input signals (total IBO). OBO is the ratio of the signal power measured at the output of an HPA to the

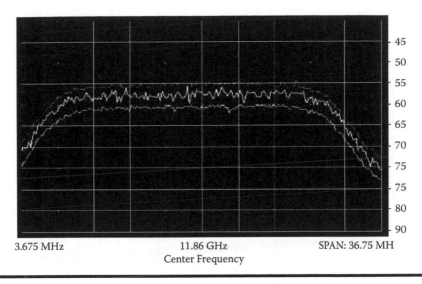

Figure 2.6   Actual plot of signal within a transponder channel.

maximum output signal power. The output back-off is expressed in dB; OBO can be positive or negative. It can be applied to a single carrier at the output to the HPA (carrier OBO) or to the ensemble of output signals (total OBO). See Appendix 3.A for additional discussion of the topic.

Focusing on the transmission channels and considering the interference and impairment issues discussed, the following values are critical to the reliable and effective operation of the transmission links [SAT200501]. (Also see Figure 2.8.):

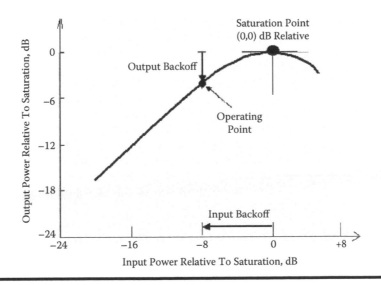

Figure 2.7   Input back off/output back off (IBO/OBO).

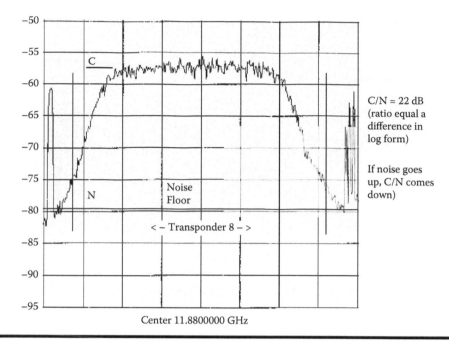

Figure 2.8   Example of carrier-to-noise ratio.

C/I (Carrier-to-interference ratio)	A measure of the quality of a signal at the receiver input. It is the ratio of the power of the carrier to the power of interference arising from manmade sources measured within a specified bandwidth (usually the modulated carrier's bandwidth). It is typically expressed in dB. The higher the ratio, the better the quality of the received signal.
C/N (Carrier-to-noise ratio)	An important measure of the quality of a modulated carrier at the receiver input. It is the ratio of the power of the carrier to the power of the noise introduced in the transmission medium, measured within a specified bandwidth (usually the modulated carrier's bandwidth). It is typically expressed in dB. The higher the ratio, the better the quality of the received carrier.
C/(N + I) (Carrier-to-noise-plus-interference ratio)	A measure of the quality of a signal at the receiver input. It is the ratio of the power of the carrier to the combined power of noise and manmade interference, measured within a specified bandwidth (usually the modulated carrier's bandwidth). It is typically expressed in dB. The higher the ratio, the better the quality of the received signal.
S/N (Signal-to-noise ratio, aka SNR)	A measure of the quality of an electrical signal, usually at the receiver output. It is the ratio of the signal level to the noise level, measured within a specified bandwidth (typically, the bandwidth of the signal). It is typically expressed in dB. The higher the ratio, the better the quality of the signal.
E_b/N_0 (Bit energy-to-noise ratio)	The ratio of bit energy per symbol to noise power spectral density, in decibels. The measure is utilized in digital modulation/transmission environments. 0 dB means the signal and noise power levels are equal, and a 3 dB increment doubles the signal relative to the noise.

References

[ANS200001] ANS T1.523-2001, Telecom Glossary 2000, American National Standard (ANS), An Outgrowth of the Federal Standard 1037 Series, *Glossary of Telecommunication Terms*, 1996.

[BLA200701] K. Blattenberger, Free Space Loss, RF Café Website, Mt. Airy, NC.

[CAS200701] L. Castanet, A. Bolea-Alamanac, M.Bousquet, "Interference And Fade Mitigation Techniques For KA And Q/V Band Satellite Communications Systems", ONERA—ElectroMagnetics and Radar Department 2 avenue E. Belin—BP 4025, 31055 Toulouse CEDEX 4, France; TESA—Laboratoire Coopératif des Télécommunications Spatiales et Aéronautiques 2, rue Charles Camichel, BP 7122, 31071 Toulouse CEDEX 7, France; SUPAERO—Aerospace Electronics and Communications Programmes Toulouse, France.

[COL200201] P. H. Cole, "A Study of Factors Affecting the Design of EPC Antennas and Readers for Supermarket Shelves," White Paper, Auto-id Centre University of Adelaide, Adelaide, SA 5005, Australia, 2002.

[DUR200501] G. J. Durnan, "Parasitic Feed Elements for Reflector Antennas," PHD Dissertation, School Microelectronic Engineering, Griffith University, Brisbane, Australia, April 2005.

[JEF199901] D. Jefferies, "Electromagnetics and Antenna," MSc in Satcoms Notes, University Of Surrey, Department of Electronic Engineering, School of Electronics and Physical Sciences, Guildford, Surrey, UK, *GU2 7XH*, 1999.

[SAT200501] Satellite Internet Inc., Satellite Physical Units and Definitions, http://www.satellite-internet.ro/satellite-internet-terminology-definitions.htm.

[USA199801] U.S. Army Information Systems Engineering Command, Fort Huachuca, Arizona, "Automated Information Systems, Design Guidance, Commercial Satellite Transmission," August 1998.

[ZYR199801] J. Zyren, A. Petrick, "Tutorial on Basic Link Budget Analysis," Application Note AN9804.1, June 1998, Intersil Corporation, Milpitas, CA.

Appendix 2A: Maxwell Equations

This section provides a brief overview of Maxwell's Equations (not the easiest topic to discuss) and is based principally on reference [JEF199901]. The basic terms were defined in Section 2.2.

All magnetic fields B form closed loops. Loops of B are linked with loops of current density J or displacement current density $(\delta/\delta t)D$. Electric field lines either begin or end on charges, or else they also form closed loops. In a radiating situation, the currents, charges, fields, and potentials are all time varying. If one assumes a sinusoidal variation with time having angular frequency ω, then a general radiation problem can be solved by Fourier superposition of the solutions at different values of the angular frequency ω. There are four of Maxwell's Equations plus a charge continuity equation that define electromagnetic theory:

- $div\, D = \rho$
- $div\, B = 0$
- $curl\, E = -(\delta/\delta t)B$
- $curl\, H = (\delta/\delta t)D + J$

and the continuity equation for charge:

- $div\, J + (\delta/\delta t)\rho = 0$

These are written formally as follows:

$$\nabla \cdot D = \rho$$

$$\nabla \cdot B = 0$$

$$\nabla \times E = -\frac{\partial B}{\partial t}$$

$$\nabla \times H = +\frac{\partial D}{\partial t}$$

$$\nabla \cdot J + \frac{\partial D}{\partial t} = 0$$

where the divergence (*div*), curl (*curl*), and gradient (*grad*) are defined in vector differential calculus. In broad terms, the gradient represents the "slope" of a scalar field along the direction of maximum change; the gradient is a vector. The divergence represents the flow out of a small volume per unit volume, and is a scalar. The curl represents the rotation of a field around a point; for a magnetic field forming closed loops, it is the limit of the size of the field multiplied by the perimeter of the loop divided by the area of the loop, as the loop shrinks to nothing. The curl is a vector because it has an associated axis of circulation or direction in space.

A high-level explanation of these equations is as follows:

■ Lines of D, electric induction, are proportional to the electric field and "diverge" away from a region containing charge density ρ. If there is a surface charge σ (coulombs per square meter), then close to the charge sheet there is an electric induction field $D = \sigma$.
■ Lines of B never diverge from anything, and form closed loops.
■ Electric field lines that form closed loops encircle a changing magnetic field. Lenz's law applies; the electric field, if it drove a current, would do so in such a way as to reduce the changing magnetic field within the loop. Electric field lines that do not form closed loops begin and end on charge, as seen from the first equation.
■ Magnetic field lines H form loops that encircle both conduction current density J and also "displacement current density" $(\delta/\delta t)D$, which is generated by time-varying electric fields. Maxwell's achievement was to realize that the term in $(\delta/\delta t)D$ was necessary; if one considers a capacitor with plates very close together, then, if the displacement current term did not generate magnetic field loops, there would be a discontinuity in the magnetic fields around the capacitor plates as one passed alternating current through the capacitor.
■ The current density J flowing out of a region ("diverging") must result in a decrease of charge within the region.

Currents and charges give rise to the fields and are called *sources*. More directly, the potentials can be calculated from the source charge and current distributions, and the fields are then derived from the potentials. In an electrostatics situation, the electric field E is given by

$$E = -grad(\phi), \text{ which begins and ends on charges.}$$

However, if there are changing magnetic fields, there is an additional contribution to the electric field forming the closed loops that circulate around the changing magnetic field lines.

The magnetic vector potential A may be used to find the magnetic field B by the relation (which is a definition of A).

$$B = curl(A)$$

To define A completely, one has to specify its divergence as well as its *curl*, and possibly an additive constant also. If one does this according to what is known as the "Lorentz Gauge," then the electric field may be calculated from

$$E = -(\delta/\delta t)A - grad(\phi)$$

Part of the source of the electric field is from the magnetic vector potential A, and part from the scalar potential ϕ. If one knows the potentials A and ϕ completely for all time and space, one can calculate the fields E and B.

A little more detailed mathematics shows that the conduction currents J give rise to the magnetic vector potential A, and the source charges ρ give rise to the scalar potential ϕ. Because there is a maximum velocity of propagation $c = 3 \times 10^8$ m/s, the potentials A and ϕ at a distance r meters from the source cannot follow changes in the source distributions until a time r/c seconds later. Considering the equation

$$E = -grad(\phi) - (\delta/\delta t)A$$

one observes that, in the far field, the potentials A and ϕ fall off as $1/r$, where r is the distance from the sources. However, applying the gradient operator to ϕ puts in a further dependence of $1/r$ on the contribution $-grad(\phi)$. Thus, the electric field E, due to the charges in the source, falls off as $1/(r^2)$ and can be neglected at large r compared to the electric field contribution $-(\delta/\delta t)A$, which falls off as $1/r$. Hence, for far-field calculations, it is true to say that only the source currents on the antenna structure need be considered.

For time-harmonic currents, because the charge continuity equation links the current density J to the source charge density ρ, the potential ϕ may be expressed in terms of the vector potential A, and therefore, there is no loss of generality in considering the far fields as being entirely due to source currents plus any preexisting electromagnetic propagating waves.

For near-field scenarios, the conduction currents on the source structures are not sufficient to use as a basis for field calculations. Consider an open-ended waveguide. Assume that the waveguide is very large in transverse dimensions compared to the wavelength. Now consider a point on the axis of the waveguide, beyond the plane at which the waveguide stops, along the z-direction. Assume that this point is closer to the point $z = 0$, which defines the exit plane, than it is to any of the current elements on the waveguide walls. If only conduction currents starting at time zero contribute to the field strength at this point, there can be no field at this observation point at a time less than the retardation time from the guide walls. Now, as one lets the guide dimensions get larger (without limit), one can show that, for any specific point on the axis of the waveguide beyond the waveguide end plane, the fields *due to the currents in the walls* are zero for finite time. If one takes the view that the waveguide is the only structure generating electromagnetic fields and that there are no preexisting propagating waves along the axis of the guide, then this result appears contradictory and nonphysical, and so the accepted theory, and nearly all the antenna calculations and simulation codes based on fields being set up *only* by the source currents, may be in error. Another way of looking at this problem is that the wave front progresses along the waveguide at

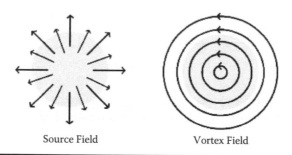

Source Field   Vortex Field

Figure 2.A1   Source and Vortex Fields.

the group velocity, which is lower than the velocity of light in the medium. Regarded as a radiating structure, the mouth of the waveguide sets off a propagating wave in free space that travels at the velocity of light. There must, therefore, be a contribution to the radiated fields from the center of the guide mouth, where the conduction currents are zero. Yet another insight may be obtained by appealing to Huygen's principle in wave optics, where each point on a propagating wave front is regarded as giving rise to an outgoing hemispherical wave front. Thus, if one considers radiation from a burst of microwaves propagating in free space at time zero, there are no source currents in the problem at all within the limits of the definition of the problem.

Most antenna calculations are made assuming that the time-harmonic radiation has persisted/ persists for all times past and future, and so the problem of what happens at the start of a radiating wave front is hidden from the analysis.

Further examination of the equations shows that A, and therefore E (in the far field), lies in the preferred direction of the current sources. (The preferred direction may be taken to be an average direction over all the source currents.) The magnetic field B, on the other hand, forms loops around the current direction, and therefore B is at right angles to E and to the preferred direction of the current sources.

Moving charge constitutes a current. In most wires, there is a near balance between mobile negative charge (electrons) and a background sea of positive charge (ions), which is stationary. It is therefore possible to have an "electrically neutral current" wherein the moving charge forming the current does not by itself provide a source of charge density ρ and, therefore, of electrostatic scalar potential ϕ. For this reason, in many antenna analyses, near-field as well as far-field radiation is assumed to be entirely determined if only the current distribution in the source is known. This is sufficient for many antenna problems. However, where there are significant discrepancies between the predictions of standard theory and the measurements on a specific antenna structure, one should look to see if there are any significant time-varying charge accumulations within the antenna conductor structure or on any local scattering objects. One should also look to see if there are any photons emitted by transitions between electron energy levels and if there are any preexisting electromagnetic waves.

Although the laws of electrodynamics are defined by Maxwell, they are most readily understood in terms of the source and vortex interpretation of Helmholtz and the field pictures of Faraday. Helmholtz has shown that vector fields may be regarded as the superposition of two different basic field types, known as source type and vortex type (see Figure 2.A1). In the illustration of a purely source field, one finds field lines of electric field originating in a region of positive charge. If these field lines terminate somewhere, it is in a region of negative charge; these source type field lines, however, never intersect nor close upon themselves. In the illustration of a purely vortex field, one finds field lines of magnetic field surrounding a wire carrying a current. These vortex field lines are always in the form of closed curves, and never have a starting point or an ending point. What one learns about electric fields from Maxwell's equations is that the electric field E can be either source type or vortex type, or a mixture of both. The regions of space providing sources of E can be charges on conductors, and those providing vortices of E are regions where there is a time-varying magnetic flux density. What one learns about magnetic fields from Maxwell's equations is that the magnetic field H can, in the absence of magnetic media, be only the vortex type. The regions of space providing vortices of H are those where there is neither an electric current nor a time-varying electric flux density [COL200201]. When electromagnetic fields propagate to a significant distance from their originating antenna, it is the property of time-varying electric fields to create surrounding vortices of the magnetic field and time-varying magnetic fields to create surrounding vortices of the electric field that may be regarded as providing the mechanism for the propagation phenomenon.

Chapter 3

Antenna Engineering Basics

This chapter provides a basic engineering overview of satellite antennas. The performance of a satellite link depends in large measure on the antenna system.

3.1 Antenna Operation

The antenna is an electrical conductor (or conductors) that radiates electromagnetic energy into space (transmission) and/or collects it from space (reception). Antennas are transducers that transfer electromagnetic energy between a transmission line and free space. A transmitting antenna behaves like an equivalent impedance that dissipates the power transmitted; the transmitter is equivalent to a generator. A receiving antenna behaves like a generator with an internal impedance corresponding to the antenna-equivalent impedance. The receiver represents the load impedance that dissipates the time averaged power generated by the receiving antenna [AMA200101]. See Figure 3.1.

Antennas are reciprocal devices. It means that they can be used both as transmitting and receiving elements; in two-way communication, the same antenna can be used for both transmission and reception (receiver protection may be used in some cases). For example, this is how the antennas on cellular phones operate. The reciprocity principle affirms that the transmitting and receiving patterns of an antenna are identical at a specified wavelength; there is no functional difference between receive and transmit modes of an antenna except that the power flow is directed inwards to the receive antenna and outwards from the transmit antenna. An antenna has the same efficiency, directivity, and polarization characteristics in receive and transmit modes. This property is called *reciprocity,* and it occurs due to the symmetry of the electromagnetic equations when the direction of time is reversed [JEF200401]. The basic characteristics of an antenna are its gain and its half-power beamwidth (HPBW), both of which are described in detail later. Some common physical antenna types include the following:

- Linear dipole fed by a two-wire line
- Linear monopole fed by a single wire over a ground plane
- Coaxial ground-plane antenna
- Loop antenna
- Uda-Yagi (or Yagi-Uda) dipole array

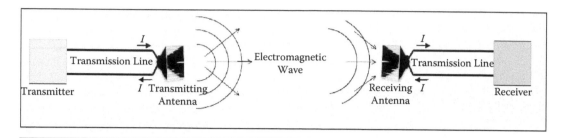

Figure 3.1 Antenna operation.

- Multiple loop antenna wound around a ferrite core
- Log-periodic array
- Parabolic (dish) antenna
- Toroidal antenna (multiple feed), including simulsat types

An antenna leaks electromagnetic energy into the surrounding space. For example, in a dipole antenna, because of the change in geometry in the transmission line, there is an abrupt change in the characteristic impedance at the transition point where the current is still continuous; hence, the dipole leaks electromagnetic energy into the surrounding space (and so, it reflects less power than the original open circuit.) The space surrounding the dipole has an electric field. At higher frequency, the current oscillates in the wires, and the field emanating from the dipole changes periodically; the field lines propagate away from the dipole and form closed loops [AMA200101]. See Figure 3.2. Note that an ideal (hypothetical) isotropic antenna radiates equally in all directions.

The aperture is the area that captures energy from the electromagnetic (radio) field that surrounds the antenna. Formally, in a directional antenna, it is the portion of a plane surface very near the antenna normal to the direction of maximum radiant intensity through which the major part of the radiation passes [ANS200001]. Examples of apertures include horns, reflectors, and radiating slots [LEE198801]. For a reflector (dish) antenna, the aperture is the size of the reflector; for a horn, it is the area of the "mouth" of the horn (aperture for other types of antennas are not described here). The larger the aperture, the stronger the signal the antenna receives or transmits. At microwave frequencies, the behavior of reflector antennas can be approximated by optical reflectors. See Table 3.1 for some basic antenna concepts [SAT200501], [EUT200701], [ANS200001].

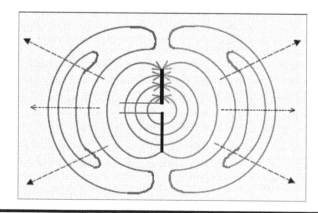

Figure 3.2 Example of radiation pattern.

Table 3.1   Basic Antenna Concepts

Adjacent satellite interference (ASI)	For a ground station, extraneous power from a signal in an adjacent satellite operating at the same band; it may be because the antenna is too small and is not able to focus properly on the specific satellite of interest or because the antenna is mispointed [ANS200001].
Antenna	A device for transmitting or receiving radio waves. In satellite communication systems, the antenna usually consists of a parabolic reflector and a feed horn (feed). In a receiving system, the reflector focuses radio waves onto the feed horn for detection and conversion into electrical signals. In a transmitting system, the reflector concentrates the radio waves emitted by the feed horn into a narrow beam aimed toward the satellite.
Antenna alignment	The proper positioning and pointing of the antenna so that it can receive the signal only from the intended satellite (thereby minimizing the ASI). The antenna uses a three-axis pointing system that one needs to adjust when pointing it to a satellite: azimuth (the magnetic compass direction at which one points the dish), elevation (the angle above the horizon at which one points the dish), and polarization (the horizontal/vertical rotation to align with the signal's polarization).
Antenna aperture	The total reflective area of an antenna over which radio waves are radiated or captured.
Antenna efficiency	The ratio (typically expressed as a percentage) of the signal strength transmitted toward (or received from) a particular direction in space by a real antenna to the signal strength that would be obtained with a theoretical reference antenna of the same size.
Antenna gain	A measure of the "amplifying" or focusing power of an antenna when transmitting to, or receiving from, a particular direction in space. The gain of an antenna is the ratio of the power radiated (or received) per unit solid angle by the antenna in a given direction to the power radiated (or received) per unit solid angle by an isotropic antenna fed with the same power. The gain is usually expressed in dBi.
Antenna illumination	On the transmit side, the radiation of electromagnetic energy from the feed horn (feed) to the surface of the parabolic reflector of a transmit antenna; on the receive side, the focusing of electromagnetic energy captured by the reflector of a receiving antenna toward the feed horn. With perfect illumination, no signal energy is lost to the surrounding terrain; in practice, there are losses that have to be taken into account.

(Continued)

Table 3.1 Basic Antenna Concepts (*Continued*)

Antenna noise temperature	The temperature of a hypothetical resistor at the input of an ideal noise-free receiver that would generate the same output noise power per unit bandwidth as that at the antenna output at a specified frequency. The antenna noise temperature depends on antenna coupling to all noise sources in its environment as well as on noise generated within the antenna. It is a measure of noise whose value is equal to the actual temperature of a passive device [ANS200001].
Beam	A unidirectional flow of radio waves concentrated in a particular direction. A term commonly used to refer to an antenna's radiation pattern by analogy with a light beam. It is most often used to describe the radiation pattern of satellite antennas. The intersection of a satellite beam with the earth's surface is referred to as the (beam's) footprint.
Beamwidth	A measure of the ability of an antenna to focus the signal energy toward a particular direction in space (e.g., toward the satellite for a ground-based transmitting antenna) or to collect it from a particular direction in space (e.g., from the satellite for a ground-based receiving antenna). The beamwidth is measured in a plane containing the direction of maximum signal strength. It is usually expressed as the angular separation between the two directions in which the signal strength is reduced to one-half of the maximum value (the −3 dB half-power points).
Boresight	The direction of maximum antenna gain. For a receiving antenna, the boresight is aligned with the satellite as accurately as possible for maximum received signal strength.
Carrier-to-receiver noise density (C/kT)	The ratio of the received carrier power to the receiver noise power density. The carrier-to-receiver noise density ratio is usually expressed in dB. C is the received carrier power in watts, k is Boltzmann's constant in joules per kelvin, and T is the receiver system noise temperature in kelvins. The receiver noise power density kT is the receiver noise power per hertz.
Dual-feed antenna	An antenna system consisting of a reflector, a support structure, and two LNBs, each equipped with a separate feed horn (feed) or sharing an integrated feed assembly. The focal point of each feed is set so that the antenna system can receive signals from two different satellite orbital positions simultaneously. The angular separation between satellite positions is usually around 2°–4°, although other angles are possible.

Table 3.1   Basic Antenna Concepts (*Continued*)

Effective isotropic radiated power (EIRP)	A measure of the signal strength that a satellite transmits toward the earth, or an earth station toward a satellite, expressed in dBW. Also, the arithmetic product of (a) the power supplied to an antenna, and (b) its gain [ANS200001].
Elevation	The angle measured in the local vertical plane between the satellite and the local horizon. It is the vertical coordinate that is used to align a satellite antenna.
F/D ratio	The ratio of an antenna's focal length to its diameter. It describes the basic geometric architecture of the antenna, which affects its physical size, design, and electrical performance. F = focal length; D = diameter.
Feed horn (feed)	A device resembling a horn that emits radio waves in a concentrated beam or collects and focuses radio waves that are incident on its aperture. In a receiving system, it collects microwave signals reflected from the surface of the antenna. In a transmitting system, it directs microwave signals onto the reflector surface for focusing into a narrow beam aimed at the satellite. The feed is mounted at the focus of the parabolic reflector. It is usually designed to match a particular antenna geometry (F/D ratio).
Focal length	The distance F from the reflective surface of an antenna to its focal point, usually measured in the horizontal plane. Incoming satellite signals are directed to the feed horn (feed), which is normally located at the focal point.
G/T Figure of merit	The ratio of the maximum gain G of a receiving antenna to the receiving system's equivalent noise temperature T. This value is usually expressed in dB/K. It is a measure of the ability of an earth station to receive a satellite signal with good quality (high carrier-to-noise ratio). In general, the G/T increases with increasing antenna diameter.
Isotropic antenna	A theoretical device that radiates energy or receives energy equally from all directions.
Multibeam	Refers to the use of multiple antenna beams on board the satellite to cover a contiguous geographical area, instead of a single wide-area beam. Multibeam architectures are often utilized in satellites operating in the Ka-band, which is characterized by narrower beamwidths with respect to the Ku-band. Single, wide-area beams predominate in the latter. This term also applies to toroidal antennas, as well as parabolic antennas that have multiple beam feed assemblies built into their structure. The former typically can accommodate as many as 35 feeds, spanning nearly 70 degrees of the orbital arc (35 satellites spaced at 2 degree intervals), while the latter typically can accommodate as many as 5 feeds, spanning up to 8 degrees of the orbital arc (5 satellites spaced at 2 degree intervals).

(Continued)

Table 3.1   Basic Antenna Concepts (*Continued*)

Off-axis	Any direction in space that does not correspond to an antenna's boresight direction.
Offset (fed) antenna	An antenna having a feed horn (feed) that is offset from the center of the reflector. It generally offers better performance than a symmetrically fed antenna because the feed system does not block the main reflector aperture.
Offset feed	An LNB that is slightly displaced with respect to the focal point of the reflector so that it receives signals originating from a different direction to that obtained with an LNB placed at the focal point. A technique used in dual-feed reception systems, which receive signals from satellites located at two different orbital locations.
Parabolic antenna	An antenna having a main reflector surface that is a paraboloid or is shaped like a paraboloid. It has the property of reflecting parallel incoming signals to a single focal point. A paraboloid is a geometric surface whose sections parallel to two coordinate planes are parabolic (parabolic: a geometric shape formed by the intersection of a cone by a plane parallel to its side), and whose sections parallel to the third plane are either elliptical or hyperbolic.
Pointing angles	The elevation and azimuth angles that specify the direction of a satellite from a point on the earth's surface.
Pointing error (antenna)	A value that quantifies the amount by which an antenna is misaligned with the satellite's position in space. This is either expressed as an angular error or as a loss in signal strength with respect to the maximum that would be achieved with a perfectly aligned antenna.
Polar mount	A mechanical support structure for an earth station antenna that permits all satellites in the geosynchronous arc to be scanned with the movement of only one axis.
Radiation pattern	A three-dimensional representation of the gain of a transmit or receive antenna as a function of the direction or radiation or reception.
Receiver noise temperature	The equivalent noise temperature of a complete receiving system, excluding contributions from the antenna and the physical connection to the antenna, referred to the receiver input.
Shaped beam	The radiation pattern of a satellite antenna that has been designed so that its footprint follows the boundary of a specified geographical area (the area of service provision) as closely as possible. Shaped beams maximize the antenna gain over the service area and reduce the likelihood of interference into systems serving other geographical areas.

Table 3.1   Basic Antenna Concepts (*Continued*)

Side lobe	Part of an antenna's radiation pattern that can detect or radiate signals in an unwanted direction (i.e., off-axis) that can produce interference into other systems or susceptibility to interference from other systems. The larger the side lobes, the more noise and interference an antenna can detect. Side-lobe levels are determined by the design of the antenna. In addition to determining the size of side lobes, antenna size also determines the angular displacement of the side lobes from the main lobe.
Spot beam	An antenna radiation pattern designed to serve a relatively small or isolated geographic area, usually with high gain.
Steerable beam	An antenna beam that can be repointed by mechanical and/or electrical means. Usually used to refer to relatively narrow satellite beams that can be steered over a part or the whole of the portion of the earth's surface that is visible from the satellite's orbital position.
Tracking	The process of continuously adjusting the orientation of an antenna so that its boresight follows the movements of the satellite about its nominal position. Used in earth stations equipped with large antennas and earth stations operating with satellites in an inclined orbit, LEOs, MEOs, and polar orbits.

From a functionality perspective, the following antenna types are common, among others [SEY200401]:

■ (Ideal) isotropic antenna (radiating equally in all directions; not available in practice and/or not useful for satellite communications)
■ Dipole antennas (e.g., half-wave dipole antenna, also known as Hertz antenna; quarter-wave vertical antenna, also known as monopole antenna)
■ Aperture antennas (e.g., parabolic reflective antenna, horn antenna, Cassegrain antenna, and lens antenna.) For commercial satellite applications, parabolic (dish) antennas are used almost exclusively; furthermore, prime-focus systems are typical, at least for the reception portion.
■ Directive-beam antenna
■ Phase-array antenna

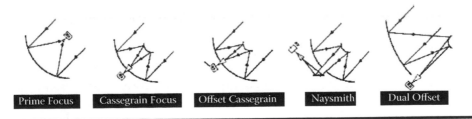

Figure 3.3   Reflective-type antenna (partial list).

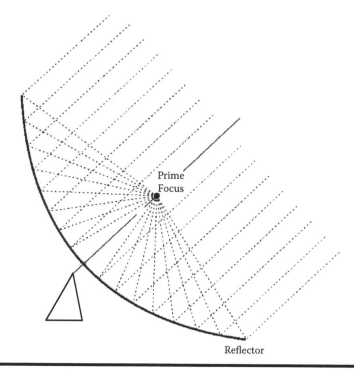

Figure 3.4   Prime focus antenna.

As noted, the most-widely used narrow-beam antennas in satellite communications are reflector antennas in which the shape is generally a paraboloid of revolution. In a parabolic antenna, when a source (i.e., the feed/feed horn) is placed at the focal point, the signal rays that reflect from the reflecting surface form a parallel beam. In a perfect antenna, the beam is purely directional; in real antennas, there will also be a side-lobe radiation. See Figure 3.3. In a satellite link, there are at least three antennas involved: (i) the uplink antenna provided by the end user for an application where the end user is transmitting in a private network or provided by the service provider at an earth station; (ii) the receiving/transmitting antenna on the satellite; and, (iii) the receiving antenna. For a given application, an engineer may need to determine the correct size of a field antenna (but the size of the space-borne antenna is established at satellite-manufacturing time.) For full-earth coverage from a geostationary satellite, the angular diameter of the earth is 17.4°, and a horn antenna is employed. Horns are also used as feeds for reflector antennas.

In a direct-feed reflector, such as on a satellite or a small earth terminal, the feed horn is located at the focus, or may be offset to one side of the focus. See Figure 3.4. Large earth station antennas have a subreflector at the focus. In the Cassegrain design, the subreflector is convex with a hyperboloidal surface, whereas in the Gregorian design it is concave with an ellipsoidal surface (in Gregorian designs the required subreflector is slightly larger than would be the case with a Cassegrain antenna). The subreflector allows the antenna feed to be located near the base of the antenna. It reduces losses because the length of the waveguide between the transmitter or receiver and the antenna feed is reduced; the system noise temperature is also reduced because the receiver looks at the cold sky instead of the warm earth. In addition, mechanical stability is improved, resulting in higher pointing accuracy [NEL200701]. See Figures 3.5 and 3.6.

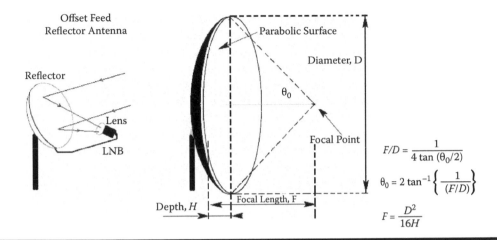

$$F/D = \frac{1}{4\tan(\theta_0/2)}$$

$$\theta_0 = 2\tan^{-1}\left\{\frac{1}{(F/D)}\right\}$$

$$F = \frac{D^2}{16H}$$

Figure 3.5   Typical satellite antenna (e.g., receive only).

As seen in these figures, with a prime-focus dish the feed is at the focal point, out in front of the parabolic reflector, so either a lossy feedline is necessary or part of the equipment needs to be placed near the feed; this design is reasonable for receiving systems because low-noise amplifiers are compact. On the other hand, high-power transmitters tend to be large and heavy; advantageously, the Cassegrain antenna allows the feed to be placed at a more convenient point, near the

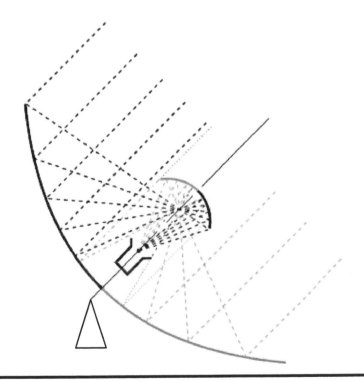

Figure 3.6   Gregorian antenna.

vertex (the center of the parabolic reflector) with the feedline coming through the center of the dish. The main advantage of the Cassegrain antenna is that the secondary reflector may be used to reshape the illumination of the reflector for better performance. A symmetric reflector that is a paraboloid is completely defined by two parameters: the focal length F and the diameter D. (The depth of the dish H is a derived parameter, where $H = D^2/16F$.) This is the reason why commercial antennas are typically described in terms of D and F/D. Typical F/D values range from 0.3 to 1.0 (as the ratio gets larger, the dish becomes flatter). Reshaping the illumination is particularly significant for very deep dishes because there are no good feeds for F/D less than ~0.3, whereas there are very efficient feeds for shallow dishes with $F/D > 0.6$. By proper shaping of the subreflector, one can use a feed to efficiently illuminate a deep dish [WAD200401].

Phased-array antennas may be used to produce multiple beams or for electronic steering; they are found on many nongeostationary satellites. Phased arrays can also be used on the ground, but due to their cost they are usually employed (only) in military systems.

3.2 Antenna Gain

As already noted, the gain of an antenna is a measure of how much of the input power is concentrated in a particular direction. It is expressed with respect to an isotropic antenna. The gain (also called the *directivity*) of the antenna is the ratio of the radiation intensity in a given direction to the radiation intensity averaged over all directions. By way of definition, the boresight is the direction of the axis of a directional antenna. The required gain of an antenna at a given wavelength is achieved by appropriately selecting the size of the antenna, as seen next. Often directivity and gain are used interchangeably (the technical difference is that directivity neglects antenna losses such as dielectric, resistance, and polarization; because these losses in most classes of antennas are normally small, they are approximately equal).

Assuming that the antenna pattern is uniform, the gain is equal to the area of the isotropic sphere divided by the sector (cross-section) area (the gain is independent of actual power output and distance at which measurements are taken).

$$G = \frac{Area\ of\ Sphere}{Area\ of\ Antenna\ pattern}$$

When the angle in which the radiation is constrained is reduced, the gain goes up. For example, using an isotropic radiating source, the gain is 0 dB (by definition). If the spatial angle is decreased to one hemisphere, the power radiated, P_{in}, would be the same but the area would be half as much, so the gain would double to 3 dB. If the angle is a quarter sphere, the gain is 6 dB (see Figure 3.7).

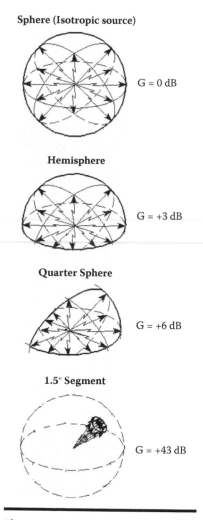

Sphere (Isotropic source)

G = 0 dB

Hemisphere

G = +3 dB

Quarter Sphere

G = +6 dB

1.5° Segment

G = +43 dB

Figure 3.7 Gain as a function of the cross-section area.

More specifically, the antenna gain in direction (θ, ϕ) is given by the differential equation [NEL200701]:

$$G(\theta, \phi) = (dP/d\Omega)/(P_{in}/4\pi)$$

where dP is the increment of the radiated output power in a solid angle $d\Omega$, and P_{in} is the total input power. The total output power can be shown to be $P_{in} = E_a^2 A/\eta Z_0$, where E_a is the average electric field over the area A of the aperture, Z_0 is the impedance of free space, and η is the net antenna efficiency. The output power over the solid angle $d\Omega$ is $dP = E^2 r^2 d\Omega/Z_0$, where E is the electric field at distance r. An isotropic antenna has a solid angle $\Omega = 4\pi$. The gain of the antenna is defined as the ratio of the peak radiation intensity to average radiation intensity:

$$D = 4\pi/\Omega$$

and therefore, a highly directive antenna has a small Ω [GAR200401].

Antenna efficiency is of interest. There is a net antenna efficiency factor already mentioned η and there are other efficiency factors, notably the "aperture efficiency" η_a. Let A be the physical area of the antenna and A_0 be the collecting area when the antenna is pointed directly at a source. A_0 may not be the same as the physical area of the aperture. The ratio of these two terms can be used to define an aperture efficiency $\eta_a < 1$, or [GAR200401]:

$$A_0 = \eta_a A$$

Hence, the aperture efficiency η_a of an antenna is the ratio of the effective radiating (for transmission or collecting for reception) area of an antenna to the physical area of the antenna. It takes into account a variety of losses. A basic relationship between A_0 and Ω is

$$A_0 \Omega = \lambda^2$$

where λ^2 is the wavelength. This relationship shows that as A_0 gets larger, Ω gets smaller. Along the boresight direction it can be shown (by the theory of diffraction) that $E = E_a A/r\lambda$. From this, it follows that the boresight gain is determined by the equation

$$G = \eta (4\pi/\lambda^2) A$$

If one specifies the gain (needed) at a given wavelength, this equation can be used to determine the required antenna area to achieve the specified gain (the gain is maximum along the boresight direction) [NEL200701]. For a reflector antenna, the area is simply the projected area. For a circular reflector of diameter D, the area is $A = \pi D^2/4$, and the gain is

$$G = \eta (\pi D/\lambda)^2 = \eta (\pi D f/c)^2$$

Note that the gain increases as the wavelength decreases or the frequency increases.

For example, an antenna with a diameter of 3 m and an efficiency of 0.55 would have a gain of 19540 at the C-band uplink frequency of 6 GHz and wavelength of 0.050 m. The gain expressed in dB is $10 \log(19540) = 42.9$ dB. This states that the power radiated by the antenna is 19540 times more concentrated along the boresight direction than for an isotropic antenna, which by definition has a gain of 1 (0 dB). At Ku band, with an uplink frequency of 14 GHz and wavelength 0.021 m, the gain for a 3-m antenna is 110774 or 50.44 dB. As stated earlier, the gain is higher at the higher frequency for the same-size antenna; for example, under the same conditions, a 6-ft C-band antenna has a 35 dBi gain, whereas the same size of antenna has a 44.5 dBi gain at Ku band. See Table 3.2; also see Figure 3.8.

Table 3.2   Gain as a function of diameter and wavelength (Top: C band; Bottom: Ku band)

Efficiency	0.55	0.55	0.55	0.55	0.55
Diameter	1	2	3	4	5
Lambda	0.05	0.05	0.05	0.05	0.05
G	2171.18	8684.74	19540.66	34738.96	54279.62
dB	33.37	39.39	42.91	45.41	47.35
Efficiency	0.55	0.55	0.55	0.55	0.55
Diameter	1	2	3	4	5
Lambda	0.021	0.021	0.021	0.021	0.021
G	12308.30	49233.22	110774.74	196932.87	307707.61
dB	40.90	46.92	50.44	52.94	54.88

The Ku band allows the use of smaller antennas than at C band. This is because the satellite effective isotopic radiated power (EIRP) at Ku band is typically about 10 dB higher than at C band, which basically compensates for the higher free-space loss incurred at the higher frequency of operation.

$$20 \log(12\ \text{GHz}/4\ \text{GHz}) = 9.5\ \text{dB}$$

(Note that the power received by an earth station antenna is the same for antennas of equal gain; however, because the antenna gain is proportional to the square of the frequency at Ku band, a significantly smaller antenna can be used to achieve the same gain [JOS200701].)

More specifically, the antenna gain can be described as two components: copolar (CP) gain and cross-polar (XP) gain [DUR200501]:

■ The CP gain is the common measurement given when referring to the gain of an antenna in the plane of interest. It defines the antenna's directivity, taking into account antenna

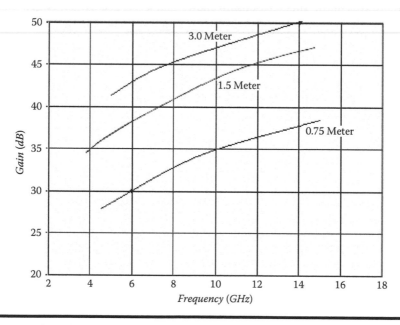

Figure 3.8   Typical satellite antenna gain.

efficiency in the principal plane of interest (typically this is the electrical field E). The CP gain of an aperture antenna is

$$G = \eta_a D = \frac{4\pi\eta_a A_p}{\lambda_o^2} \qquad \text{(parabolic antenna)}$$

$$A_p = \frac{\pi D^2}{4}$$

η_a = *aperture efficiency* (range of 50 to 60%)

$G = 20 \log_{10} D + 20 \log_{10} F + 17.8$ (parabolic antenna 55%, aperture efficiency)
  here
    G = gain over isotropic antenna, in dBi
    F = frequency in GHz
    D = parabola diameter in meters

■ The XP gain refers to the antenna gain in the plane perpendicular with (orthogonal to) the principal plane of interest (typically the magnetic field H). For maximum signal energy transfer by a transmitting/receiving antenna system, the electric field of both antennas must be aligned. In an ideal case, if the electric field planes of the transmitting and receiving antennas are perpendicular to each other, then no energy transfer occurs. In practice, this is not precisely the case, and cross-polarized energy transfer occurs. The XP performance is usually specified in comparison with the main-lobe copolar gain.

Now we expand on the issue of antenna efficiency. Antenna taper is defined as the variation in the electric field across the antenna diameter. The gain pattern of a reflector depends on how the antenna is illuminated by the electromagnetic fields of interest. The taper is chosen to maximize the illumination efficiency, which is defined as the product of aperture taper efficiency and spillover efficiency. Illumination efficiency is a measure of the nonuniformity of the field across the aperture caused by the tapered radiation pattern. It reaches a maximum value for an optimum combination of taper and spillover [ARO200601]. The antenna efficiency η_a is a factor that includes all reductions from the maximum gain. It can be expressed as a percentage, or in dB. Several types of "loss" must be accounted for in the efficiency term η_a [ANT200601]:

  i. Illumination efficiency, which is the ratio of the directivity of the antenna to the directivity of a uniformly illuminated antenna of the same aperture size
  ii. Phase error loss or loss due to the aperture being not a uniform phase surface
  iii. Spillover loss (for reflector antennas), which reflects the energy spilling beyond the edge of the reflector into the back lobes of the antenna
  iv. Mismatch loss, derived from the reflection at the feed port due to impedance mismatch (especially important for low-frequency antennas)
  v. RF losses between the antenna and the antenna feed port (or measurement point)

The efficiency η_a includes the preceding items (i) and (ii). It does not result in any loss of power radiated but affects the gain and pattern; it depends on the electric field distribution over the antenna aperture. η_a is nominally 0.6–0.8 for a planer array and 0.13–0.8 (with a typical value of 0.50–0.65) for a parabolic antenna. Items (iii), (iv), and (v) represent the radio frequency (RF) or power losses. The efficiency varies and generally gets lower with wider bandwidths. The gain equation is optimized for small angles; this explains why the efficiency also gets lower for wider beamwidth antennas [ANT200601].

As implied by this discussion, the aperture efficiency of a feed-and-reflector combination η_a is comprised of five components: (i) the illumination efficiency (aka "taper efficiency") η_t, (ii) the phase efficiency η_p, (iii) the spillover efficiency η_s, (iv) the XP efficiency η_x, and (v) the surface efficiency η_{sf} ($\eta_{sf} = \exp(-(4\pi\sigma/\lambda)2)$; e.g., for $\sigma = \lambda/16$, $\eta_{sf} = 0.5$). Then one has

$$\eta_a = \eta_t \times \eta_s \times \eta_p \times \eta_x \times \eta_{sf}$$

One can also write

$$\eta_a = \eta_t \times \eta_s \times \eta_{bl} \times \eta_{misc} \times \eta_{sf}$$

with

η_{bl} = blockage efficiency
η_{misc} = diffraction, phase, match, loss efficiency

As noted, the illumination efficiency is a measure of the nonuniformity of the field across the aperture caused by the tapered radiation pattern. Because the illumination is less toward the edges, the effective area is less than the geometric area of the reflector. The illumination efficiency is given by the following equation [CHE199901]:

$$\eta t = \frac{\left| \int_0^R g(r)dr \right|^2}{\int_0^R |g(r)|^2 \, dr},$$

where $g(r)$ is the aperture field. This expression has a maximum value of 1 when the aperture illumination is uniform, that is, $g(r) = 1$; for a theoretically uniform illumination, the electric field is constant across the antenna diameter and, therefore, the aperture taper efficiency η_a is 1.

The illumination efficiency can also be written in terms of the electric field pattern of the feed $E(\theta)$:

$$\eta_t = 2\cot^2 \frac{\theta_0}{2} \cdot \frac{\left| \int_0^{\theta_0} E(\theta)\tan(\theta/2)d\theta \right|^2}{\int_0^{\theta_0} |E(\theta)|^2 \sin(\theta)d\theta},$$

where θ_0 is the angle subtended by the edge of the reflector at the focus and $\cot(z) = 1/\tan(z)$. When a feed illuminates the reflector, only a proportion of the power from the feed will intercept the reflector, the remainder being the spillover power. This loss of power is quantified by the spillover efficiency η_s, which is given by the formula [CHE199901]:

$$\eta_S = \frac{\int_0^{\theta_0} |E(\theta)|^2 \sin(\theta)d\theta}{\int_0^{\pi} |E(\theta)|^2 \sin(\theta)d\theta}$$

Note that the illumination and the spillover efficiencies are countervailing; as the edge taper increases, the spillover decreases (and so η_s increases), whereas the illumination or taper efficiency η_t decreases. The trade-off between η_s and η_t has an optimum point, as indicated by the product $\eta_s \times \eta_t$ in Figure 3.9. The maximum of $\eta_s \times \eta_t$ occurs for an edge taper of about −11 dB, and has a value of about 80 percent.

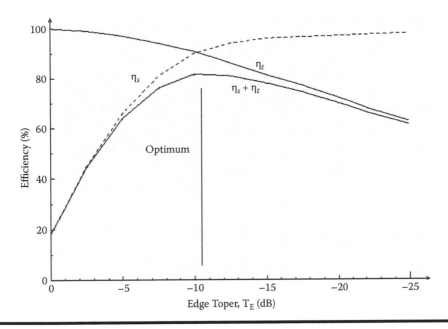

Figure 3.9   Illumination efficiency and spillover efficiency as a function of edge taper.

Many antenna designs (e.g., prime focus and Cassegrain) require structures to hold the feed and/or subreflector in place, and these structures block part of the aperture, impacting the aperture blockage efficiency. The blockage affects the beam *shape* in addition to the aperture efficiency. Some designs employ off-axis mechanics to reduce or eliminate blockage (but this can cause other unwanted effects). The efficiency due to blockage is [GAR200401]

$$\eta_{bl} = (1 - \text{area blocked/total area})^2$$

For a small blocked-to-total ratio, this is

$$\eta_{bl} \sim (1 - 2 \times \text{area blocked/total area})$$

For a prime focus antenna, spillover efficiency is the fraction of radiated power intercepted by the reflector; power not intercepted is lost. For a Cassegrain design, the feed is at the reflector, and the radiation has to hit the subreflector. Typically, $0.7 < \eta_s < 0.97$ [GAR200401].

The following specifications can be achieved for a typical satellite antenna with optimum edge taper (0.316) assuming a parabolic illumination [ARO200601]:

- Aperture taper efficiency = 0.92
- Spillover efficiency = 0.90
- Overall illumination efficiency = 0.83 (= 0.92 × 0.90)
- Beam efficiency = 0.95
- Half-power beamwidth (HPBW) = 1.75° (e.g., 65.3° × λ/D)
- First-side lobe below peak = −22.3 dB

Define the total radiation efficiency $\eta^* = P/P_{in}$ associated with various losses such as spillover, ohmic heating, phase nonuniformity, blockage, surface roughness, and cross-polarization. The net antenna efficiency η is then

$$\eta = \eta_a \, \eta^*$$

For a typical antenna, $\eta = 0.55$, but, as noted, in some cases can be as high as 0.65.

The boresight gain G can be expressed in terms of the antenna beam solid angle Ω_A that contains the total radiated power. It is the solid angle through which all the power would be concentrated if the gain were constant and equal to its maximum value [NEL200701]. The antenna gain is reduced (impacted) by the radiation efficiency η^*, as seen by this equation:

$$G = \eta^* \, (4\pi / \Omega_A)$$

The directivity does not include radiation losses.

3.3 Half-Power Beamwidth

The half-power beamwidth ($HPBW$) is the angular separation between the half-power points on the antenna radiation pattern where the gain is one-half the maximum value. At the half-power points to either side of the boresight direction, the gain is reduced by a factor of 2, which equates to 3 dB. The half-power (-3 dB) beamwidth is a measure of the directivity of the antenna. It is defined as the beamwidth of the main CP gain lobe at the -3 dB point (see Figure 3.10).

For a reflector antenna, the $HPBW$ decreases with decreasing wavelength and increasing diameter. Here, it may be expressed as [NEL200701]:

$$HPBW = \alpha = k\lambda/D$$

where k is a factor that depends on the shape of the reflector and the method of illumination (for a typical antenna, $k = 70$ if α is in degrees, or 1.22 if α is in radians). For example, for a 2-m antenna, the $HPBW$ at 6 GHz is approximately 1.75°. At 14 GHz, the it is approximately 0.75°. See Table 3.3 for some examples.

Note that, by algebraic manipulation, the antenna gain may be expressed directly in terms of the $HPBW$:

$$G = \eta \, (\pi k / \alpha)^2$$

Table 3.3 Half-power beamwidth (Top: C band; Bottom: Ku band).

Efficiency	0.55	0.55	0.55	0.55	0.55
Diameter	0.75	1.5	3	4	5
Lambda (C)	0.05	0.05	0.05	0.05	0.05
G	1221.29	4885.17	19540.66	34738.96	54279.62
dB	30.87	36.89	42.91	45.41	47.35
HPBW	4.67	2.33	1.17	0.88	0.70
Efficiency	0.55	0.55	0.55	0.55	0.55
Diameter	0.75	1.5	3	4	5
Lambda (Ku)	0.021	0.021	0.021	0.021	0.021
G	6923.42	27693.68	110774.74	196932.87	307707.61
dB	38.40	44.42	50.44	52.94	54.88
HPBW	1.96	0.98	0.49	0.37	0.29

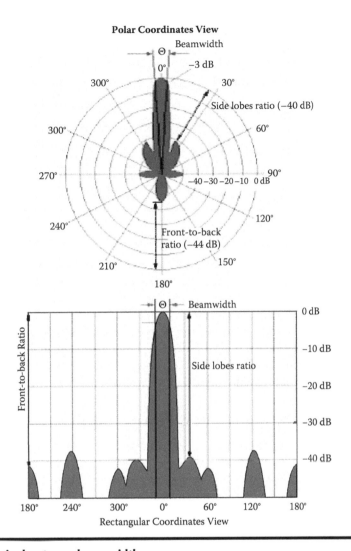

Figure 3.10   Typical antenna beamwidth.

For example, for the typical values of $\eta = 0.55$ and $k = 70°$, one obtains the approximation

$$G = 27{,}000/(\alpha°)^2$$

where $\alpha°$ is expressed in degrees. See Figure 3.11. One can see how the size of the beam established the gain. For example, for a satellite antenna with a circular spot beam of diameter 1°, the gain is 27,000 (44.3 dB.) For a Ku-band downlink at 12 GHz, the required antenna diameter for the satellite-borne antenna is determined from the equation given earlier:

$$G = \eta \, (\pi D/\lambda)^2 = \eta \, (\pi D \, f/c)^2 = 27{,}000$$

Solving for D one gets $D = 1.75$ m. As another example, a horn used to provide full earth coverage (angular diameter of 17.4°) needs a gain of 89.2 or 19.5 dB. Using the same efficiency factor of 0.70, the horn diameter for a C-band downlink antenna operating at a frequency of 4 GHz is 0.27 m. In summary, see Figure 3.12 for some basic approximation formulas for satellite antennas.

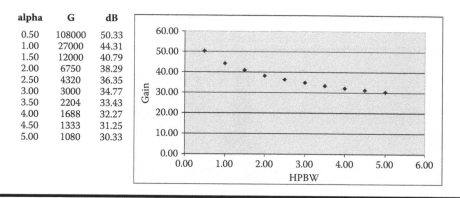

alpha	G	dB
0.50	108000	50.33
1.00	27000	44.31
1.50	12000	40.79
2.00	6750	38.29
2.50	4320	36.35
3.00	3000	34.77
3.50	2204	33.43
4.00	1688	32.27
4.50	1333	31.25
5.00	1080	30.33

Figure 3.11   Relation between *HPBW* and *G*.

For maximum antenna gain, high efficiency and very narrow beamwidth illumination of the reflector (by the feed) needs to occur; narrow illumination can be achieved by nonuniform illumination (e.g., steeper edge tapers result in lower side lobes and wider main lobes). Cassegrain-based antennas do a good job at achieving narrow beamwidth. Beams that are too large in the transmit direction can cause interference to satellites in the vicinity of the intended satellite (specifically, ASI); hence, there is regulation that defines what the beamwidth needs to be. Some mitigation approaches involve the use of spread-spectrum techniques (Code Division Multiple Access) and/or the dynamic use of signal processing methods that capture and then cancel the unwanted signal from the main intended signal.

3.4 Effective Isotropic Radiated Power (EIRP)

Define the power flux density (PFD) Φ as the radiated power P per unit area S, namely, $\Phi = P/S$. (Keep in mind that $P = \eta^* P_{in}$, where P_{in} is the input power and η^* is the radiation efficiency, and $S = d^2 \Omega_A$, where d is the slant range to the center of coverage and Ω_A is the solid angle containing the total power). Φ is the signal power received over a surface area of $1m^2$, expressed in dBW/m²; it

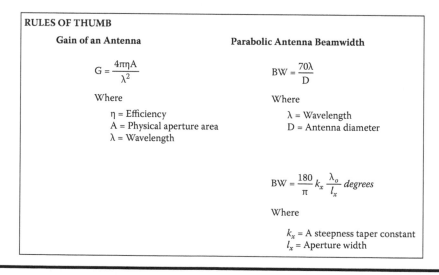

Figure 3.12   Basic approximation formulas for satellite antennas.

Table 3.4 Key definitions related to flux density

Flux density	The fundamental magnetic force field. "Flux" means to flow (around a current-carrying conductor, for example), and "density" refers to its use with an enclosed area and Faraday's Law to determine induced voltage. Also called the *induction field*. The unit of flux density is a tesla, volt–second per square meter per turn (the unit of magnetic flux density is the gauss; there are 10,000 gauss per tesla).
Power flux density (PFD) Φ (aka power density)	The power crossing unit area normal to the direction of wave propagation. Measured in units of watts per square meter (W/m²).
Power spectral density (PSD)	Amount of power per unit (density) of frequency (spectral) as a function of the frequency. A PSD function is a real-valued continuous function of frequency, presented with the frequency on the horizontal axis and density on the vertical axis where the signal is compared with itself. An equivalent definition is the Fourier transform of the autocorrelation function.
Spectral power flux density (SPFD)	SPFD is PFD per unit bandwidth. Its units are W/m²/Hz or Jansky (1 Jansky = 10⁻²⁶ W/m²/Hz implying −260 dBWm⁻²Hz⁻¹).
Spectral power flux density at the antenna aperture	The spectral power flux density at the antenna aperture is calculated as follows [IEEE200001]: $$sfd = \frac{Pr}{A_e} = \frac{Pr}{\lambda^2 \frac{G}{4\pi}} = Pr - 10\text{Log}(\lambda^2) - G + 10\text{Log}(4\pi)$$ where Pr = interference power level into receiver A_e = effective antenna aperture λ = wavelength G = antenna gain

quantifies the strength of a radio wave at the reception point of an earth–space link. See Table 3.4 for some key related definitions and concepts.

EIRP is a measure of the signal strength that a satellite transmits toward the earth, or an earth station toward a satellite. It is the power radiated equally in all directions, which would produce a power flux density equivalent to that of the actual antenna beam. *EIRP* often expressed in decibels relative to 1W, namely, as dBW ("decibels above one watt"). *C*-band satellites typically have an *EIRP* at about 35–40 dBW. The Ku-band satellites typically have an EIRP at about 50 dBW, but the majority of the existing U.S. GEO satellites provide only around 45 dBW. Some Ku-band satellites have spot beams that concentrate the power into small areas with 52–55 dBW levels.

After some algebraic manipulation [NEL200701],

$$\Phi = EIRP/4\pi d^2$$

and

$$EIRP = GP_{in}$$

The last equation states that the *EIRP* is the product of the antenna gain of the transmitter and the power applied to the input terminals of the antenna (the antenna efficiency is absorbed in the definition of gain). For a practical satellite, the *EIRP* depends on the antenna's location, design, and other factors.

Satellite coverage (footprint) is the geographical area where satellite signals can be transmitted or received with a sufficient quality when using appropriately sized earth stations. It is usually described in the form of footprints displaying satellite antenna gain-to-noise-temperature (see Section 3.5), EIRP, or by other quantities such as the antenna size required for good-quality reception of a particular service [SAT200501]. As see in Figure 3.13, one makes use of a collection of concentric (topological) footprints, each representing a particular satellite's EIRP value. For typical satellites, EIRP levels tend to fall away from the center of the footprint pattern in descending values: a typical footprint map, for example, might show a boresight point strength of 37 dBW with concentric lines indicating 36 dBW, 35, 34, etc., toward the outer fringes. These values do not take into account the path loss incurred between the satellite and the receiving antenna, but are important indicators of available signal strength [JOS200701]. C-band satellites typically transmit signal levels ranging from around 35 to 40 dBW, with the strongest signals found at the center of the satellite coverage beam. Depending on the receiving location, a typical communications antenna aperture varies from 2.0 to 11 m in diameter (2.0–3.7 m are typical for commercial applications). Ku-band satellites transmit signal levels typically ranging from 47 to 52 dBW. Because of the higher power and frequency, a receiving antenna as small as 0.3 m in diameter can be used to receive the Ku-band satellite signals. Heavy downpours (e.g., as may be

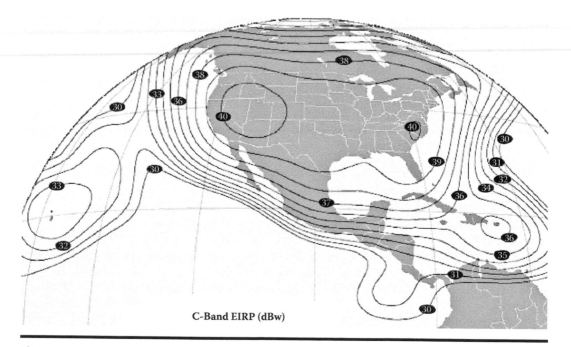

Figure 3.13   Example of *EIRP* for a geosynchronous satellite.

Table 3.5 Maximum typical allowable uplink EIRP values (earth station EIRP limitations)

C band (FCC 25.209)	
EIRP PSD	*Off-axis angle*
29–25.log θ dBi	$1.0° \leq \theta \leq 7.0°$
+8 dBi	$7.0° < \theta \leq 9.2°$
32–25.log θ dBi	$9.2° < \theta \leq 48°$
−10 dBi	$48 < \theta \leq 180°$
Ku band (FCC 25.222)	
EIRP PSD	*Off-axis angle*
15–25.logθ–10 log (N)dBW/4 kHz	$1.25° \leq \theta \leq 7.0°$
−6 –10 log (N)dBW/4 kHz	$7.0° < \theta \leq 9.2°$
18–25.log θ–10 log (N) dBW/4 kHz	$9.2° < \theta \leq 48°$
−24 –10 log (N) dBW/4 kHz	$48 < \theta \leq 180°$
Ka band (FCC 25.138)	
EIRP PSD	*Off-axis angle*
18.5–25.log θ dBW/40 kHz	$2.0° \leq \theta \leq 7.0°$
−2.63–10 log (N)dBW/40 kHz	$7.0° < \theta \leq 9.23°$
21.5–25.log θ –10 log (N) dBW/40 kHz	$9.23° < \theta \leq 48°$
−10.5–10 log (N)dBW/40 kHz	$48 < \theta \leq 180°$

experienced in Southeast Asia or the Caribbean) can lower the level of a Ku-band satellite signal by 20 dB, severely degrading the quality of the signals or even temporarily interrupting the reception entirely. Earth stations that operate with an inclined-orbit satellite (due perhaps to near-end-of-life conditions for the satellite) will experience greater Doppler uncertainty and range variation than stations operating with a normal noninclined orbit satellite. Additionally, the worst-case performance of the spacecraft coverages, which occurs at least once per day, may become significant and could require additional earth station *EIRP* margins.

The uplink *EIRP* of a Ku-band antenna is typically in the range of 60–75 dBW. The typical maximum allowable uplink EIRP values for commercial applications are shown in Table 3.5. Also, to protect a satellite system from excessive unbalancing among the carriers transmitted into each transponder, it is generally required that a proper stability of the earth station's EIRP be guaranteed; that is, the *EIRP* of any carrier transmitted in the direction of the satellite system, measured in a continuous 24-hr period, should not vary by more than 0.5 dB. Most satellites require a flux density of −92 dBW/square meter for normal operations and between −80 and −60 dBW/square meter for emergency operation. Stations with uplink power control should not exceed authorized satellite flux densities at the spacecraft by more than 1 dB at any time.

It is typically required that the uplink antenna-pointing stability is such that environmental conditions, both internal and external to the uplink earth station, will not cause sufficient antenna movement to produce more than a ±1 dB change in operational flux density at the satellite under clear-sky conditions.

3.5 Antenna Gain-to-Noise-Temperature (*G/T*)

The system temperature T (also known as T_{sys}) is a measure of the total noise power and includes contributions from the antenna and the receiver. The ratio G/T provides a characterization of antenna performance, a figure of merit, where G is the antenna gain in decibels at the receive frequency, and T is the equivalent noise temperature of the receiving system in Kelvins [ANS200001]. It is an important figure of merit because it is independent of the reference point where it is calculated, even though the gain and the system temperature individually are different at different points.

A link budget commonly refers to the complete gain and loss equation from the transmitter, through the ambient medium (air, cable, waveguide, fiber, etc.) and through to the receiver. For the RF link budget, the required antenna values are the EIRP, "figure of merit" G/T, antenna size, antenna efficiency, gain G, location (longitude/latitude), rain zone, desired availability, and desired (digital) throughput (give the modulation and FEC scheme), among other variables. Although the complete equation would incorporate many (additional) terms, the following high-level equation and block diagram shown are often used to illustrate the link budget [BLA200701]:

$$P_{rx} = P_{tx} + G_{tx} + G_{rx} - A_{fs} - A_m$$

where

P_{rx} = received power at detector (dBW)
P_{tx} = transmitter output power (dBW)
G_{tx} = transmitter antenna gain (dBi)
G_{rx} = receiver antenna gain (dBi)
A_{fs} = free-space attenuation (dB)
A_m = miscellaneous attenuation (radome, rain, etc.)

The C-band low noise block downconverter (LNB) noise temperature is typically 25 K, whereas Ku-band LNBs have a noise temperature of 65 K. When this difference in noise temperature is used in G/T calculations, it results in a −4.1 dB performance loss on Ku band. This topic is covered in Chapter 4.

3.6 Antenna Taper

As noted, an antenna's taper is the variation in electric field across the antenna diameter. The gain pattern of a reflector antenna depends on how the antenna is illuminated by the feed. The total antenna solid angle containing all of the radiated power, including side lobes, is calculated as follows [NEL200701]:

$$\Omega_A = \eta^* (4\pi/G) = (1/\eta_a)(\lambda^2/A)$$

where η_a is the aperture taper efficiency, and η^* the radiation efficiency associated with losses, defined earlier. Let Ω_M be the solid angle for the main lobe. Then the beam efficiency is defined as

$$\varepsilon = \Omega_M/\Omega_A$$

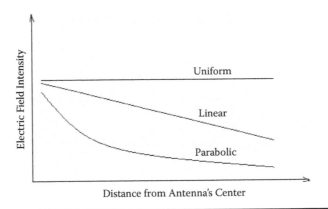

Figure 3.14 Various antenna tapers.

The efficiency value of η_a is calculated from the electric field distribution in the aperture plane; the (beam) efficiency value of ε is calculated from the antenna radiation pattern. If the feed is designed to cause the electric field to decrease with distance from the center, then the aperture taper efficiency decreases; however, the proportion of power in the main lobe increases. In general, maximum aperture taper efficiency occurs for a uniform distribution, but the maximum beam efficiency occurs for a highly tapered distribution, as the observations that follow make it clear [NEL200701] (also see Figure 3.14):

■ For uniform illumination, the *HPBW* is

$$58.4° \times \lambda/D$$

where the first side lobe is 17.6 dB below the peak intensity in the boresight direction. Here, the main lobe contains about 84 percent of the total radiated power, and the first side lobe contains about 7 percent.

■ If the electric field amplitude has a parabolic distribution falling to zero at the reflector's edge, then the aperture taper efficiency is 0.75 but, advantageously, the fraction of power in the main lobe increases to 98 percent. The *HPBW* is

$$72.8° \times \lambda/D$$

where the first side lobe is 24.6 dB below the peak intensity. Therefore, although the aperture taper efficiency is less, more power is contained in the main lobe as indicated by the larger *HPBW* and lower side-lobe intensity.

■ If the electric field decreases to a fraction C of its maximum value (C being called the *edge taper*), the reflector will not collect all the radiation from the feed; hence, there will be energy spillover, and the resulting efficiency is approximately $1 - C^2$. However, as the spillover efficiency decreases, the aperture taper efficiency increases. The taper is chosen to maximize the illumination efficiency, which reaches a maximum value for an optimum combination of taper and spillover. For a typical antenna, the optimum edge taper C is about 0.316, or −10 dB (that being $20 \log C$). With this edge taper and a parabolic illumination, the aperture taper efficiency is 0.92, and the spillover efficiency is 0.90 (as seen in Figure 3.9). The *HPBW* is

$$65.3° \times \lambda/D$$

where the first side lobe is 22.3 dB below peak. Therefore, the overall illumination efficiency is 0.83. The beam efficiency is about 95 percent.

In summary, and noting that reflector antennas are generally used to produce narrow beams for earth stations and for geostationary satellites, the efficiency of the antenna is optimized by the method of illumination and selection of edge taper.

3.7 Antenna Patterns

Satellite antenna patterns have a significant effect on the utilization of the satellite orbit. To achieve efficient use of the orbital arc, it is necessary to control the degree of interference between different satellite networks. By imposing regulatory guidelines for the radiation characteristics of both satellite and earth station antennas, the designs can be influenced in such a way as to minimize this interference level [RAD199701]. Electromagnetic energy propagates as waves; waves combine both constructively and destructively to form a diffraction pattern that manifests itself in the main lobe and side lobes of the antenna. Therefore, another measure of an antenna's performance is the side-lobe level. It is a measure of the directivity of minor lobes, particularly the first minor lobe, when compared with the main lobe. High side-lobe levels are a major contributor to noise in receiving terminals. Typically for a reflector antenna, this should exceed −10 dB. A design goal for the reflector/feed assembly is to keep the side-lobe level as low as possible (e.g., −15 dB is better than −10 dB.) To optimize performance, an antenna designer needs to have the reflector illumination at the edge of the dish to be significantly less than the power at the center of the dish (this difference being the edge taper). A typical goal is to have the power illuminating the edge of the dish to be 10 dB less than the power at the center of the dish [DUR200501].

The physical optics integration technique is the most widely used technique for the analysis of reflector antennas. This technique provides accurate results in the main beam region and the first few near-in side lobes. The geometrical theory of diffraction (GTD) is then used to account for diffraction and to calculate the wide-angle side-lobe radiation pattern [RAD199701]. The radiation pattern of an antenna is important in the region of the main lobe as well as side lobes. Copolar (copol, CP) refers to the case in which one considers the incident (and scattered) electromagnetic waves with the same polarization state. Cross-polar (cross-pol, or X-pol or XP) refers to the case in which one considers the incident (and scattered) electromagnetic waves with the orthogonal polarization states.

Figure 3.15 shows the concept of antenna side lobes in the copolarization frequency and in the cross-polarization frequency.

Within the main lobe of an axis-symmetric antenna, the gain $G(\theta)$ in a direction θ with respect to the boresight direction can be approximated by the expression [NEL200701]:

$$G(\theta) = G - 12\ (\theta/\alpha)^2$$

where G is the boresight gain, and α is the *HPBW* (the angular separation between the half-power points on the antenna radiation pattern, where the gain is one-half the maximum value). Note that at $\theta = \alpha/2$, where by definition the gain is reduced by a factor of 2, $G(\theta) = G - 12\ (1/2)^2 = G - 3$; that is, G is reduced by 3 dB as we would expect (the details of the antenna, including its shape and illumination, are embodied in the value of the HPBW angle α).

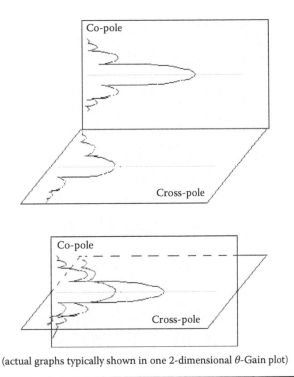

(actual graphs typically shown in one 2-dimensional θ-Gain plot)

Figure 3.15   Antenna side lobes in the copolarization frequency and in the cross-polarization frequency.

By regulation (to minimize the interference between adjacent satellites in the geostationary arc with a 2° spacing), for transmitting antennas with $D/\lambda > 100$, the gain of the side lobes must fall within the envelope 29–25 log θ, as noted in Table 3.5. See Figure 3.16 for an example [AND200101].

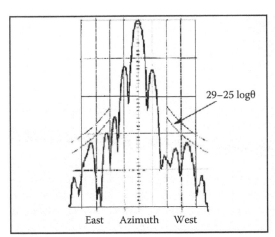
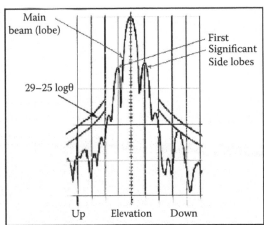

Courtesy: Andrew Corporation

Figure 3.16   Gain of side lobes.

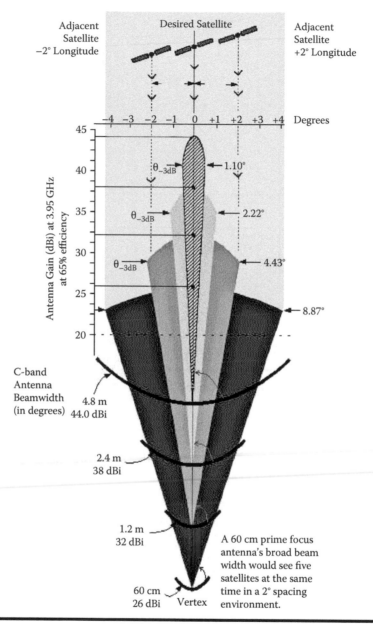

Figure 3.17   Selectivity of C-based antennas as a function of size.

For receiving antennas, the series of graphs in Figures 3.17–3.19 provide a perspective of the selectivity of an antenna based on its size and wavelength of operation. As one can see, a 60-cm C-band antenna could potentially receive five satellites within its main beam if the satellites are separated by 2° in longitude. If this "neighborhood" was comprised of C-band-supporting satellites, then severe ASI interference would result. On the other hand, a Ku antenna can be smaller: the graph shows the performance of various antenna apertures, and one can see that 0.3–0.6 m dish are possible for reception because of the reduction in parabolic antenna beam width at the

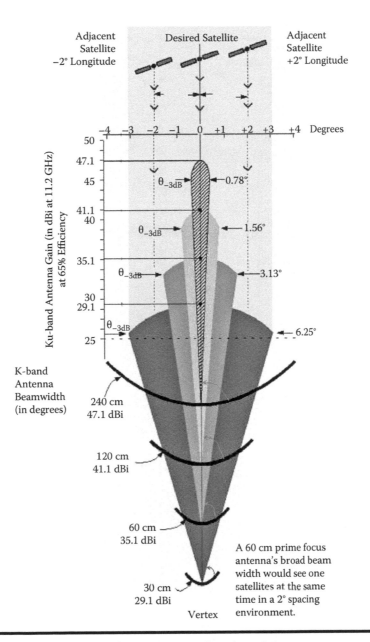

Figure 3.18   Selectivity of Ku-based antennas as a function of size.

higher frequency bands. Note, however, that the increased susceptibility to absorption and rain requires that a certain size (gain margin) be used to achieve a desired availability as part of the link budget calculation. To address the effects of rain fade, Ku/Ka-band system designers typically use a larger antenna than what would be required under clear-sky conditions. This increase in antenna aperture gives the system several decibels of margin so that the receiving system can operate in moderate rainstorms. For example, for this higher margin, a Ka system may use 0.7–1 m in diameter for direct-to-home (DTH) applications (note that the outages can also be caused by

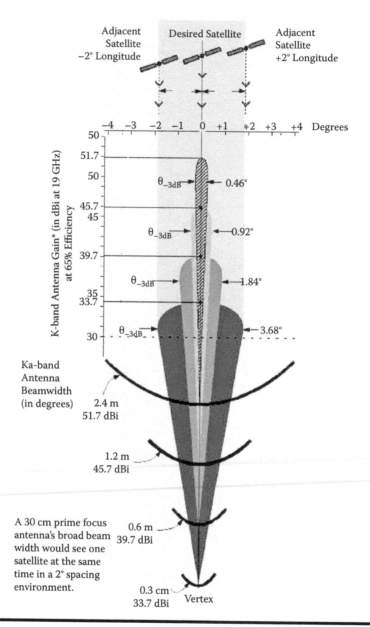

Figure 3.19 Selectivity of Ka-based antennas as a function of size.

intense sandstorms because the presence of any atmospheric particulate can have an adverse effect on satellite reception).

As stated, at angles away from the axis of the main lobe, the antenna response/behavior is embodied in the side lobes; side-lobe characteristics have important impacts on interference received from adjacent satellites or interference caused by the antenna in question to adjacent satellites. The angular position and amplitude of side lobes varies significantly between antennas and is in large measure determined by the size (the smaller the size, the larger the lobes) and the illumination approach.

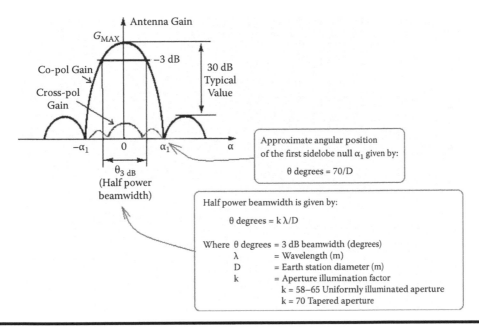

Figure 3.20 Co-pol/cross-pol gain for an antenna.

- The co-polar (CP) performance of a reflector system is its gain plot as a function of Θ from the boresight position in dBi. It describes the geometric spread of the main signal (polarization under consideration) at a given frequency of operation (and polarization). See Figure 3.20 for a generalized example.
- The cross-polar (XP) performance of a reflector system is primarily determined by the feed radiation pattern. Reflector antenna system geometries (i.e., dual offset or circularly symmetric) can be designed that minimize or practically eliminate the reflector-generated XP contribution, the XP performance of the system being primarily determined by the feed horn pattern. For a reflector antenna system, the geometry-generated XP component can be minimized by using a circularly symmetric system or a compensated dual-offset system. The contribution and limitations of the cross-polarized radiation are then determined by the feed horn and the components of the feed chain. Recent advances in the design of waveguide components and feed horns, accompanied by precision machining (at least for satellite-based antennas), have improved the expected XP performance of feed chains [RAD199701].

The CP component is an "unwanted" element of the transmitted energy from an uplink. The cross-polar discrimination (XPD) represents the ratio of the signal power received (or transmitted) by an antenna on one polarization (the polarization of the desired signal) to the signal power received (transmitted) on the opposite polarization. The XPD value is a function of the antenna, feed, and Ortho Mode Transducer (OMT). It is expressed as −dB. In linear, orthogonal reuse satellite transmission systems, CP is always present and never wanted. Some antenna designs are better at reducing the XPD than others, typically an offset-fed antenna of 0.6 F/D has a high XP component that causes the interference in the opposite or (orthogonal) transponder. When an uplink is first brought into service, the satellite operator needs to test the station to ensure that the

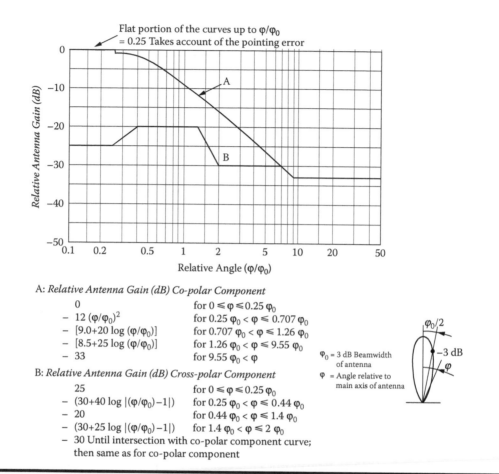

A: *Relative Antenna Gain (dB) Co-polar Component*

0	for $0 \leq \varphi \leq 0.25\, \varphi_0$
$-12\,(\varphi/\varphi_0)^2$	for $0.25\,\varphi_0 < \varphi \leq 0.707\,\varphi_0$
$-[9.0 + 20 \log (\varphi/\varphi_0)]$	for $0.707\,\varphi_0 < \varphi \leq 1.26\,\varphi_0$
$-[8.5 + 25 \log (\varphi/\varphi_0)]$	for $1.26\,\varphi_0 < \varphi \leq 9.55\,\varphi_0$
-33	for $9.55\,\varphi_0 < \varphi$

$\varphi_0 = 3$ dB Beamwidth of antenna

φ = Angle relative to main axis of antenna

B: *Relative Antenna Gain (dB) Cross-polar Component*

25	for $0 \leq \varphi \leq 0.25\, \varphi_0$
$-(30 + 40 \log |(\varphi/\varphi_0) - 1|)$	for $0.25\,\varphi_0 < \varphi \leq 0.44\,\varphi_0$
-20	for $0.44\,\varphi_0 < \varphi \leq 1.4\,\varphi_0$
$-(30 + 25 \log |(\varphi/\varphi_0) - 1|)$	for $1.4\,\varphi_0 < \varphi \leq 2\,\varphi_0$
-30 Until intersection with co-polar component curve; then same as for co-polar component	

Figure 3.21   Example of ITU copol/cross-pol side-lobe gain envelope for antenna (early example).

XP is as low as possible and does not cause any interference. Choosing a good-quality and well-designed antenna/feed/OMT package will (a) save money, and (b) provide a better and cleaner environment for satellite communications [BAR200101].

The International Telecommunication Union (ITU) has published generic antenna masks that need to be supported for the transmit compliance. Figure 3.21 depicts one early example of ITU CP/XP side-lobe gain envelope for antennas. Newer guidelines have been published; different antenna "regions" have different requirements (see Figure 3.22).

3.7.1 Copolar Side-Lobe Guidelines

It is generally required that the off-axis transmit antenna gain of the earth station for a CP frequency at an angle θ measured between the main beam electrical boresight and the direction considered needs to be controlled. For circularly polarized satellites, the CP gain should not be higher than the values shown in Table 3.6 (antenna CP gain over the angular range shown; θ is the angle in degrees from the axis of the main lobe.)

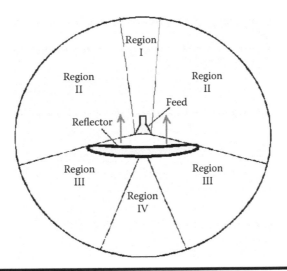

Figure 3.22 ITU antenna regions used in recommendations.

3.7.2 Cross-Polar Side-Lobe Guidelines

The off-axis transmit antenna cross-polarized gain of the earth station for all antenna for an XP frequency needs to be controlled. For circularly polarized satellites, the XP gain is expected not to exceed the levels defined in Table 3.7 (antenna XP gain over the angular range shown.)

3.7.3 FCC Side-Lobe Guidelines

The Federal Communications Commission (FCC) has issued guidelines for antenna operations in the United States. The required antenna performance standards are included in the FCC rules under Part 25.209 (a) and (b). The issue is mainly concerned with interference into adjacent satellites. The performance standard at issue concerned off-axis angles between 1 and 7°, that is, the earth station's antenna gain must be within the following envelope [RYA199901]:

$$G = 29-25 \log (\theta) \text{ dBi}$$

where, again, θ is the angle in degrees from the axis of the main lobe.

Table 3.6 Antenna transmit copolar off-axis gain (C-band)

Copolar gain in direction θ	Off-axis angle θ
29–25.log θ dBi	$1.0° \leq \theta \leq 7.0°$
+8 dBi	$7.0° \leq \theta \leq 9.2°$
32–25.log θ dBi	$9.2° \leq \theta \leq 48°$
−10 dBi	$48° \leq \theta \leq 180°$

Table 3.7 C-, Ku-, and Ka-band antenna transmit cross-polar off-axis gain

Cross polar gain indirection θ	Off-axis angle θ
19–$25.\log \theta$ dBi	For $2.5° \le \theta \le 7°$
-2dBi	For $7° \le \theta \le 9.2°$

FCC §25.212 specifies the power density requirements for the routine licensing of narrowband emissions for small-diameter earth stations. At Ku-band, any antenna with equivalent diameter of 1.2 m or greater will be routinely licensed if, for analog transmissions, the maximum bandwidth is 200 kHz and the input power density into the antenna does not exceed −8 dBW/4 kHz, and, for digital narrowband or wideband transmissions, the maximum input power density does not exceed −14 dBW/4 kHz (see §25.212 (c)).

At C band, any antenna with equivalent diameter of 4.5 m or greater will be routinely licensed if, for analog SCPC carriers with bandwidths up to 200 kHz, the maximum power density does not exceed 0.5 dBW/4 kHz and, for narrow and wideband digital transmissions, the maximum power density does not exceed −2.7 dBW/4 kHz (see §25.212(d)). These two rules combined create a maximum EIRP power density restriction.

Appendix 3B contains four charts illustrating these limits for the four combinations of analog/digital and C/Ku-band operations. Also refer back to Table 3.5.

3.8 Coverage Area of Satellite-Based Antenna

The general theory of operation of satellite and ground-based antennas is the same; however, space-borne antennas are more complex because they need to survive the harsh environment of space, and are limited in physical size and weight [ARO200601]. The gain of a satellite-based antenna is designed to provide a specified area of coverage on the earth. The area of coverage within HPBW is

$$S = d^2 \Omega$$

where d is the slant range to the center of the footprint, and Ω is the solid angle of a cone that intercepts the half-power points, which may be expressed in terms of the angular dimensions of the antenna beam. The boresight gain can be approximated in terms of this solid angle by the relation [NEL200701]:

$$G = \eta' \, (4\pi / \Omega) = (\eta'/K)(41{,}253/\alpha°\beta°)$$

where $\alpha°$ and $\beta°$ are in degrees; η' is an efficiency factor that depends on the HPBW (although η' is different from the net efficiency η, in practice these two efficiencies are approximately equal for a typical antenna taper); and K is a factor that depends on the shape of the coverage area; for example, for a square or rectangular area of coverage $K = 1$, whereas for a circular or elliptical area of coverage, $K = \pi/4$. If $\eta' = (\pi K/4)^2\eta$ and $K = 1$, one can see that the two values are comparable.

3.9 Satellite-Based Antennas and Shaped Beams

It was mentioned that, in theory, a satellite-based antenna could cover about one-third of the earth's surface; however, the area of desired coverage often has an irregular shape, for example, a specific country or continent. Traditionally, service providers have created the desired coverage by means of a beam-forming network. In this arrangement, superposition of multiple individual circular beams produces the specified shape. Each beam is driven by its own feed and illuminates the reflector area of the satellite antenna and, by appropriately exciting the beam-forming network, the specified areas of coverage are illuminated. The shaped reflector is a relatively new approach to the same issue. Here, there is a single feed that illuminates a reflector with an undulating shape that defines the required region of coverage. The advantages of the shaped reflector are lower spillover loss, lower signal losses, and a reduction in mass with resulting lower overall costs.

References

[AMA200101] Amanogawa, *Antennas,* Digital Maestro Series, 2001.

[AND200101] Pattern Optimization Procedure for Gregorian Optic Earthstation Antennas, SP50037-A, 14 March 2001, Andrew Corporation, Orland Park, IL.

[ANS200001] ANS T1.523-2001, Telecom Glossary 2000, American National Standard (ANS), an outgrowth of the Federal Standard 1037 series, *Glossary of Telecommunication Terms,* 1996.

[ANT200601] Antenna Introduction/Basics, ANITA (Antarctic Impulsive Transient Antenna) Department of Physics and Astronomy at the University of Hawaii and Manoa, RF Hardware Reference Library, excerpts from the Military EW Handbook.

[ARO200601] A. Aroh, "A Study on the Principles of the Engineering and Technology of The Satellite Professional's Product(s)/Service(s)," 25 Feb 2006, Society of Satellite Professionals International, c/o The New York Information Technology Center, New York.

[BAR200101] M. Bartlett, Satellite FAQ, S+AS Limited, 6 The Walled Garden, Wallhouse, Torphichen, West Lothian, EH48 4NQ, SCOTLAND, July 2001.

[BLA200701] K. Blattenberger, Free Space Loss, RF Café Website, Mt. Airy, NC.

[CHE199901] J. N. Chengalur, Y. Gupta, K. S. Dwarakanath, Lecture Notes, Low Frequency Radio Astronomy, National Centre for Radio Astrophysics (NCRA), Pune, June 21–July 17, 1999. Tata Institute of Fundamental Research, Pune University Campus, Ganeshkhind, Pune, India.

[DUR200501] G. J. Durnan, Parasitic Feed Elements for Reflector Antennas, Ph.D. dissertation, School Microelectronic Engineering, Griffith University, Brisbane, Australia, April 2005.

[EUT200701] Eutelsat, Glossary, Paris, France.

[FOC200701] Staff, *An Introduction to Earth Stations,* Focalpoint Consulting, St. Gaudens, France, 2007.

[GAR200401] D. E. Gary, Radio Astronomy Lecture Notes, NJIT, Newark, NJ.

[IEEE200001] IEEE 802.16.2, Recommended Practices to Facilitate the Coexistence of Broadband Wireless Access (BWA) Systems, Revision 2, 2000-02-24 (Revision of IEEE 802.16c-99/02r1).

[JEF200401] D. Jefferies, "Microwaves: Satcoms Applications," MSc in Satcoms Notes, University Of Surrey, Department of Electronic Engineering, School of Electronics and Physical Sciences, Guildford, Surrey, UK, 18th March 2004.

[JOS200701] Promotional Material, Joseph & Sons, LLC, Alexandria, VA.

[LEE198801] Y. T. Lo, S. W. Lee, *Antenna Handbook: Theory, Applications, and Design,* Van Nostrand Reinhold, New York, 1988.

[NEL200701] R. A. Nelson, "Antenna and Antenna Array Fundamentals," 2007.

[RAD199701] Radiocommunications Agency, Space Services Unit, "A Study into the Assessment of FSS Satellite Antenna Reference Radiation Pattern Envelopes", New King's Beam House, London, June 1997.

[RYA199901] K. Ryan, "FCC Licencing of Non-Compliant Antennas," Comsearch, March 20, 1999. Unpublished Memorandum.
[SAT200501] Satellite Internet Inc., Satellite Physical Units and Definitions.
[SEY200401] J. S. Seybold, "Antennas", IEEE Melbourne COM/SP AP/MTT, November 9, 2004.
[SIL200201] J. Sills, "Improving PA Performance with Digital Predistortion," *CommsDesign*, Oct. 2, 2002.
[WAD200401] P. Wade, "Multiple Reflector Dish Antennas," W1GHZ ©2004, Shirley, MA.

Appendix 3A: Amplifier Preemphasis

The issue of amplifier power is important for proper operation of a satellite system. Sufficient power is needed to be able to close the link (as covered in Chapter 4). Backoff is needed to avoid intermodulation problems. Preemphasis is needed to deal with group delay considerations (covered in Chapter 3). This appendix provides a quick tutorial on amplifications, based primarily on reference [SIL200201] by J. Sills.

Three types of HPAs are found in satellite applications, as follows [FOC200701]:

■ Solid-state power amplifiers (SSPA). HPAs that use solid-state components (e.g., transistors) to amplify the RF power level at its input. Solid-state amplifiers typically have more linear characteristics than other types of amplifier, but they are limited to lower power levels (however, in recent years, they have achieved fairly high levels in the 2–3 KW range).

■ Traveling wave tube amplifiers (TWTA). HPAs that use a complex series of magnets, cavities, and an electron beam encased in a vacuum tube to amplify the power of the input signal. TWTAs can provide more power than SSPAs, but they are not normally operated at their full output power level because of nonlinear effects that adversely affect the transmission quality. They are usually operated at several decibels below their maximum output power level so that the device's behavior is approximately linear. This is especially true if the TWTA is amplifying multiple carriers, which are susceptible to mutual interference caused by device nonlinearities.

■ Klystron Power Amplifier (KPA). HPAs that use a series of interconnected resonant cavities to successively amplify the input signal. They produce similar output power levels to TWTAs and, for equal output powers, can be the more economical choice. However, their bandwidth is limited to around 100 MHz or less, and consequently, they may not be suitable for earth stations that need to be able to transmit over a wide frequency range.

Figure 3.A.1 depicts the typical power response of HPAs.

Figure 3.A.2 (top) shows the input–output block diagram of a power amplifier, where $v_i(t)$ and $v_0(t)$ are the input and output signals, respectively. The output can be expressed as

$$v_0(t) = g(|v_i(t)|^2)v_i(t) = g_0(|v_i(t)|^2)e^{j\,g_\phi^{(|v_i(t)|^2)}}v_i(t)$$

where

$$g_0(|v_i(t)|^2) \text{ and } g_\phi^{(|v_i(t)|^2)}$$

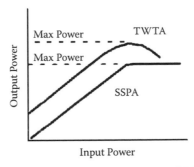

Figure 3.A.1   HPA power response.

are the amplifier's amplitude-transfer (AM-to-AM) and phase-transfer (AM-to-PM) characteristics. The amplitude and phase values can be plotted as a function of input power; see Figure 3.A.3 for an example (the transfer characteristics can be measured by applying an input pulse to the power amplifiers (PA) and measuring the output amplitude and phase).

The amplifier's linear characteristic is defined by

$$\overline{g}_a(|v_i(t)|^2) = G$$

where G corresponds to a given gain. The linear region is defined as that set of inputs for which

$$g_a = \overline{g}_a \text{ and } g_\phi = \overline{g}_\phi$$

For the example in Figure 3.A.3, the upper limit of the linear region falls between +5 and +10 dBm; the amplifier exhibits a 35-dB gain; beyond that, the amplifier cannot sustain this gain.

The 1-dB compression point is defined as the input power at which the amplifier's output power is 1 dB below the linear response. In the example highlighted in Figure 3.A.3, the 1-dB compression point occurs at +20 dBm. With the input power at +20 dBm, the output power is +54 dBm, which is 1 dB less than the +55-dBm output power required for linear operation. The saturation point corresponds to the input level that results in the largest output. In this case, the saturation point is at +25 dBm.

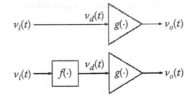

Figure 3.A.2   PA operation. Top: block diagram of PA's input and output. Bottom: diagram of a cascaded digital predistorter plus a PA.

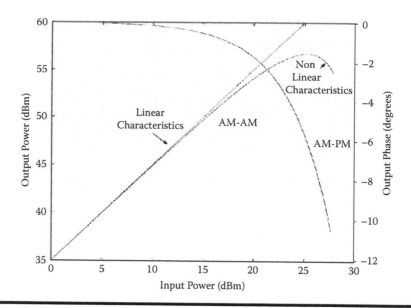

Figure 3.A.3   Diagram of AM–AM and AM–PM characteristics.

A predistortor preceding the amplifier can be used to linearize the amplifier. Figure 3.A.2 (bottom) shows the predistortion PA cascade, where $v_d(t)$ is the predistortion output.

Expanding the equation given at the start of this discussion gives

$$v_0(t) = g(|f(|v_i(t)|^2)v_i(t)|^2)f(|v_i(t)|^2)$$

By definition, the PA is linearized when

$$G = g(|f(|v_i(t)|^2)v_i(t)|^2)f(|v_i(t)|^2) = \bar{g}_a(|v_i(t)|^2)$$

For input levels in the linear range of the amplifier, the predistortor applies a gain of 0 dB and has no effect on the signal. In the nonlinear range, it applies a gain to either amplify or attenuate the input signal. The value of the gain depends on the input level. For example, with an input of +20 dBm, the predistortor applies a gain of approximately 2 dB. The amplifier input is now +22 dBm, which results in the desired +55-dBm output.

The predistortor must also linearize the amplifier's phase response. Figure 3.A.3 shows that the phase distortion is approximately 2° at the 1-dB compression point, and nearly 6° at saturation. It introduces a phase shift that is equal and opposite to that of the amplifier. The two-tone analysis that follows gives clear evidence that phase linearization cannot be neglected.

PAs can be characterized by their response to inputs constructed from two sinusoidal tones because that drives intermodulation. Consider the two-tone input

$$v_i(t) = ve^{j\omega_1 t} + ve^{j\omega_2 t}$$

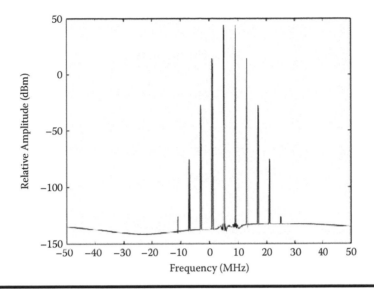

Figure 3.A.4   Spectrum of PA output.

The input can be expressed using a trigonometric identity as

$$v_i(t) = 2v \cos(\omega_m t)e^{j\omega t}$$

Here, the PA output is the product of three terms:

$$(1)\ g_a(|v_i(t)|^2);\ (2)\ e^{jg_\phi(|v_i(t)|^2)};\ \text{and}\ (3)\ v_i(t)$$

that is,

$$v_0(t) = g(|v_i(t)|^2)v_i(t) = g_a(|v_i(t)|^2)e^{jg\phi(|v_i(t)|^2)}v_i(t)$$

The spectrum of the output is equal to the spectral convolution of these three terms. Figure 3.A.4 depicts the spectrum of the two-tone input signal, where $\omega_1 = 5$ MHz, and $\omega_2 = 9$ MHz.

The phase response can be expressed as a Taylor's series:

$$e^{jg_\phi(|v_i(t)|^2)} = e^{j\alpha 4v^2 \cos^2 \omega_m t}$$

$$= 1 + j\alpha 4v^2 \cos^2 \omega mt = \frac{(j\alpha 4v^2 \cos^2 \omega_m t)^2}{2!} + \frac{(j\alpha 4v^2 \cos^2 \omega_m t)^3}{3!} + \cdots$$

$$= C_0 + C_{1 \cos 2\omega_m t} + C_{2 \cos 4\omega_m t} + C_{3 \cos 6\omega_m t} + \cdots$$

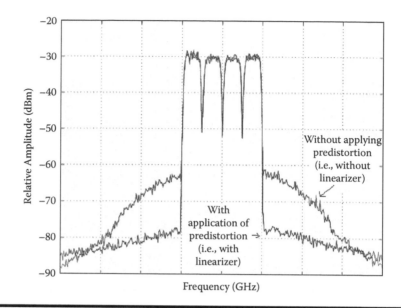

Figure 3.A.5 **PA with and without predistortion. (Courtesy CommsDesigns, www.commsdesigns.com)**

where

$$C_0 = 1 + \sum_{m-1}^{\infty} \frac{(j4\alpha v^2)^m}{m!} \binom{2m}{m} \frac{1}{2^{2m}}$$

and

$$C_k = 1 + \sum_{m-k}^{\infty} \frac{(j4\alpha v^2)^m}{m!} \binom{2m}{m-k} \frac{1}{2^{2m}}, k = 1, 2, \dots$$

It is clear here that the phase response has an infinite number of spectral components, but the amplitude of these components falls off rapidly as shown in Figure 3.A.4.

Say, for example, that the peak-to-average ratio (PAR) is 3 dB (the average power of the two-tone signal being +17 dBm, and the peak being +20 dBm). Thus, in this example, the average power was backed off by 3 dB so that the peak power did not exceed the 1-dB compression point of the amplifier. Multicarrier communication signals can exceed 9-dB PAR and therefore require greater backoff. Unfortunately, as backoff increases, efficiency decreases. For an ideal strong nonlinearity, the minimum backoff required for linear operation is equal to the PAR. In a more realistic model that includes a weak nonlinearity, the backoff needs to be increased even further. Digital predistortion can be used to linearize the weak nonlinear behavior of the amplifier and reduce backoff, thereby increasing efficiency. Figure 3.A.5 shows a typical PA output with and without predistortion.

Appendix 3B: FCC Rules on EIRP Density

This appendix contains four charts illustrating the FCC EIRP density limits for four combinations of analog/digital and C/Ku-band operations [RYA199901].

Ku-Band Analog			
	E/S Antenna Gain (dBi) CFR 25.209(a)(1)	RF Power Density (dBW/4kHz) CFR 25.212(c)	EIRP Density (dBW/4kHz)
1	29.0	−8	21.0
2	21.5	−8	13.5
3	17.1	−8	9.1
4	13.9	−8	5.9
5	11.5	−8	3.5
6	9.5	−8	1.5
7	8.0	−8	0.0
−1	29.0	−8	21.0
−2	21.5	−8	13.5
−3	17.1	−8	9.1
−4	13.9	−8	5.9
−5	11.5	−8	3.5
−6	9.5	−8	1.5
−7	8.0	−8	0.0

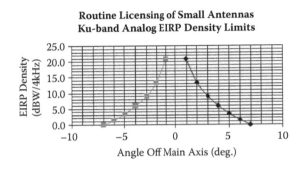

**Routine Licensing of Small Antennas
Ku-band Analog EIRP Density Limits**

Figure 3.B.1   Routine licensing of small antennas: Ku-band analog EIRP density limits.

Ku-Band Digital			
	E/S Antenna Gain (dBi) CFR 25.209(a)(1)	RF Power Density (dBW/4kHz) CFR 25.212(c)	EIRP Density (dBW/4kHz)
1	29.0	−14	15.0
2	21.5	−14	7.5
3	17.1	−14	3.1
4	13.9	−14	−0.1
5	11.5	−14	−2.5
6	9.5	−14	−4.5
7	8.0	−14	−6.0
−1	29.0	−14	15.0
−2	21.5	−14	7.5
−3	17.1	−14	3.1
−4	13.9	−14	−0.1
−5	11.5	−14	−2.5
−6	9.5	−14	−4.5
−7	8.0	−14	−6.0

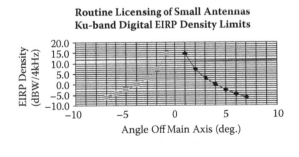

**Routine Licensing of Small Antennas
Ku-band Digital EIRP Density Limits**

Figure 3.B.2   Routine licensing of small antennas: Ku-band digital EIRP density limits.

C-Band Analog			
	E/S Antenna Gain (dBi)	RF Power Density (dBW/4kHz)	EIRP Density (dBW/4kHz)
	CFR 25.209(a)(1)	CFR 25.212(c)	
1	29.0	0.5	29.5
2	21.5	0.5	22.0
3	17.1	0.5	17.6
4	13.9	0.5	14.4
5	11.5	0.5	12.0
6	9.5	0.5	10.0
7	8.0	0.5	8.5
−1	29.0	0.5	29.5
−2	21.5	0.5	22.0
−3	17.1	0.5	17.6
−4	13.9	0.5	14.4
−5	11.5	0.5	12.0
−6	9.5	0.5	10.0
−7	8.0	0.5	8.5

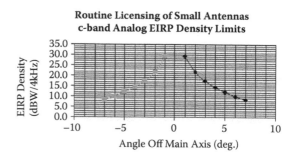

Figure 3.B.3   Routine licensing of small antennas: C-band analog EIRP density limits.

C-Band Digital			
	E/S Antenna Gain (dBi)	RF Power Density (dBW/4kHz)	EIRP Density (dBW/4kHz)
	CFR 25.209(a)(1)	CFR 25.212(d)	
1	29.0	−2.7	26.3
2	21.5	−2.7	18.8
3	17.1	−2.7	14.4
4	13.9	−2.7	11.2
5	11.5	−2.7	8.8
6	9.5	−2.7	6.8
7	8.0	−2.7	5.3
−1	29.0	−2.7	26.3
−2	21.5	−2.7	18.8
−3	17.1	−2.7	14.4
−4	13.9	−2.7	11.2
−5	11.5	−2.7	8.8
−6	9.5	−2.7	6.8
−7	8.0	−2.7	5.3

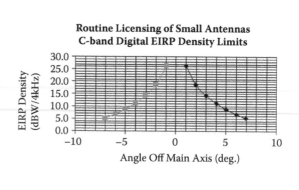

Figure 3.B.4   Routine licensing of small antennas: C-band digital EIRP density limits.

Chapter 4

Modulation and Multiplexing Techniques

This chapter covers the critical topics of modulation and channel multiplexing (multiple access). These functions (in addition to forward error correction [FEC], discussed in Chapter 5) drive, in large measure, the performance of the satellite link, the effective channel throughput, and the service availability that one is able to obtain over the satellite link. This chapter only presents a basic introduction to these topics and focuses pragmatically only on commercial satellite applications. Each of these topics would require a textbook to begin to cover the discipline in a more complete manner.

4.1 Modulation

Radio signals are electromagnetic waves that propagate in space with an underlying sinusoidal pattern defining the wave's instantaneous amplitude/frequency/phase value, as covered in Chapter 2. To transmit a content-bearing signal over a radio frequency (RF) link, a system needs to support three main stages [AGI200101]:

1. A carrier is generated at the transmitter. A carrier signal is a single-frequency signal that is used to carry the intelligence (data). There are three characteristics of a sinusoidal carrier signal that can be changed over time: amplitude, phase, and frequency. (Note, however, that phase and frequency are just different ways to view or measure the same signal change.)
2. The carrier is modulated with the information to be transmitted. Modulation is the process of overlaying a signal that has (some encoded) intelligence over an appropriate underlying carrier, thereby enabling the long-distance transport and reception of the intelligent signal. Any reliably detectable change in signal characteristics can carry information. When the underlying signal is modulated, it is able to carry a message over a distance; in general, the intelligent signal would otherwise be unfit for direct baseband transmission.*

---

* Note, however, that some short-haul applications, such as wireline Local Area Networks, do not employ modulation but transmit the signal in baseband mode for up to 100 m.

3. At the receiver, the signal modifications or changes are detected and demodulated at this point; the modulating signal (intelligence or data) is recovered.

Modulation techniques fall into three categories or combinations thereof:

■ Amplitude modulation (AM) (as in the case of AM radio)
■ Frequency modulation (FM) (as in the case of FM radio)
■ Phase modulation (PM) (typical of data and satellite communications)

AM changes only the magnitude of the signal; PM changes only the phase of the signal; FM is similar to PM, but here the frequency is the controlled parameter rather than the relative phase. Amplitude and phase modulation can also be used together. When the intelligent signal is digital, one talks about *digital modulation*. AM is then known as amplitude shift keying (ASK), FM is then known as frequency shift keying (FSK), and PM is then known as phase shift keying (PSK). Contemporary (satellite) communications make use of digital modulation, hence the focus of the discussion in this chapter is on digital techniques; however, we start out with a basic overview of modulation in general.

4.1.1 Analog Frequency Modulation

FM uses the instantaneous frequency of a modulating signal (voice, music, data, etc.) to directly vary the frequency of a carrier signal. It is the most common analog modulation technique used in mobile communications systems. In FM, the frequency of a rapidly varying signal is modulated by a slowly varying signal; the amplitude of the modulating carrier is kept constant, whereas its frequency is varied by the modulating message signal. The modulation index, β, is used to describe the ratio of maximum frequency deviation of the carrier to the maximum frequency deviation of the modulating signal [BLA200701]. For example, if maximum frequency (fmax) causes a maximum deviation of $1 \times$ fmax in the carrier, then $\beta = 1/1 = 1$; if maximum frequency (fmax) causes a maximum deviation of $5 \times$ fmax in the carrier, then $\beta = 5/1 = 5$; and so on.

The basic mathematical description of the modulation of the frequency of a sinusoidal carrier by another sinusoidal signal is given in the following formula.

Let the carrier be

$$x_c(t) = X_c \times \cos(\omega_c t)$$

and the modulating signal be

$$x_m(t) = \beta \times \sin(\omega_m t)$$

Then, the FM-modulated signal is

$$x(t) = X_c \times \cos[\omega_c t + \beta \cdot \sin(\omega_m t)]$$

with

$$\beta = \Delta\omega/\omega_m$$

Narrowband FM is defined as the condition where β is small enough to make all terms after the first two in the series expansion of the FM equation negligible; this is the case for

$$\beta = \Delta\omega/\omega_m < 0.2$$

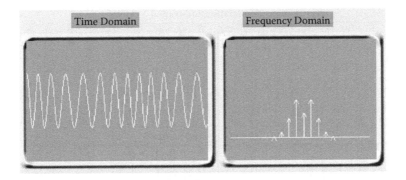

Figure 4.1   Time/frequency view of FM.

The occupied bandwidth (BW) is a measure of the frequency spectrum covered by the signal in question. The following approximation can be used for BW in the narrowband case:

$$BW \sim 2\omega_m$$

Wideband FM is defined as a case when a significant number of sidebands have nonnegligible amplitudes. Here, an approximation for BW is

$$BW \sim 2\Delta\omega$$

J. R. Carson showed that a reasonable approximation for BW is

$$BW \sim 2\,(\Delta\omega + \omega_m) = 2 \times \omega_m\,(1 + \beta)$$

A plot of the resulting signal in the time domain (amplitude versus time) is shown in Figure 4.1. An expansion of the formula shows that the resulting signal consists of a large number of frequency components, or sidebands. This is in contrast to the case of AM, where only two sidebands are created. The sidebands are spaced apart by the modulation frequency, X_m, and centered about the carrier frequency, X_c. This issue has implication in digital modulation as we discuss later. The frequency domain portion of Figure 4.1 shows the carrier and sideband components as a plot of amplitude versus frequency; only the most significant components are shown. An infinite number of higher-order components have negligible amplitude. In fact, the amplitude of each component is given by a Bessel function of the appropriate component order; the amplitude of the n-th sideband is $J_n(\beta)$ [AGI200701].

To further illustrate FM, in the examples of Figure 4.2, the carrier frequency is 11 times the value of the modulation frequency; the dashed lines represent the modulation envelope, and the solid lines represent the modulated carrier [BLA200701]. Figure 4.3 depicts the modulation process when the modulating signal is digital.

4.1.2 Analog Amplitude Modulation

AM uses the instantaneous amplitude of a modulating signal (voice, music, data, etc.) to directly vary the amplitude of a carrier signal. Here, the amplitude of a high-frequency carrier signal is varied in proportion to the instantaneous amplitude of the modulating message signal. The

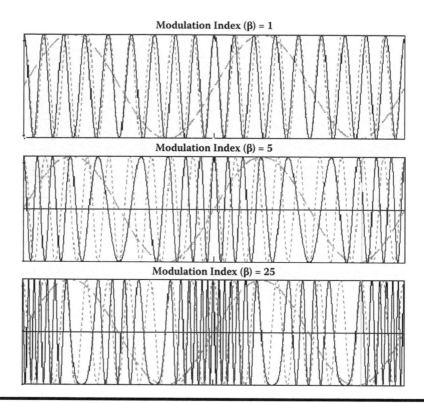

Figure 4.2   FM-modulated analog signal as a function of the modulation index.

modulation index, m, is used to describe the ratio of maximum voltage to minimum voltage in the modulated signal. It is defined as

$$m = (V_{max} - V_{min})/(V_{max} + V_{min})$$

If the modulating signal is equal in magnitude to the carrier, then $m = 1$, and the modulated signal varies from a (scaled) maximum of unity down to zero (see Figure 4.4). When $m = 0$, no modulation of

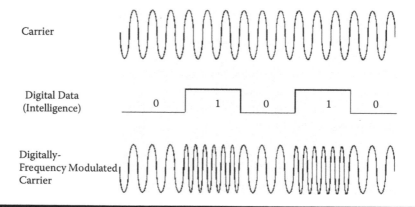

Figure 4.3   FM modulation where the modulating signal is digital.

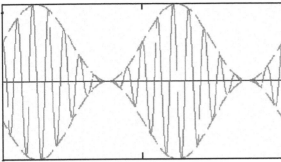

$m = (2-0)/(2+0) = 1.0$

100% Modulation: Here, the maximum voltage (Vmax) is 2 V and the minimum (Vmin) is 0 V.

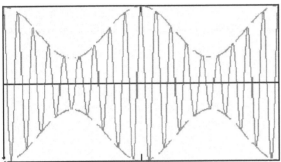

$m = (3-1)/(3+1) = 0.5$

50% Modulation: Here, the maximum voltage (Vmax) is 3 V and the minimum (Vmin) is 1 V.

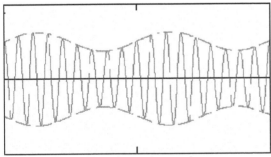

$m = (1.25-0.75)/(1.25+0.75) = 0.25$

25% Modulation: Here, the maximum voltage (Vmax) is 1.25 V and the minimum (Vmin) is 0.75 V.

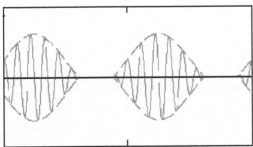

$m = (2.5-(-0.5))/(2.5+(-0.5)) = 1.5$

150% Modulation: Here, the maximum voltage (Vmax) is 2.5 V and the minimum (Vmin) is −0.5 V.

Figure 4.4 AM-modulated analog signal as a function of the modulation index.

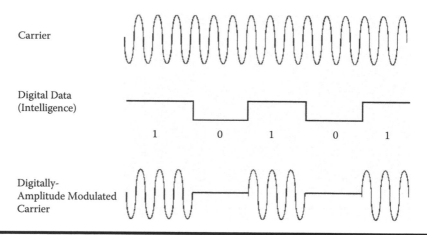

Carrier

Digital Data
(Intelligence)

1   0   1   0   1

Digitally-
Amplitude Modulated
Carrier

Figure 4.5   **AM modulation where the modulating signal is digital.**

the carrier is performed. If *m* is greater than 1, the carrier is actually cut off for some period of time, and unwanted harmonics are created at the transmitter output. In the frequency domain, the carrier frequency is flanked on both sides by mirror image copies of the modulating signal, as follows [BLA200701]:

$$\omega_{M1} = \omega_c \pm \omega_{m1}$$
$$\omega_{M2} = \omega_c \pm \omega_{m2}$$

Let the carrier be

$$x_c(t) = X_c \times \sin\,(\omega_c t)$$

and the AM signal be

$$x_m(t) = X_m \times \sin\,(\omega_m t)$$

Then

$$x(t) = X_c \times [1 + m \times \sin\,(\omega_m t)] \times \sin\,(\omega_c t)$$

In the examples of Figure 4.4, the carrier frequency is nine times the modulation frequency. The dashed lines represent the modulation envelope, and the solid lines represent the modulated carrier [BLA200701]. Figure 4.5 depicts the modulation process when the modulating signal is digital.

4.1.3 *Phase Modulation*

Phase modulation is basically similar to FM, as seen in Figure 4.6; the phase of the carrier is changed based on the modulating signal. Figure 4.7 depicts the modulation process when the modulating signal is digital.

4.1.4 *Digital Modulation and Constellations*

Digital modulation is the process by which an analog carrier wave is able to carry a digital signal. Basic modulation can support a bit per baud (also referred to as signal change or symbol); higher-order modulation allows one to encode several bits per baud (per symbol).

Modulated signal has the form:

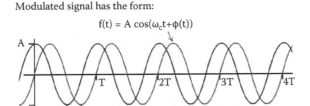

	Phase Modulation	Frequency Modulation
$x_c(t) =$	$X_c\cos(\omega_c t + \beta\cos(\omega_i t))$	$X_c\cos(\omega_c t + \beta\sin(\omega_i t))$
Modulation index $\beta =$	$K_p a$	$\dfrac{\Delta f}{f_i} = K_f a$

K_p, K_f Constants
   A Maximum amplitude of modulating signal, $x_m(t) = a \cdot \cos(\omega_i t)$
   Δf Peak frequency deviation
   f_i Frequency of the information signal (Hz)

Figure 4.6   Phase modulation.

ASK involves increasing the amplitude (power) of the wave in step with the digital signal (low = 0, high − 1). FSK changes the frequency in step with the digital signal; different bits are represented by different frequencies that can then be detected by a receiver (see Figure 4.8). PSK is similar to FSK: the approach here is to change the phase of the carrier in step with the digital message. PSK/FSK modulation is more resistant to noise than ASK (because typical noise usually changes the amplitude of the signal), and it is able to support higher data rates (i.e., higher bits

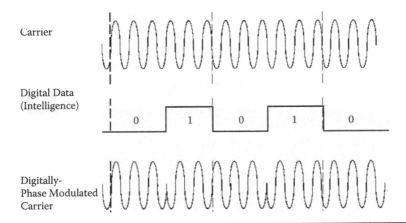

Figure 4.7   PM modulation where the modulating signal is digital.

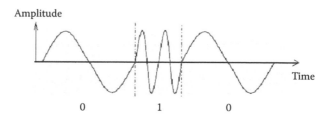

Figure 4.8   Frequency shift keying.

per symbol); however, intermodulation due to linearities in the system (e.g., amplifiers) can be problematic. As noted, PSK is very common in satellite communications.

There are several different types of PSK modulation schemes; these include the following:

- *Two-phase shift keying* (2-PSK), also known as binary phase shift keying (BPSK): Here, 1 bit is processed to produce a single-phase change; each symbol "carries" 1 bit. This "older technology" is rarely used at this time.
- *Four-phase shift keying* (4-PSK), also known as quadrature phase shift key (QPSK): Here 2 bits are processed to produce a single-phase change; each symbol "carries" 2 bits. This is a very typical, simple but robust form of digital modulation; it is employed in the Digital Video Broadcast–Satellite (DVB-S) standard, which was especially in use in the past 10 years. (Signals that are separated by 90° are known to be orthogonal to each other, or in quadrature; signals that are in quadrature do not interfere with one another.)
- *Eight-phase shift keying* (8-PSK): Here, 3 bits are processed to produce a single-phase change; this means that each symbol "carries" 3 bits. This is now the current direction in satellite modulation technique; for example, it is employed in the Digital Video Broadcast–Satellite–Second Generation (DVB-S2) standard, being field deployed in the recent past.
- *Sixteen-phase shift keying* (16-PSK): Here, 4 bits are processed to produce a single-phase change; this means that each symbol consists of 4 bits. Noise issues begin to become critical at this stage of modulation complexity.
- *Sixteen-quadrature amplitude modulation* (16-QAM): Here, 4 bits are processed to produce a single vector that maps to a signal change. QAM can also operate with 32, 64, and 256 states, each of which is defined by a specific amplitude and phase value. As the number of states increases, however, the generation and detection of symbols becomes more complex; naturally, as the number of states per symbol increases, the total data rate increases. (These modulation schemes all occupy the same channel bandwidth after filtering.) Satellite communications typically only use QAM-16. QAM's generation and detection of symbols are more complex than in a QPSK because, as noted, in QAM both the phase and the amplitude changes have also to be detected.

For BPSK, each symbol indicates two different states; this supports 1 bit per symbol (phase shift of 0° = 0; phase shift of 180° = 1). (See Figure 4.9.) QPSK uses four phase values: 45, 135, 225, and 315°; now 2 bits per symbol can be transmitted. Each symbol's phase is compared relative to the previous symbol (this is called *differential modulation**); hence, if there is a phase shift of 45°, the

---

* Differential modulation is modulation in which the choice of the significant condition for any signal element is dependent on the significant condition for the previous signal element. This is in contrast to coherent modulation in which the phase of the carrier is discretely modulated in relation to the phase of a reference signal; this reference signal must be present at both the modulator and the demodulator. Differential modulation is generally the norm.

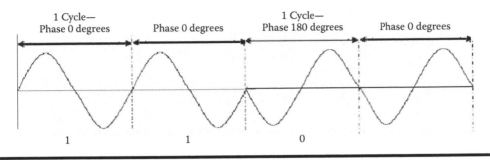

Figure 4.9   Phase shift keying (illustrative).

bits "00" are represented; if there is a phase shift of 135°, the bits "01" are encoded [SHO200401]. See Figure 4.10 and also Table 4.1. 8-PSK follows a similar format, as seen in Table 4.2.

As noted, ASK and PSK can be combined to create QAM, where both the phase and amplitude are changed. Figure 4.11 depicts the signal changes in a 16-QAM environment. QAM is the representation and transmission of digital information by encoding bit sequences of fixed, specified lengths (e.g., 4, 5, 6, 7, or 8 bits), and representing these bit sequences as a function of (a) the amplitude of an analog carrier, (b) a phase shift of the analog carrier with respect to the phase that represented the preceding bit sequence, where the permissible phase shift is an integral multiple of $\pi/2$ radians (90°, or one-quarter unit interval) or (c) both. The name *quadrature* originates from the stipulation that a phase shift, when required, must be an integral multiple of $\pi/2$ radians, that is, one-quarter of a cycle (unit interval, or baud) [ANS200001]. A representative QAM table is shown in Table 4.3.

There is no limit, in theory, to the data rate that may be supported by, or associated with, a given baud rate in a perfectly stable, noiseless transmission environment. In practice, the governing factors are the amplitude (and consequently, phase) stability and the amount of noise present in both the terminal equipment and the transmission medium (carrier frequency or communication channel) involved. The permitted relative (incremental) phase shift does not necessarily have to be a multiple of $\pi/2$ radians. (The term *quadrature*, however, does not apply if any other minimum phase shift were specified or permitted.) When using a modulation technique with higher bit rates, such as 64-QAM, an adequate link signal-to-noise ratio (SNR) is needed to overcome any interference and maintain an acceptable bit error rate (BER) (also called bit error ratio).

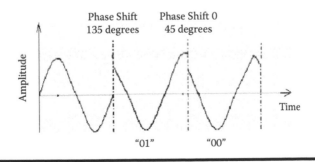

Figure 4.10   Quadrature phase shift keying (illustrative).

Table 4.1 Quadrature phase shift keying

Symbol	Phase shift (degrees)
00	45
01	135
11	225
10	315

Because it becomes difficult to depict the signal variations graphically as shown in previous figures when the number of bits per baud (bits per signal change, or bits per symbol) becomes large, a graphical method based on *constellations* is employed in general. Constellation diagrams are also used to graphically represent the quality as well as the distortion of a digital signal. One method to view the amplitude and phase is with a polar diagram, where

- The carrier signal becomes a frequency and phase reference.
- The modulating signal can be expressed as a magnitude and a phase.

The phase is relative to a reference signal, typically the carrier; the magnitude is either an absolute or relative value. It is also common to describe the signal vector by its rectangular coordinates of I (in-phase) and Q (quadrature). On a polar diagram, the I axis lies on the 0° phase reference, and the Q axis is rotated by 90°. The signal vector's projection onto the I axis is its "I" component, and its projection onto the Q axis is its "Q" component. In Figure 4.12, the phase is represented as the angle, and the magnitude is represented as the distance from the center. I/Q diagrams are useful because they reflect the way digital communications signals are created. In the transmitter, I and Q signals are mixed using a local oscillator (LO), with a 90° phase shifter placed in one of the LO paths; then they are summed to a composite output signal. Eventually, after passing through a transmission channel, the composite signal arrives at the receiver input. The input signal is mixed with the LO signal at the carrier frequency in two forms: one is at a 0° phase shift, and the other, with a 90° phase shift. The composite input signal is thus separated into in-phase, I, and quadrature, Q, components. Digital modulation is easy to accomplish with I/Q modulators because digital modulation maps the data to a number of discrete points on the I/Q plane; these

Table 4.2 Eight-point phase shift keying

Bit sequence represented	Relative phase shift radians (degrees)
000	0 (0°)
001	$\pi/4$ (45°)
010	$\pi/2$ (90°)
011	$3\pi/4$ (135°)
100	π (180°)
101	$5\pi/4$ (225°)
110	$3\pi/2$ (270°)
111	$7\pi/4$ (315°)

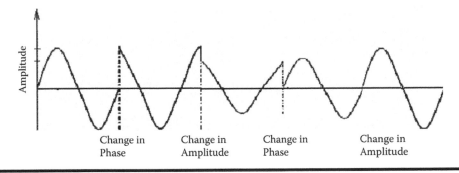

Figure 4.11 QAM (illustrative).

are the constellation points. As the signal moves from one point to another, simultaneous amplitude and phase modulation results [AGI200101].

Figure 4.13 depicts the concept of constellations in general, whereas Figure 4.14 depicts in more detail the constellation point/values described thus far; the top left is 4-PSK (QPSK), top right is 8-PSK, bottom right is 16-PSK, and bottom right is 16-QAM. In 16-state QAM, there are four *I* values and four *Q* values; this results in a total of 16 possible states for the signal. The symbol rate is one-fourth of the bit rate; therefore, this modulation format supports a more spectrally efficient transmission than BPSK, QPSK, or 8-PSK. Other constellations are possible (but not generally used in satellite communication). For example, in 32-QAM, there are six *I* values and six *Q* values, resulting in a total of 36 possible states; this equates to too many states for a power of two (32). Hence, the four corner symbol states, which require the highest power to transmit, are omitted, reducing the amount of peak power the transmitter has to generate. The current practical limits are approximately 256-QAM (16 *I* values and 16 *Q* values), but there is work underway to extend the limits to 512- or 1024-QAM. In these constellations, the symbols are very close together, and, thus, more subject to errors due to noise and distortion. Figure 4.15 depicts transition states.

Two other schemes are of (some) interest. 16-APSK (amplitude phase shift keying) and 32-APSK are high-efficiency 16-ary and 32-ary coded modulation schemes optimized for nonlinear satellite channels (nonlinear channels are typical satellite environments). APSK is, prima facie, an attractive modulation format for digital transmission over nonlinear satellite channels due to its

Table 4.3 QAM signal values

Bit sequence represented	Normalized carrier amplitude	Relative phase shift radians (degrees)
000	1/2	0 (0°)
001	1	0 (0°)
010	1/2	$\pi/2$ (90°)
011	1	$\pi/2$ (90°)
100	1/2	π (180°)
101	1	π (180°)
110	1/2	3 $\pi/2$ (270°)
111	1	3 $\pi/2$ (270°)

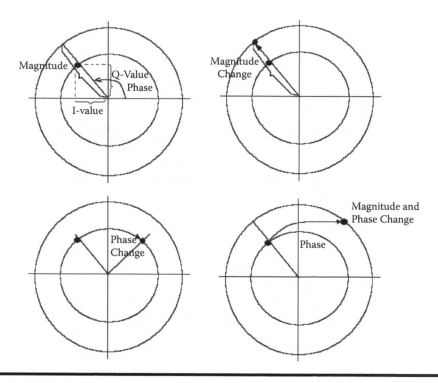

Figure 4.12   Different state transitions in polar form.

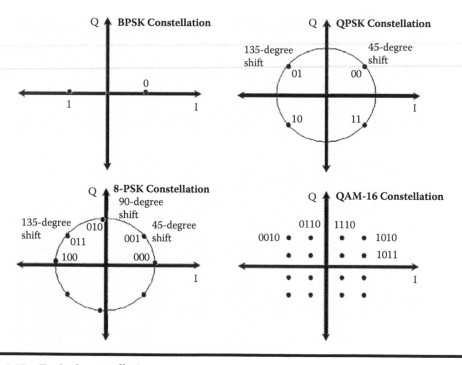

Figure 4.13   Typical constellations.

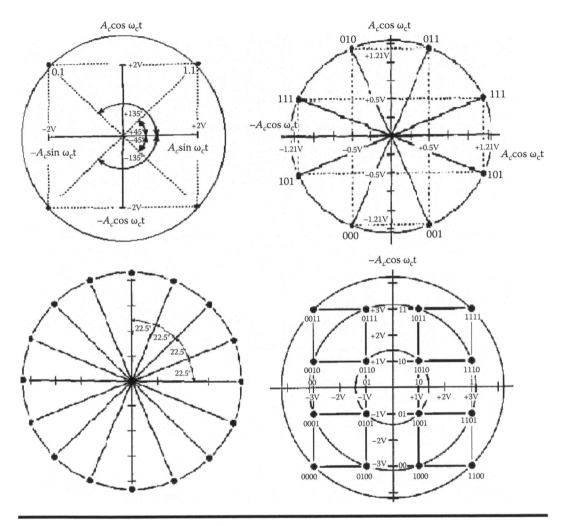

Figure 4.14   Constellation points/values.

power and spectral efficiency combined with its inherent robustness against nonlinear distortion. For these reasons APSK has been included in the new DVB-S2 [DEG200601].

M-APSK constellations are composed of n_R concentric rings, each with uniformly spaced PSK points. The signal constellation points x are complex numbers, drawn from a set X, given by

$$
X = \begin{cases}
r_1 e^{j\left(\frac{2\pi}{n_1}i + \theta_1\right)} & i = 0,\dots,n_1 - 1, \quad (\text{ring } 1) \\[2ex]
r_2 e^{j\left(\frac{2\pi}{n_2}i + \theta_2\right)} & i = 0,\dots,n_2 - 1, \quad (\text{ring } 2) \\[2ex]
\vdots & \\[1ex]
r_{n_R} e^{j\left(\frac{2\pi}{n_R}i + \theta_{n_R}\right)} & i = 0,\dots,n_{n_R} - 1, \quad (\text{ring } n_R)
\end{cases}
$$

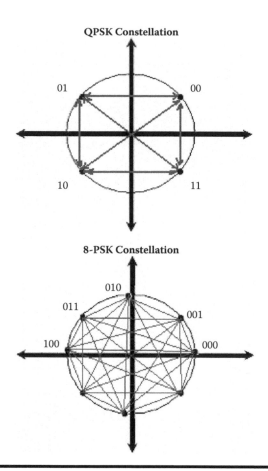

Figure 4.15   Signal transitions.

where n_l represents the number of points corresponding to the l-th ring, and r_l and μ_l are the radius and the relative phase shift corresponding to the l-th ring, respectively. Some refer to these modulations as "$n_1 + \ldots + nn_R$-APSK." Figure 4.16 depicts the 4 + 12-APSK and 4 + 12 + 16-APSK modulations with quasi-Gray mapping. In particular, for next generation broadband systems, the constellation sizes of interest are $|X| = 16$ and 32, with $n_R = 2$ and 3 rings, respectively. One can also define the phase shifts and the ring radii in relative terms rather than in absolute terms; this removes one dimension in the optimization process, yielding a practical advantage. Let $\phi_l = \Theta_l - \Theta_1$ for $l = 1, \ldots, n_R$ be the phase shift of the l-th ring with respect to the inner ring. One also defines $\rho_l = r_l - r_1$ for $l = 1, \ldots, n_R$ as the relative radii of the l-th ring with respect to r_1. In particular, $\rho_1 = 0$ and $\rho_1 = 1$. Figure 4.17 shows the symbol error rate (SER) for APSK as a function of ϕ. Note that the SER is less than would be the case with 16 QAM.

Note: A number of other modulation schemes are available for wireless, television broadcast, telephony, and other applications; these, however, are not as common in commercial satellite communications and hence are not covered here.

At the high-point constellations, the transmitting and receiving equipment become more complex, and the signal is more susceptible to errors caused by noise and distortion. Error rates of higher-order QAM systems degrade more rapidly than QPSK, as seen later. FEC methods can

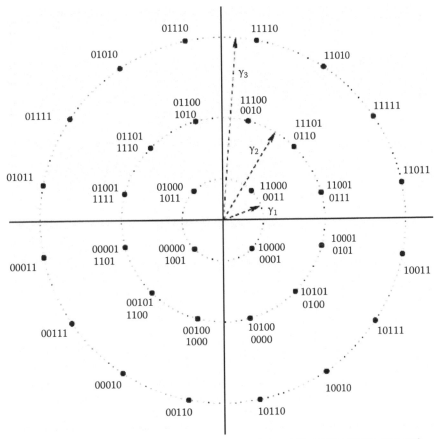

Parametric description and pseudo-Gray mapping of 16 and 32-APSK constellations with $n_1 = 4$, $n_2 = 12$, $\phi_2 = 0$ and $n_1 = 4$, $n_2 = 12$, $n_3 = 16$, $\phi_2 = 0$, $\phi_3 = \pi/16$ respectively. For the first two rings: mapping below corresponds to 4+12-APSK, mapping above to 4+12+16-APSK.

Figure 4.16   16-APSK (amplitude phase shift keying) and 32-APSK. Courtesy [DEG200601]

address some of these issues; however, tradeoffs are ever present. FEC adds bits to the payload that can be employed to recover the signal in case of bit impairment. For example, a 1/4 code adds 3 bits (= 4 − 1) every 1 bit of payload, a 5/6 code adds 1 bit (= 6 − 5) every 5 bits of payload, and a 9/10 code adds 1 bit (= 10 − 9) every 9 bits of payload. FEC is critical to systems that operate with a low SNR and suffer from distortion or other channel impairments. Modern FEC algorithms can reduce BER by 3 to 5 orders of magnitude. For example, an uncorrected BER of 10^{-3} becomes a corrected BER of 10^{-8}. FEC codes must combine power efficiency and low BER floor with flexibility and simplicity to allow for high-speed implementation.

The two most often used coding techniques for error correction are convolutional encoding (Viterbi and turbo codes) and block codes (Reed–Solomon encoding). Convolutional encoding can be used in isolation, but it is increasingly used in combination with Reed–Solomon encoding in order to improve the overall error correction performance of the link, or bit error rate. These techniques are complex mathematical functions that are applied to the required data within the modulator. These coding techniques, especially convolutional encoding, are often referred to as FEC because they do not need a return path from the receiver to the transmitter in order to function

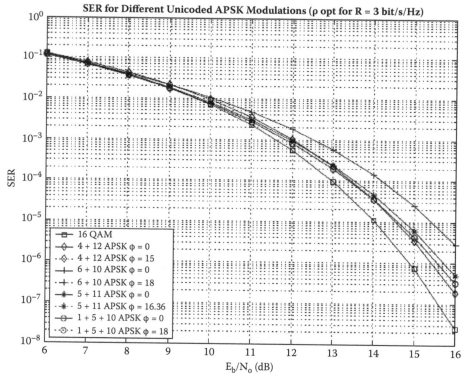

Upper bound on the uncoded symbol error probability for several APSK modulations.
Note that the continuous line and the dashed line are indistinguishable
because they are superimposed.

Figure 4.17   Performance of 16-APSK (amplitude phase shift keying) and 32-APSK. Courtesy [DEG200601]

correctly (i.e., they work in the "forward" direction only, and the receiver handles any errors without further assistance from the transmitter). A commonly used and efficient decoding technique for convolutional codes is known as *Viterbi* after its inventor. Convolutional encoding is sometimes referred to as Viterbi coding, although strictly speaking this term applies to the decoding process only. Reed–Solomon coding is a specialized coding system that performs well in correcting short bursts of errors, such as those seen in satellite communications at the output of Viterbi decoders. This is why convolutional encoding and Reed–Solomon encoding are often used in tandem, with the Reed–Solomon code used as an "outer" code and the convolutional code as the "inner" code with Viterbi decoding. The Reed–Solomon coding is specified as RS(*n*, *k*), where n is the total number of bits in the word, and k is the actual data bits in the word. For example, an RS(255,223) code takes 255 bits, including redundant data, to transmit 223 bits of information [FOC200701].

Practical, simple, and powerful coding designs for binary modulations now include turbo codes and low-density parity-check (LDPC) codes, both of which were defined in the recent past. Starting with trellis-coded modulation (TCM), the approach had been to consider channel code and modulation as a single entity, jointly designed and demodulated/decoded. Schemes have been published in the literature, where turbo codes are successfully merged with TCM. A new pragmatic paradigm has crystallized under the name of bit-interleaved coded modulation (BICM), where good results are obtained with a standard nonoptimized code. An additional advantage of BICM is its inherent flexibility, an appealing feature for broadband satellite communication systems, where a large set of spectral efficiencies is needed [DEG200601]. This topic is discussed in detail in Chapter 5, but for now the concepts introduced in this paragraph suffice at this high level of the discussion.

4.1.5 E_b/N_0 (E_b over N_0)

E_b/N_0 is a commonly used parameter to compare digital systems. It is a signal-to-noise ratio; specifically, it is the ratio of bit energy to noise power spectral density. See Table 4.4 for related definitions [MAT200701].

Table 4.4   Related definitions

Bit rate	Bit rate = (symbol rate) × (number of bits sent per symbol)
	That is, the symbol rate is the bit rate divided by the number of bits that can be transmitted with each symbol. If 1 bit is transmitted per symbol (e.g., in BPSK), then the symbol rate would be the same as the bit rate. If 2 bits are transmitted per symbol (e.g., in QPSK), then the symbol rate would be half of the bit rate.
E_b/N_0 (dB)	The ratio of bit energy per symbol to noise power spectral density, in decibels. 0 dB means the signal and noise power levels are equal, and a 3 dB increment doubles the signal relative to the noise.
E_s/N_0 (dB)	The ratio of signal energy per symbol to noise power spectral density, in decibels.

$$E_s/N_0 = (T_{sym}/T_{samp}) \times SNR$$
$$E_s/N_0 = E_b/N_0 + 10\log_{10}(k) \text{ in dB}$$

where

E_s = signal energy (J)
E_b = bit energy (J)
N_0 = noise power spectral density (W/Hz)
T_{sym} = symbol period parameter of the block in E_s/N_0 mode
k = number of information bits per input symbol (in this context)
T_{samp} = sample time of the block, in seconds.

In a communication system, k might be influenced by the size of the modulation alphabet or the code rate of an error-control code. For example, if a system uses a rate-1/2 code and 8-PSK modulation, then the number of information bits per symbol (k) is the product of the code rate and the number of coded bits per modulated symbol: (1/2) $\log_2(8) = 3/2$. In such a system, three information bits correspond to six coded bits, which in turn correspond to two 8-PSK symbols.

Input signal power (W)	The mean square power of the input symbols.
Number of bits per symbol	The number of bits in each input symbol.
SNR (dB)	The ratio of signal power to noise power, in decibels.
Symbol period (s)	The duration of a channel symbol, in seconds.

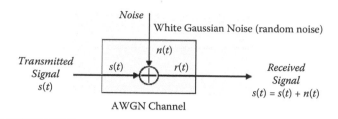

Figure 4.18 Example of noisy channel.

E_b is a measure of the bit energy, that is P_{avg}/R_b, where R_b is the bit rate; N_0 is the noise density, that is, the total noise power in the frequency band of the signal divided by the bandwidth of the signal, P_n/B_n, with P_n = noise power (the units here in Joules) and B_n = noise bandwidth. N_0 is measured in W/Hz, and it is the noise power in 1 Hz of bandwidth. Figure 4.18 depicts a model of a noisy channel.

Obviously, if the power increases, E_b/N_0 increases; if the noise increases, E_b/N_0 decreases. As we saw in Chapter 2, noise power is computed using Boltzmann's equation, that is,

$$N = kTB$$

where
  k is Boltzmann's constant = 1.380650×10^{-23} J/K,
  T is the effective temperature in Kelvin, and
  B is the channel bandwidth.

One can derive the relationship between E_s/N_0 and *SNR* as follows:

$$E_s/N_0 \text{ (dB)} = 10 \log_{10}((S \cdot T_{sym})/(N/B_n))$$

$$= 10 \log_{10}((T_{sym} F_s) \cdot (S/N))$$

$$= 10 \log_{10}(T_{sym}/T_{samp}) + SNR \text{ (dB)}$$

where
  T_{sym} = signal's symbol period
  T_{samp} = signal's sampling period
   S = input signal power, in Watts
   N = noise power, in Watts
   B_n = noise bandwidth, in Hertz
   F_s = sampling frequency, in Hertz

Note that $B_n = F_s = 1/T_{samp}$.
  One can show that

$$\text{Probability of error} = P(e) = \frac{1}{2} erfc \left(\frac{E_b}{N_0} \right)^{1/2}$$

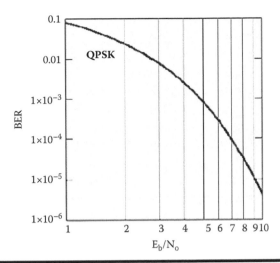

Figure 4.19 BER versus E_b/N_0.

where *erfc*, called the complimentary error function, describes the cumulative probability curve of a Gaussian distribution, namely,

$$efrc(x) = (2/\sqrt{\pi})\int_x^\pi e^{-u^2}\,\mathrm{du}$$

BER is related to E_b/N_0. For example, for QPSK in an additive white Gaussian noise (AWGN) channel, the BER is

$$\mathrm{BER} = \frac{1}{2}\,erfc(\sqrt{E_b/N_0})$$

Figure 4.19 shows this relationship in the case of QPSK. Note that an approximation for $erfc\sqrt{(E_b/N_0)}$ is $(1/\sqrt{\pi})\frac{\exp(-E_b/N_0)}{\sqrt{(E_b/N_0)}}$ when $E_b/N_0 \geq 4(\sim 6\ \mathrm{dB})$ [MAR200201].

To improve BER performance beyond what the basic E_b/N_0 provides (or to optimize operation at a given fixed E_b/N_0), one needs to employ various FEC techniques. Figure 4.20 depicts the general behavior of a modulation scheme used in conjunction with FEC. As can be seen in Figure 4.21, curves become steeper as the code's performance improves: gradients become such that a small improvement in E_b/N_0 has a major impact in BER. (Note that modern FEC algorithms are getting very close to the Shannon limit.)

Figures 4.22 and 4.23 further depict the improvements due to FEC. Note that in an analog environment the measures of interest are C/N (carrier power to noise power ratio) and C/N_0, as we saw in Chapter 2. These two measures are employed in the same way E_b/N_0 is used in digital environments. C/N is the carrier power in the entire usable bandwidth, C/N_0 is carrier power per unit bandwidth. One can convert E_b/N_0 to C/N using the equation

$$\frac{C}{N} = \frac{E_b}{N_0} \times \frac{R_b}{B}$$

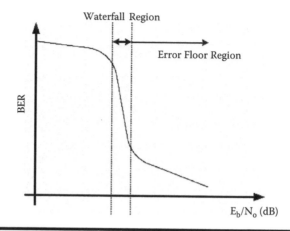

Figure 4.20 Basic behavior of FEC.

where
   B = channel bandwidth.
   R_b is the bit rate.

When all the terms are in dB, then one has

$$\frac{C}{N} = \frac{E_b}{N_0} + R_b - B$$

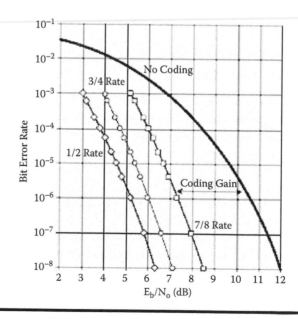

Figure 4.21 Improvements due to FEC.

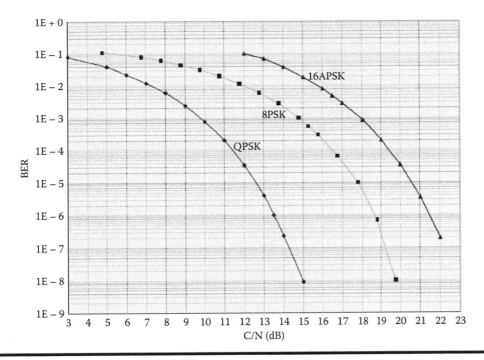

Figure 4.22   Uncoded error rate performance for QPSK, 8-PSK, and 16-APSK.

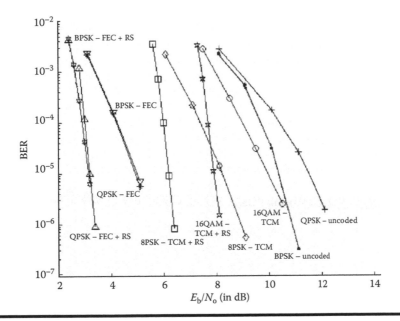

Figure 4.23   FEC-improved error rate performance for a number of schemes.

Table 4.5   Example of signal issues in digital modulation

Amplitude imbalance	The different gains of the I and Q components of a signal. In a constellation diagram, amplitude imbalance shows one signal component expanded, and the other one, compressed.
Phase error	The difference between the phase angles of the I and Q components referred to 90°. A phase error is caused by an error of the phase shift of the I/Q modulator. The I and Q components are in this case not orthogonal to each other after demodulation.
Interferers	Spurious sinusoidal signals occurring in the transmission frequency range and superimposed on the QAM signal at some point in the transmission path. After demodulation, the interferer is contained in the baseband form of low-frequency sinusoidal spurious signals. The frequency of these signals corresponds to the difference between the frequency of the original sinusoidal interference and the carrier frequency in the RF band. In the constellation diagram, an interferer shows in the form of a rotating pointer superimposed on each signal status. The constellation diagram shows the path of the pointer as a circle around each ideal signal status.

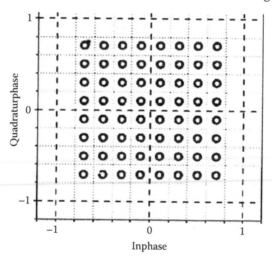

Carrier suppression or leakage	A special type of interference in which its frequency equals the carrier frequency in the RF channel. Carrier leakage can be superimposed on the QAM signal in the I/Q modulator. In the constellation diagram, carrier leakage shows up as a shifting of the signal states corresponding to the DC components of the I and Q components.
Additive Gaussian noise	Noise that disturbs the digitally modulated signal during analog transmission, for instance, in the analog channel. Additive superimposed noise normally has a constant power density and a Gaussian amplitude distribution throughout the bandwidth of a channel. If no other error is present at the same time, the points representing the ideal signal status are expanded to form circular "clouds."

Table 4.5 Example of signal issues in digital modulation (*Continued*)

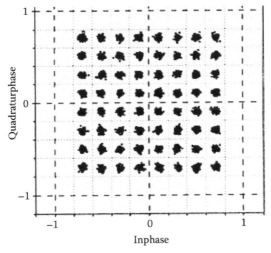

Phase jitter or phase noise	Noise caused by transponders in the transmission path or by the I/Q modulator. It may be produced in carrier recovery, a possibility that is to be excluded here. In contrast to the phase error described previously, phase jitter is a statistical quantity that affects the I and Q path equally. In the constellation diagram, phase jitter shows up by the signal states being shifted about their coordinate origin.

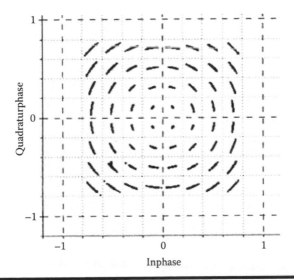

Table 4.5, based on [BLO200701], depicts some key impairments that impact demodulation. If the input signal is distorted or greatly attenuated, the receiver can no longer recover the symbol clock, demodulate the signal, or recover the information. In some cases, a symbol will fall far enough away from its intended position for it to cross over to an adjacent position; the in-phase (*I*) and quadrature (*Q*) level detectors used in the demodulator would misinterpret such a symbol as being in the wrong location, causing bit errors. QPSK is not as efficient as, say, 8-PSK or 16-QAM, but the states are much farther apart, and the system can tolerate a lot more noise before suffering symbol errors [AGI200101].

4.1.6 Filters and Roll-Off Factors

Modulation may give rise to signals in bands other than the assigned/intended band. Occupied bandwidth (BW), introduced earlier, is a measure of how much frequency spectrum is covered by the signal in question; the units are in Hertz, and the measurement of occupied BW generally implies a power percentage or ratio. Figure 4.24 shows a block diagram of a typical direct conversion transmitter; notice the low-pass filters (LPFs). The digital baseband low-pass filters eliminate out-of-band signal components from the baseband signal, but these components may reappear along the way as a result of intermodulation distortion generated within the transmitter, for example, in the downstream power amplifier, such as the satellite transponder. This tendency to regain undesirable signal components is called *spectral regrowth*. Figure 4.25 depicts this graphically. The regrowth of digital data power spectra due to amplifier nonlinearity is a known problem. Adjacent channel power ratio (ACPR) is a popular measure of spectral regrowth in digital transmitters and is often a design specification [CHE200701].

Consider driving a transmitter with baseband I/Q (in-phase (I) and quadrature (Q)) signals that, when plotted against each other, trace out a circular trajectory:

$$I = A \times \cos(\omega_m t)$$

and

$$Q = A \times \sin(\omega_m t)$$

Ideally, the RF signal would be

$$x(t) = A \times [\cos(\omega_m t) \times \cos(\omega_c t) - \sin(\omega_m t) \times \sin(\omega_c t)] = A \times \cos[(\omega_m + \omega_c) \times t]$$

However, nonlinearities in the transmitter introduce harmonics of ω_m above and below the carrier fundamental frequency. The harmonics distort the baseband signal. Analysis of the spectrum at the output of a soft saturating passband amplifier with a finite number of pure sinusoidal inputs generates explicit expressions for the magnitudes of output intermodulation products in terms of the amplitudes of the sinusoidal inputs and of the nonlinear amplifier characteristic expressed as a Taylor's series function of its input. The frequency-domain expression for amplifier output shows that the spectral regrowth depends on the cubic coefficient of the Taylor's series of the amplifier nonlinearity as well as the combination of input phase modulation and amplitude ripple [AMO199701].

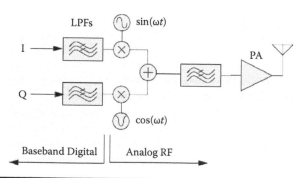

Figure 4.24   Typical (generic) transmitter.

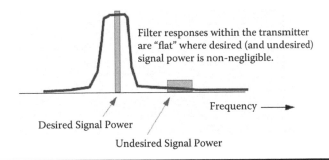

Figure 4.25   Spectral regrowth.

Given these observations on the possibility for spectral regrowth, a key question is whether a modulated signal fits "neatly" inside a transponder, or whether there may be a leakage of signal into another transponder. Figure 4.26 depicts the relative power spectral density (in W/Hz, measured in dB) of a digitally modulated carrier using BPSK and QPSK without applying any filtering (the x-axis shows normalized frequency $(f - f_c)$/(bit rate), where f_c is the carrier frequency; the y-axis is the relative level of the power density with respect to the maximum value at carrier frequency f_c). Two issues are worth considering:

1. The width of the principal lobe of the spectrum of the modulated signal, which affects the required/utilized bandwidth of the channel
2. The secondary side lobes, which affect adjacent carriers by generating interference when the spectral decay with respect to frequency is not as sharp as one would hope

Figure 4.26 shows that QPSK is "better" than BPSK, having a lower main lobe. This is because, in general, constellation diagrams show that transition to new states could result in large amplitude

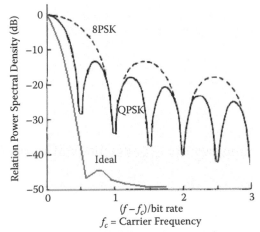

Normalized frequency is the difference between a frequency point f and the carrier frequency f_c with respect to the bit rate R_c modulating the carrier

Figure 4.26   Power sidelobes generated by the modulation process.

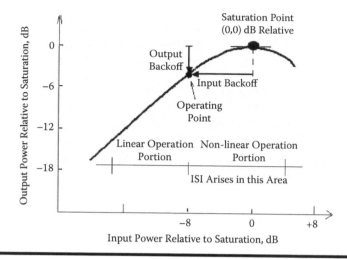

Figure 4.27   Nonlinear channel. ISI = Intersymbol interference.

changes (e.g., going from 010 to 110 in 8-PSK), and a signal that changes amplitude over a very large range will exercise amplifier nonlinearities, which cause distortion products. In continuously modulated systems, large signal changes will cause "spectral regrowth," or wider modulation sidebands (a phenomenon related to intermodulation distortion). The problem lies in nonlinearities in the circuits. If the amplifier and associated circuits were perfectly linear, the spectrum (spectral occupancy or occupied bandwidth) would be unchanged [AGI200101].

As stated, any fast transition in a signal, whether in amplitude, phase, or frequency, results in a power-frequency spectrum that requires a wide occupied bandwidth. Any technique that helps to slow down transitions narrows the occupied bandwidth. To deal with this spectral regrowth issue and limit unwanted interference to adjacent channels, filtering is used at the transmitter side. In addition to the intrinsic phenomenon of spectral spreading of the modulation sidelobes, an attempt to transmit at maximum power, either at the remote terminal or at the satellite, produces amplitude and phase distortions. To avoid operating in the nonlinear region of the amplifier (also as covered in Appendix 3.A), one typically needs to back off (from using) the maximum transponder power, as shown in Figure 4.27.

The rest of this section focuses on filtering. Filtering allows the transmitted bandwidth to be reduced without losing the content of the digital data, thereby improving the spectral efficiency of the signal. The most common filtering techniques are raised cosine filters, square-root raised cosine filters, and Gaussian filters. Filtering smooths transitions (in I and Q) and, as a result, reduces interference, because it reduces the tendency of one signal or one transmitter to interfere with another in a frequency division multiple access (FDMA) system. On the receiver end, reduced bandwidth improves sensitivity because more noise and interference are rejected.

It should be noted, however, that some tradeoffs must be taken into consideration: Carrier power cannot be limited (clipped) without causing the spectrum to spread out once again; as narrowing the spectral occupancy was the reason for the filtering to be inserted in the first place, the designer must select the choice carefully [AGI200101]. One tradeoff is that some types of filtering cause the trajectory of the signal (the path of transitions between the states) to overshoot in a number of cases. This overshoot can occur in certain types of filters (such as Nyquist filters).

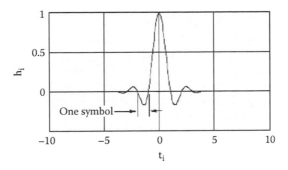

Figure 4.28   Nyquist or raised cosine filter.

This overshoot path represents carrier power and phase. For the carrier to take on these values, it requires more output power from the transmitter amplifiers; specifically, it requires more power than would be necessary to transmit the actual symbol itself. Another consideration is intersymbol interference (ISI). ISI is the interference between adjacent symbols often caused by system filtering, dispersion in optical fibers, or multipath propagation in radio systems. This occurs when the signal is filtered enough so that the symbols blur together, and each symbol affects those around it. This is determined by the time-domain response or impulse response of the filter.

There are many different types of filtering. As noted, the most common are[*]

- Raised cosine
- Square-root-raised cosine
- Gaussian filters

4.1.6.1   Nyquist/Raised Cosine Filter

Figure 4.28 shows the impulse or time-domain response of a raised cosine filter, a class of Nyquist filters. Nyquist filters have the property that their impulse response rings[†] at the symbol rate. The time response of the filter goes through zero with a period that exactly corresponds to the symbol spacing. Adjacent symbols do not interfere with each other at the symbol times because the response equals zero at all symbol times, except the center (desired) one. Nyquist filters heavily filter the signal without blurring the symbols together at the symbol times; this is important for transmitting information without errors caused by ISI. Note that ISI does exist at all times except at the symbol (decision) times. Usually, the filter is split, half being in the transmit path and half in the receiver path; in this case, root Nyquist filters (commonly called *root-raised cosine*) are used in each part such that their combined response is that of a Nyquist filter.

The raised cosine filter is an implementation of a low-pass Nyquist filter; its spectrum exhibits odd symmetry, about $1/2T$, where T is the symbol period of the communications system. The

---

[*] The rest of this subsection is based on Reference [AGI200101].

[†] That is, they have the impulse response of the filter cross through zero.

frequency-domain description of the raised cosine filter is a piecewise function characterized by two values: α, the roll-off factor, and T, specified as

$$H(f) = \begin{cases} 1.0, & |f| \le \frac{1-\alpha}{2T} \\ \frac{1}{2}\left[1 + \cos\left(\frac{\pi T}{\alpha}\left[|f| - \frac{1-\alpha}{2T}\right]\right)\right], & \frac{1-\alpha}{2T} < |f| \le \frac{1+\alpha}{2T} \\ 0, & |f| > \frac{1+\alpha}{2T} \end{cases} \quad \text{with } 0 \le \alpha \le 1$$

The impulse response of such a filter, defined in terms of the normalized sinc function, is given by

$$h(t) = \text{sinc}\left(\frac{t}{T}\right)\frac{\cos\left(\frac{\pi \alpha t}{T}\right)}{1 - \frac{4\alpha^2 t^2}{T^2}}$$

The roll-off factor, α, is a measure of the excess bandwidth of the filter, that is, the bandwidth occupied beyond the Nyquist bandwidth of $1/2T$. If we denote the excess bandwidth as Δf, then

$$\alpha = \frac{\Delta f}{\left(\frac{1}{2T}\right)} = \frac{\Delta f}{R_S/2} = 2T\Delta f$$

where $R_s = 1/T$, and T, the symbol period. (Symbol energy expands into the $[-T, T]$ interval in a time-amplitude domain.)

As can be seen in Figure 4.29, the time-domain ripple level increases as α decreases (from 0 to 1). This shows that the excess bandwidth of the filter can be reduced, but only at the expense of an elongated impulse response. As α approaches 0, the roll-off zone becomes infinitesimally narrow; hence, the filter converges to an ideal or brick-wall filter. When $\alpha = 1$, the nonzero portion of the spectrum is a pure raised cosine.

The bandwidth of a raised cosine filter is most commonly defined as the width of the nonzero portion of its spectrum, that is,

$$BW = \frac{1}{2}R_S(1+\alpha)$$

A typical cosine roll-off factor is $\alpha = 0.25$.

4.1.6.2 Transmitter–Receiver Matched Filters

In some situations, filtering is desired at both the transmitter and receiver. See Figure 4.30. Filtering in the transmitter reduces the adjacent-channel power radiation of the transmitter and, thus, its potential for interfering with other transmitters. Filtering at the receiver reduces the effects of broadband noise and also interference from other transmitters in nearby channels. To get zero ISI, both the filters are designed such that the combined result of the filters and the rest of the system is a full Nyquist filter. Potential differences can cause problems; these may occur because the transmitter and receiver are often manufactured by different suppliers. If the design is performed correctly, the results are optimal data rate, efficient radios, and reduced effects of interference and noise.

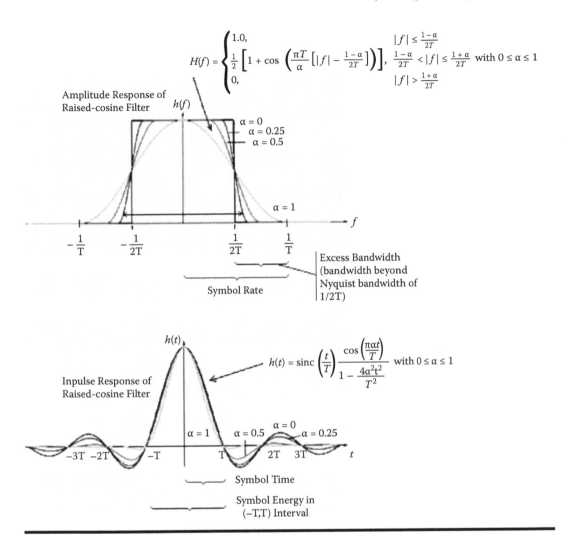

$$H(f) = \begin{cases} 1.0, & |f| \le \frac{1-\alpha}{2T} \\ \frac{1}{2}\left[1 + \cos\left(\frac{\pi T}{\alpha}\left[|f| - \frac{1-\alpha}{2T}\right]\right)\right], & \frac{1-\alpha}{2T} < |f| \le \frac{1+\alpha}{2T} \quad \text{with } 0 \le \alpha \le 1 \\ 0, & |f| > \frac{1+\alpha}{2T} \end{cases}$$

Amplitude Response of Raised-cosine Filter

$h(f)$

α = 0
α = 0.25
α = 0.5

α = 1

f

$-\frac{1}{T}$ $-\frac{1}{2T}$ $\frac{1}{2T}$ $\frac{1}{T}$

Excess Bandwidth (bandwidth beyond Nyquist bandwidth of 1/2T)

Symbol Rate

$h(t)$

$$h(t) = \text{sinc}\left(\frac{t}{T}\right)\frac{\cos\left(\frac{\pi\alpha t}{T}\right)}{1 - \frac{4\alpha^2 t^2}{T^2}} \quad \text{with } 0 \le \alpha \le 1$$

Inpulse Response of Raised-cosine Filter

α = 1   α = 0.5   α = 0   α = 0.25

−3T −2T −T   T   2T   3T t

Symbol Time

Symbol Energy in (−T,T) Interval

Figure 4.29   Raised cosine filter.

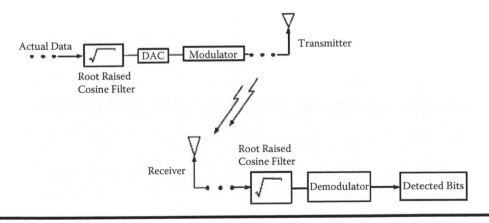

Actual Data

Root Raised Cosine Filter

DAC   Modulator

Transmitter

Receiver

Root Raised Cosine Filter

Demodulator   Detected Bits

Figure 4.30   Transmitter–receiver-matched filters.

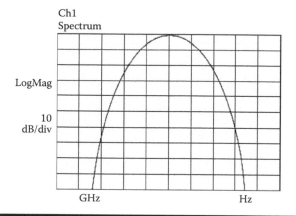

Figure 4.31   Gaussian filter.

4.1.6.3   Gaussian Filter

System architects must decide just how much of the ISI can be tolerated in a system and combine that with noise and interference. The Gaussian filter (see Figure 4.31) is a Gaussian shape in both the time and frequency domains. Its effects in the time domain are relatively short, and each symbol interacts significantly (that is, causes ISI) with only the preceding and succeeding symbols. This reduces the tendency for particular sequences of symbols to interact, which makes amplifiers easier to build and more efficient.

4.1.6.4   Filter Bandwidth Parameter Alpha

Now we return to the raised cosine filter. The sharpness of a raised cosine filter is described by α (see Figure 4.32); this provides a measure of the occupied bandwidth of the system:

$$\text{Occupied bandwidth} = \text{symbol rate} \times (1 + \alpha)$$

with symbol rate $= 1/2T = (1/2)\ R_s$.

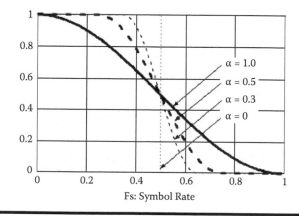

Figure 4.32   Filter bandwidth parameters "α."

Sometimes, α is called the "excess bandwidth factor," as it indicates the amount of occupied bandwidth that will be required in excess of the ideal occupied bandwidth (which would be the same as the symbol rate). If the filter had a perfect (brick wall) characteristic, with sharp transitions and $\alpha = 0$, the occupied bandwidth would be

$$\text{Occupied bandwidth} = \text{symbol rate} \times (1 + 0) = \text{symbol rate.}$$

In a perfect world, the occupied bandwidth would be the same as the symbol rate, but this is not practical. An $\alpha = 0$ is impossible to implement. At the other extreme, consider a broader filter with an $\alpha = 1$, which is easier to implement. The occupied bandwidth will then be

$$\text{Occupied bandwidth} = \text{symbol rate} \times (1 + 1) = 2 \times \text{ symbol rate}$$

An $\alpha = 1$ uses twice as much bandwidth as an $\alpha = 0$. In practice, it is possible to implement an α below 0.2 and make good, compact, practical devices, though some video systems use an α as low as 0.10. Typical values range from 0.35 to 0.5.*

4.1.6.5 *Filter Bandwidth Effects*

A designer seeks to maximize the power and bandwidth efficiency of satellite terminals, particularly for very small aperture terminals (VSATs), where a key limitation to communication capacity is a nonlinear transmitting power amplifier. It is highly undesirable to waste the RF spectrum by using channel bands that are too wide. Therefore, filters are used to reduce the occupied bandwidth of transmission. Figure 4.33 depicts the use of filters in an actual satellite chain. The data pulse stream is band limited for bandwidth efficiency; however, the effect of filtering (in the time domain) is to cause envelope excursions outside the linear region of the power amplifier characteristic. The traditional approach is to reduce the average power at the input of the amplifier such that the maximum signal excursions are within the linear range of the amplifier characteristic. This reduction is termed *power back-off*, as defined earlier, and can be significant for bandwidth-efficient systems. Figure 4.34 shows the necessary back-off for QPSK for different roll-off factors. For example, for a roll-off factor of 25 percent, the excursions of the envelope are 5.4 dB above the average power level, increasing to 7.68 dB for a 10 percent roll-off [AMB200301].

Consider again a QPSK signal, and refer to Figure 4.35. If the radio has no transmitter filter (as is the case on the left of the graph), the transitions between states are instantaneous. The center figure is an example of a signal at an $\alpha = 0.75$; the figure on the right shows the signal at an $\alpha = 0.375$. The filters with α of 0.75 and 0.375 smooth the transitions and narrow the frequency spectrum required. Different filter α's also affect transmitted power. In the case of the unfiltered signal, the maximum or peak power of the carrier is the same as the nominal power at the symbol states; no extra power is required due to the filtering.

Consider the example of a $\pi/4$ DQPSK signal. If an $\alpha = 1.0$ is used, the transitions between the states are gradual; less power is needed to handle these transitions. Using an $\alpha = 0.5$, the transmitted bandwidth decreases from 2 to 1.5 times the symbol rate. This results in a 25 percent improvement in occupied bandwidth. The smaller α takes more peak power because of the

---

* The corresponding term for a Gaussian filter is BT (bandwidth time product). The occupied bandwidth cannot be stated in terms of BT because a Gaussian filter's frequency response does not go identically to zero, as does a raised cosine. Common values for BT are 0.3 to 0.5.

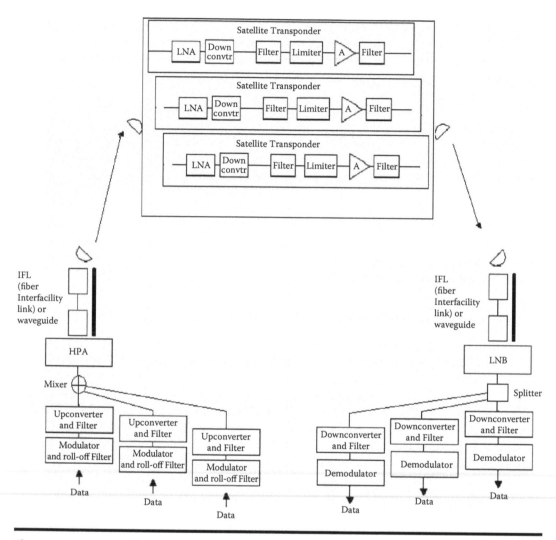

Figure 4.33 **Use of filters in actual satellite transmit/receive chain.**

overshoot in the filter's step response. This produces trajectories that loop beyond the outer limits of the constellation. At an $\alpha = 0.2$, about the minimum of most radios at press time, there is a significant excess power beyond that needed to transmit the symbol values themselves. A typical value of excess power needed at $\alpha = 0.2$ for QPSK with Nyquist filtering is approximately 5 dB. This is more than three times as much peak power, caused by the filter used to limit the occupied bandwidth.

These principles apply to QPSK and the varieties of QAM, such as 16-QAM, 32-QAM, 64-QAM, and 256-QAM. Not all signals behave in exactly the same way, and exceptions include FSK, MSK, and others with constant envelope modulation. The power of these signals is not affected by the filter shape.

Narrow filters with sufficient accuracy and repeatability are somewhat difficult to build. Smaller values of α increase ISI because more symbols can contribute; this tightens the requirements on

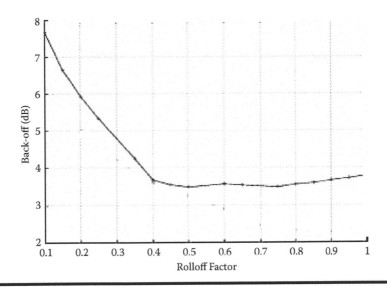

Figure 4.34   QPSK back-off.

clock accuracy. These narrower filters also result in more overshoot and, therefore, more peak carrier power. The power amplifier must then accommodate the higher peak power without distortion. Larger amplifiers cause more heat and electrical interference, because the RF current in the power amplifier will interfere with other circuits. In summary, spectral efficiency is highly desirable, but there are penalties in cost, size, weight, and complexity. A quick discussion of amplifier issues was provided in Appendix 3.A.

4.1.7   Predistortion

Predistortion methods in the uplink station can be used to minimize the effect of transponder nonlinearity. As seen earlier, power amplifiers (PAs) are inherently nonlinear, and when operated near saturation, cause intermodulation products that interfere with adjacent and alternate channels. This interference affects the adjacent channel leakage ratio (ACLR). Linearization techniques allow PAs to operate efficiently and at the same time maintain acceptable ACLR levels (in effect, linearization affords the use of a lower-cost more efficient PA in place of a higher-cost less efficient PA). PAs in the field today are predominately linearized by some form of feed-forward technology. In recent years, however, designers have started to employ digital predistortion; compared to

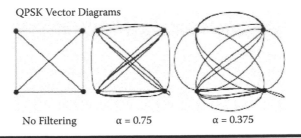

Figure 4.35   Effect of different filter bandwidths.

feed-forward architectures, designs based on digital predistortion approaches enjoy higher efficiency at lower cost [SIL200201].

In a group of waves that have slightly different individual frequencies, the group delay time is the time required for any defined point on the envelope (i.e., the envelope determined by the additive resultant of the group of waves) to travel through a device or transmission facility. Group delay is the rate of change of the total phase shift with respect to angular frequency, $d\theta/d\omega$, through a device or transmission medium, where θ is the total phase shift and ω is the angular frequency equal to $2\pi f$, where f is the frequency. Preemphasis (also known as predistortion) can be employed to deal with group delay. Preemphasis is a system process designed to increase, within a band of frequencies, the magnitude of some (usually higher) frequencies with respect to the magnitude to other (usually lower) frequencies in order to improve the overall signal-to-noise ratio by minimizing the adverse effects of such phenomena as attenuation differences, or saturation of the PA [ANS200001].

4.2 Multiple Access Schemes

Multiplexing, the ability to support multiple users over a single facility (e.g., a radio channel, a satellite band, a transponder, etc.), is as fundamental to communications as modulation. Multiple access protocols are channel allocation schemes that have desirable performance characteristics.

Five principal types of multiple access schemes are,

- Space division multiple access (SDMA; also known as space division multiplexing)
- Frequency division multiple access (FDMA; also known as frequency division multiplexing)
- Time division multiple access (TDMA; also known as time division multiplexing)
- Code division multiple access (CDMA; also known as code division multiplexing)
- Random access. The use of algorithms that support randomized transmission in a distributed system without central control

TDMA is currently the most commonly used approach in satellite communications.

At the highest level of the multiple access protocols classification, there are two categories: conflict-free protocols and contention protocols. Conflict-free protocols are those ensuring that a transmission, whenever initiated, is a successful one—that is, it will not be interfered with by another transmission. In these protocols, the channel is allocated to the users without any overlap between the portions of the channel allocated to different users. The channel resources can be viewed from a time, frequency, or mixed time–frequency perspective. The channel is allocated to the users either by static or dynamic techniques. It can be "divided" at the physical layer (of the open systems interconnection reference model) by giving the entire frequency range (bandwidth) to a single user for a fraction of the time, as done in TDMA, or giving a fraction of the frequency range to every user all of the time, as done in FDMA, or providing every user access to the bandwidth at varying frequencies or time "minislots" by utilizing spread spectrum mechanisms, such as CDMA. Dynamic allocation assigns the channel based on demand such that a user who happens to be idle uses only little of the channel resources, leaving the majority to the other more active users. Such an allocation can be done by statistical multiplexing at the data link layer (of the open systems interconnection reference model), for example, using Frame Relay or Asynchronous Transfer Mode protocols, or by reservation schemes (where the users first announce their intent to transmit, and all those who have so announced will transmit before new users have a chance to announce their intent to transmit) [ROM198901]. One of the drawbacks of conflict-free protocols is that idle users do consume (are assigned) a portion of the channel resources unless a more complex data link layer mechanism is used; this becomes a major issue when the number of potential users in the system is large. In systems with contention schemes, a transmitting user is not guaranteed to be instantaneously successful (retransmissions may be needed); however, these are more bandwidth efficient; these protocols are discussed later.

Space division multiplexing (also know as *frequency reuse*) is the ability to use the same frequencies repeatedly across a system, but where there is geographical and/or physical separation. SDMA is a method of allowing multiple users to share a single communications channel by arranging the users so that they are not in each other's communications range. It is a method whereby a number of entities use a single service capability by having portions of physical space within the service facility dedicated for their individual use. For example, the radio FM band and the over-the-air TV band are reused in localities that are 75 or more miles apart. Cellular phone frequencies also make use of this technique. The use of a C-band frequency by two earth stations using two C-band satellites is another example of space division multiplexing.

TDMA is a communications technique that uses a common channel (multipoint or broadcast) for communications among multiple users by allocating unique time slots to different users. TDMA is used extensively in satellite systems as well as in many other environments. TDMA is a multiple access technique whereby users share a transmission medium by being assigned and using (each user at a time and for a limited time interval) a number of time division multiplexed slots (subchannels); several transmitters use the same overall channel for sending several bit streams [ANS200001]. In TDMA, a channel (of a given bandwidth—typically, a transponder) is shared by all the active remote stations, but each is only permitted to transmit in predefined short bursts of time (slots) allocated by a predetermined sequencing mechanism, thus sharing the channel among all the remote stations by dividing it over time (hence, time division). Hence, in TDMA, the entire bandwidth is used by each user for a fraction of the time. See Figure 4.36 for a pictorial view of this scheme.

FDMA makes use of frequency division to provide multiple and simultaneous transmissions to a single transponder. Frequency-division multiplexing (FDM) is the approach of deriving two or more simultaneous, continuous channels from a transmission medium by assigning a separate portion of the available frequency spectrum to each of the individual channels [ANS200001]. Hence, in FDMA, the bandwidth of the available spectrum is divided into separate channels, each individual channel frequency being allocated to a different active remote station for transmission. FDMA splits the available frequency band into smaller fixed frequency channels, and each transmitter or receiver uses a separate frequency. In FDMA, a fraction of the frequency bandwidth is allocated to every user all the time. Transmitters are narrowband or frequency limited. A narrowband transmitter is used along with a receiver that has a narrowband filter so that it can demodulate the desired signal and reject unwanted signals, such as interfering signals from adjacent radios [AGI200101]. See Figure 4.36 for a pictorial view of this scheme. In satellite communications, FDM is used to define the 12 (24 with polarization) transponders; however, the use of FDMA *within* a transponder is on the decline (with TDMA being the leading approach, as noted). (Some call this arrangement frequency-time division multiple access [FTDMA].)

In static channel allocation strategies, channel allocation is predetermined (typically, at system design time) and does not change during the operation of the system. For both the FDMA and the TDMA protocols, due to the static and fixed assignment, parts of the channel can be idle even though some other users in the system may have data to transmit. As noted, dynamic channel

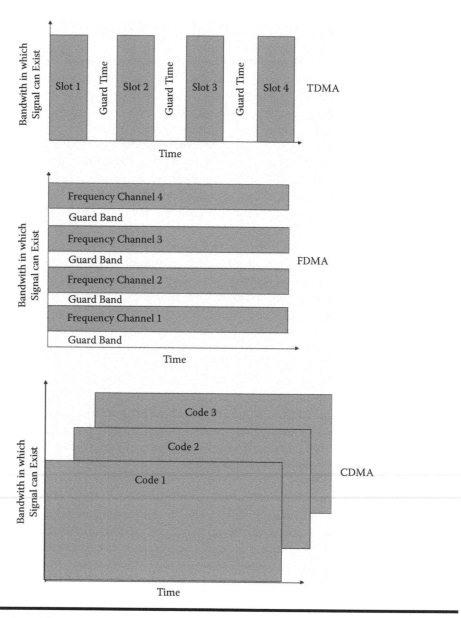

Figure 4.36 Pictorial view of multiple access schemes.

allocation protocols attempt to overcome this by changing the channel allocation, based on the current demands of the users.

CDMA is a coding scheme where multiple channels are independently coded for transmission over a single wideband channel. In some communication systems, CDMA is used as an access method that permits carriers from different stations to use the same transmission equipment by using a wider bandwidth than the individual carriers. On reception, each carrier can be distinguished from the others by means of a specific modulation code, thereby allowing for the reception of signals that were originally overlapping in frequency and time. Thus, several transmissions

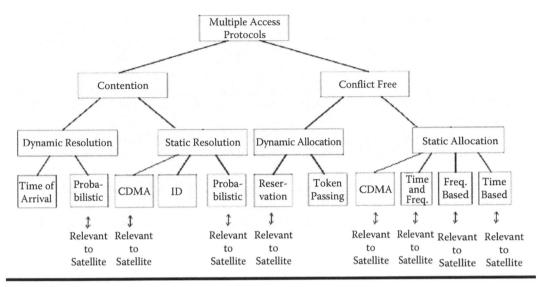

Figure 4.37 **Taxonomy of multiple access protocols.**

can occur simultaneously within the same bandwidth, with the mutual interference reduced by the degree of orthogonality of the unique codes used in each transmission [ANS200001]. CDMA is an access method where multiple users are permitted to transmit simultaneously on the same frequency. The channel (probably derived via FDM methods), however, is larger than would otherwise be the case. Channelization is added in the form of coding. In a CDMA system, all users access the same bandwidth, but they are distinguished (separated) from each other by a uniquely different coding parameter. Each user is assigned a code that is utilized to transform the user's signal into a spread-spectrum-coded version of the user's data stream, typically using direct sequence spread spectrum (digital) techniques. The receiver then uses the same spreading algorithm and code parameter to transform the spread-spectrum signal back into the original user's data stream. Users timeshare a higher-rate digital channel by overlaying a higher-rate digital sequence on their transmission. A different sequence is assigned to each terminal so that the signals can be discerned from one another by correlating them with the overlaid sequence [AGI200101]. See Figure 4.36 for a pictorial view of this scheme. CDMA is not widely used in commercial satellite applications (although a few cases do exist).

As noted, Random Access uses algorithms that support randomized transmission in a distributed system without central control. One example is contention resolution at the media access control (MAC) layer of wireless networks, where nearby radios seek to access their local radio channel. In the satellite context, this approach is typically used in VSAT environments. Figure 4.37 provides a taxonomy of access schemes [ROM198901]. Conflict-free approaches were already discussed earlier.

Static conflict-free protocols such as the FDMA and TDMA do not utilize the shared channel very efficiently, especially when the system is lightly loaded or when the loads of different users are asymmetric. This is one of the motivations for random access schemes. Most random access schemes (contention-based multiple access protocols) are noncentralized protocols where there is no single node coordinating the activities of the others (however, all nodes behave according to the same set of rules). Contention schemes do not automatically guarantee successful transmission. The protocol must prescribe a way to resolve conflicts once they occur, so all messages are eventually transmitted successfully.* The resolution process does consume resources, and is one of the major differences among the various contention protocols. If the probability of interference is small, such as might be the case with bursty (data) users, taking the chance of having to resolve the interference compensates for the resources that have to be expanded to ensure freedom of conflicts. In contention schemes, idle users do not transmit and thus do not consume any portion of the channel resources.

When contention-based multiple access protocols are used, the necessity arises to resolve the conflicts, whenever they occur. Both static and dynamic resolutions exist. Static resolution means that the actual behavior is not influenced by the dynamics of the system. A static resolution can be based, for example, on user IDs or any other fixed priority assignment, meaning that whenever a conflict arises the first user to finally transmit a message will be the one with, say, the smallest ID (this is done in some tree-resolution protocols). A static resolution can also be probabilistic, meaning that the transmission schedule for the interfering users is chosen from a fixed distribution that is independent of the actual number of interfering users, as is done in Aloha-type protocols and the Carrier Sense Multiple Access (CSMA) protocol used in classical Ethernet systems. Dynamic resolution, namely, taking advantage and tracking system changes, is also possible in contention-based protocols. For example, resolution can be based on time of arrival giving highest (or lowest) priority to the oldest message in the system. Alternatively, resolution can be probabilistic but such that the statistics change dynamically according to the extent of the interference. Estimating the multiplicity of the interfering packets and the exponential back-off scheme of the Ethernet standard fall into this category. As noted earlier, these types of protocols are used in VSAT environments.

The Aloha family of protocols is the richest family of multiple access protocols. Its popularity is due first of all to seniority, as it is the first random access technique introduced. Second, many of these protocols are relatively simple so that their implementation is straightforward. The Aloha family of protocols belongs to the contention-type or random retransmission protocols in which the success of a transmission is not guaranteed in advance. The reason is that whenever two or more users transmit on the shared channel simultaneously, a collision occurs, and the data cannot be received correctly. This being the case, packets may have to be transmitted and retransmitted, until eventually they are correctly received. Transmission scheduling is therefore the focal concern of contention-type protocols. Because of the great popularity of Aloha protocols, analyses have been carried out for a very large number of variations. The variations present different protocols for transmission and retransmission schedules as well as adaptation to different circumstances and channel features. This chapter covers a few of these variations.

The pure Aloha protocol is the basic protocol in the family of the Aloha protocols. It considers a single-hop system, with an infinite population generating packets of equal length T according to a Poisson process, at the rate of λ packets per second. The channel is error free, without capture: whenever a transmission of a packet does not interfere with any other packet transmission, the transmitted packet is received correctly, whereas if two or more packet transmissions overlap in time, a collision is caused, none of the colliding packets is received correctly, and they have to be retransmitted. The users whose packets collide with one another are called the colliding users. At the end of every transmission, each user knows whether the transmission was successful or a collision took place. The pure Aloha protocol is very simple. It states that a newly generated packet is transmitted immediately, hoping for no interference by others. Should the transmission be

---

* The rest of this section is based on reference [ROM198901].

unsuccessful, every colliding user, independently of the others, schedules its retransmission to a random time in the future. This randomness is required to ensure that the same set of packets does not continue to collide indefinitely.

The slotted Aloha variation of the Aloha protocol is simply that of pure Aloha with a slotted channel. The slot size equals *T*—the duration of packet transmission. Users are restricted to start transmission of packets only at slot boundaries. Thus, the vulnerable period is reduced to a single slot. In other words, a slot will be successful if and only if exactly one packet was scheduled for transmission sometime during the previous slot. The throughput is, therefore, the fraction of slots (or probability) in which a single packet is scheduled for transmission.

References

[AGI200101] Agilent Technologies, "Digital Modulation in Communications Systems—An Introduction," Application Note 1298, March 14, 2001, Doc. 5965-7160E, Santa Clara, CA.

[AGI200701] Staff, "Interactive Frequency Modulation Model," Agilent Technologies, Inc. Headquarters, Santa Clara, CA.

[AHA200501] Staff, Terms and Definitions, AHA/Comtech Telecommunications Corporation, Moscow, ID, 2005.

[AMB200301] A. Ambroze, M. Tomlinson and G. Wade, "Magnitude Modulation for Small Satellite Earth Terminals using QPSK and OQPSK," Proceedings of IEEE ICC2003 conference in Alaska, USA, May 2003. Faculty of Technology, University of Plymouth, United Kingdom.

[AMO199701] F. AMOROSO, R. A. Monzingo, "Analysis Of Data Spectral Regrowth From Nonlinear Amplification," IEEE, ICPWC'97, 1997, pg 142 ff.

[ANS200001] ANS T1.523-2001, Telecom Glossary 2000, American National Standard (ANS), An outgrowth of the Federal Standard 1037 series, *Glossary of Telecommunication Terms*, 1996.

[BLA200701] K. Blattenberger, "Free Space Loss," RF Café Website, Mt. Airy, NC.

[BLO200701] Staff, "QAM Defined," Blonder Tongue Laboratories, Inc., Old Bridge, NJ, 2007.

[BRE200501] D. Breynaert, M. d'Oreye de Lantremange, "Analysis of the Bandwidth Efficiency of DVB-S2 in a Typical Data Distribution Network," Newtec, CCBN2005, Beijing, March 21–23 2005.

[CHE200701] J. Chen, "Extracting and Using J-models to Estimate ACPR in Direct Conversion Transmitters," Candence White Paper, San Jose, CA.

[COM199801] CM701/DT7000 Reed–Solomon (DVB Version), Option Card, ComStream Corporation, A Spar Company, San Diego, CA, 1998.

[DEG200601] R. De Gaudenzi, A. G. Fàbregas, A. Martinez, "Turbo-coded APSK modulations design for satellite broadband communications," European Space Agency (ESA-ESTEC), Noordwijk, The Netherlands; Institute for Telecommunications Research, University of South Australia, Australia; Technische Universitat Eindhoven, Eindhoven, The Netherlands, *International Journal of Satellite Communications and Networking*, Volume 24, Issue 4, Pages 261–281, Published Online, 19 May 2006.

[EDW199001] G. Edwards, "Forward Error Correction Encoding and Decoding," Stanford Telecom Application Note 108, 1990. Sunnyvale, California.

[FOC200701] Staff, An Introduction to Earth Stations, Focalpoint Consulting, St Gaudens, France, 2007.

[HEN200201] H. Hendrix, "Viterbi Decoding Techniques for the TMS320C54x DSP Generation," Texas Instruments, Application Report SPRA071A—January 2002. Texas Instruments, Dallas.

[JOU199901] M. K. Juonolainen, Forward Error Correction in INSTANCE, Cand Scient Thesis, 1/2/1999, University of Oslo, Department of Informatics.

[KOR200701] I. Koren and C. M. Krishna, "Fault-Tolerant Systems," Morgan-Kaufman Publishers, San Francisco, CA, 2007, Department of Electrical and Computer Engineering, University of Massachusetts, Amherst, MA.

[LIT200101] L. Litwin, "Error control coding in digital communications systems," RF Design, Jul 1, 2001.

[LUB200201] M. Luby, L. Vicisano, J. Gemmell, L. Rizzo, M. Handley, J. Crowcroft, "The Use of Forward Error Correction (FEC) in Reliable Multicast," Request for Comments: 3453, December 2002.

[MAR200201] G. Maral, M. Bousquet, *Satellite Communications Systems: Systems, Techniques and Technology*, 2002, Wiley, New York.

[MAT200701] The MathWorks, Inc., Novi, MI.

[MOR200401] A. Morello and V. Mignone, "DVB-S2—Ready for lift off," Digital Video Broadcasting, EBU Technical Review – October 2004, RAI Radiotelevisione Italiana.

[RAD200501] Radyne ComStream Staff, "DVB-S2 and the Radyne ComStream DM240," White Paper, WP017, Rev. 1.3, January 2005, Radyne ComStream, Inc., Phoenix, AZ.

[ROM198901] R. Rom, M. Sidi, *Multiple Access Protocols, Performance and analysis,* Springer-Verlag, New York, June 1989.

[SHO200401] S. W. Ho, *Adaptive Modulation (QPSK, QAM)*, Intel Whitepaper, 2004, Order Number 303788-001.

[SIL200201] J. Sills, "Improving PA Performance With Digital Predistortion," CommsDesign, Oct 02, 2002,

[TOR199801] A. Torres, V. Demjanenko, Inclusion of Concatenated Convolutional Codes in the ANSI T1.413 Issue 3, Contribution to Standards Committee T1-Telecommunications, Plano, Texas T1E1.4/98-301R1, November 30-December 4, 1998, VoCAL Technologies Ltd.

[TRE200401] TrellisWare Technologies Staff, "FlexiCodes: A Highly Flexible FEC Solution," April 2004, TrellisWare Technologies, Inc. San Diego, CA.

Chapter 5

Error Correction Techniques

This chapter covers the critical topic of forward error correction (FEC), briefly introduced in Chapter 4. Along with modulation, FEC drives, in large measure, the performance of the satellite link, the effective channel throughput, and the service availability that one is able to obtain over the satellite link. This chapter only presents a basic introduction to this topic, and focuses pragmatically only on commercial satellite applications.

5.1 FEC Basics

The FEC is the key subsystem required to achieve high performance in satellite links given the typical presence of high levels of noise and interference. It is a system of error control for data transmission where the receiving device has the capability to detect and correct, in a simplex mode (one-way communications channel), any character or code block that contains fewer than a predetermined number of symbols in error; it is accomplished by adding bits to each transmitted character or code block using a predetermined algorithm [ANS200001]. In other words, FEC codes can detect and correct a limited number of errors without retransmitting the data stream. FEC refers to the ability to overcome both erasures (losses) and bit-level corruption. There are two basic types of FEC codes:

■ Block codes. In these algorithms, the encoder processes a block of message symbols and then outputs a block of code word symbols. Reed–Solomon (RS) codes belong to the family of block codes. They first appeared in the literature in 1960. The input to an FEC encoder is a certain number k of equal-length source symbols. The FEC encoder generates a certain number of encoding symbols that are of the same length as the source symbols. The chosen length of the symbols can vary upon each application of the FEC encoder, or it can be fixed. These encoding symbols are placed into packets for transmission. The number of encoding symbols placed into each packet can vary on a per-packet basis, or a fixed number of symbols (often one) can be placed into each packet. Also, enough information is placed in each packet to identify the particular encoding symbols carried in that packet. Upon receipt of packets containing encoding symbols, the receiver feeds these encoding symbols into the corresponding

FEC decoder to recreate an exact copy of the *k* source symbols. Ideally, the FEC decoder can recreate an exact copy from any *k* of the encoding symbols [LUB200201].

■ Convolution codes. In these algorithms, instead of processing message symbols in discrete blocks, the encoder works on a continuous stream of message symbols and simultaneously generates a continuous encoded output stream. These codes get their name because the encoding process can be viewed as the convolution of the message symbols and the impulse response of the encoder [LIT200101].

The performance improvement that occurs when using an error control coding is often measured in terms of coding gain. Suppose an uncoded communications system achieves a given bit error rate (BER) at a signal-to-noise rate (SNR) of 35 dB. Imagine that an error control coding scheme with a coding gain of 3 dB was added to the system. This coded system would be able to achieve the same BER at the even lower SNR of 32 dB. Alternatively, if the system was still operated at an SNR of 35 dB, the BER achieved by the coded system would be the same that the uncoded system achieved at an SNR of 38 dB. The power of the coding gain is that it allows a communications system to either maintain a desired BER at a lower SNR than was possible without coding, or achieve a higher BER than an uncoded system could attain at a given SNR [LIT200101].

Research activities in FEC techniques in the past ten years have given rise to new theoretical approaches. Modern approaches include parallel or serially concatenated convolutional codes, product codes, and low-density parity-check codes (LDPCs)—all using "turbo" (i.e., recursive) decoding techniques [MOR200401]. Figure 5.1 shows the progression of the work in the recent past. In looking at a diagram that plots BER versus E_b/N_0, there is typically an initial steep reduction in error probability as E_b/N_0 increases (this area called the *waterfall region*), followed by a

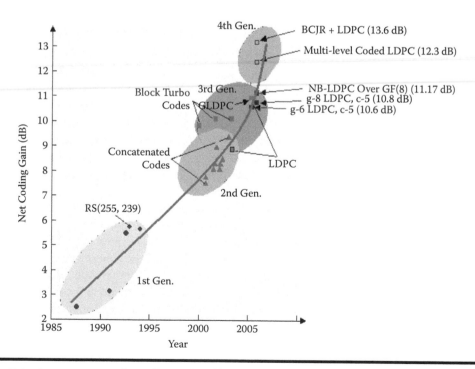

Figure 5.1 **Improvements in coding gain offered by FEC as a function of time. (Figure modeled after T. Mizuochi, IEEE JSTQE May/June 2006)**

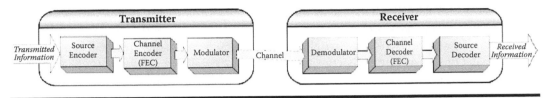

Figure 5.2 Basic diagram of FEC encoding.

region of shallower reduction (this is called the *error floor region*). Figure 5.2 depicts a basic diagram of FEC encoding.

FEC codes take a group of k data packets and generate $h = n - k$ parity packets. The total block size n consists of the k data and h parity packets. Once the parity packets have been computed, the block may be transmitted. The block format may vary according to application. Typically, one has to be concerned with the following parameters:

> k = number of data packets per block
> h = number of parity packets per block, where $h = n - k$
> n = number of total packets per block
> R = percentage of total bandwidth used for redundancy
> T_{FEC} = FEC latency

The desirable properties of an FEC algorithm include the following [TRE200401]:

- Good threshold performance. The waterfall region of an FEC algorithm's BER curve should occur at as low an E_b/N_0 as possible. This will minimize the energy expense of transmitting information.
- Good floor performance. The error floor region of an FEC algorithm's BER curve should occur at a BER that is as low as possible. For communication systems employing an automatic repeat request (ARQ) scheme (such as used in High-level Data-Link Control (HDLC)-like protocols), this may be as high as 10^{-6}, whereas most broadcast communications systems require 10^{-10} performance, and storage systems and optical fiber links require BERs as low as 10^{-15}.
- Low-complexity code constraints. To allow for low-complexity decoders, particularly for high-throughput applications, the constituent codes of the FEC algorithm should be simple. Furthermore, to allow the construction of high-throughput decoders, the code structure should be such that parallel decoder architectures with simple routing and memory structures are possible.
- Fast decoder convergence. The decoder of an FEC algorithm code should converge rapidly (i.e., the number of iterations required to achieve most of the iteration gain should be low). This will allow the construction of high-throughput hardware decoders (or low-complexity software decoders).
- Code rate flexibility. Most modern communications and storage systems do not operate at a single code rate. For example, in adaptive systems, the code rate is adjusted according to the available SNR so that the code overheads are minimized. It should be possible to fine-tune the code rate to adapt to varying application requirements and channel conditions. Furthermore, this code rate flexibility should not come at the expense of degraded threshold or floor performance. Some systems demand code rates of 0.95 or above, but this is difficult for most FEC algorithms to achieve.

- Block size flexibility. One factor that all FEC algorithms have in common is that their threshold and floor performance is maximized by maximizing their block size. However, it is not always practical to have blocks of many thousands of bits. Therefore, it is desirable that a FEC algorithm still performs well with block sizes as small as only 100 or 200 bits. Also, for flexibility, it is very desirable that the granularity of the block sizes be very small, ideally just 1 bit.
- Modulation flexibility. In modern communication systems employing adaptive coding and modulation, it is essential that the FEC algorithm easily support a broad range of modulation schemes.

Table 5.1 has a basic glossary of terms based on a number of references, including [AHA200501].

Table 5.1 Basic glossary of FEC terms

Automatic repeat request (ARQ)	An error control mechanism for data transmissions. Lost data is detected with sequence numbers and timeouts. The communication about which Protocol Data Units (PDUs) are received or lost is accomplished with acknowledgments. If the acknowledgment back to the sender is negative, lost PDUs are retransmitted. The two most common strategies for retransmission are [JOU199901] Go-back-N scheme goes back to the lost PDU and restarts transmission from the lost PDU. In selective repeat, only the lost packets are retransmitted.
Block codes	Encoding algorithms where the encoder processes a block of message symbols and then outputs a block of code word symbols. Reed–Solomon (RS) codes belong to the family of block codes.
Coding rate	The ratio of input data bits to bits transmitted, for example, 1/2, 7/8, etc.
Convolution codes	Encoding algorithms where, instead of processing message symbols in discrete blocks, the encoder works on a continuous stream of message symbols and simultaneously generates a continuous encoded output stream.
Cyclic code	A subclass of linear codes used for encoding and for syndrome computation; they can be easily implemented using shift registers, and they can be decoded using practical means.
Cyclic redundancy code (CRC)	A subclass of linear codes used for detection of errors in a computer data. An error-detection scheme that uses redundant encoded bits (parity bits) and appends them to the digital signal. The received signal is decoded, and PDUs with errors are discarded. For example, an ARQ system can be combined with CRC if an error correction is required [JOU199901].
Data coding	Operating on data with an algorithm to accomplish encryption, error correction, compression, or some other feature.
Error correction code (ECC)	Allows data that is being read or transmitted to be checked for errors and, when necessary, corrected "on the fly."
Forward error correction (FEC)	A technique where extra redundant information is added to a data stream. This information is added so that errors that may occur during the transmission can be corrected by the receiver without interaction from the transmitter. If FEC techniques were not used, and an error occurred, the receiver would have to request for retransmission that would decrease bandwidth and throughput.

Table 5.1   Basic glossary of FEC terms (*Continued*)

Low-density parity-check codes (LDPC)	FEC mechanism used in DVB-S2. LDPCs are block codes defined by a sparse parity-check matrix. This sparseness admits a low-complexity iterative decoding algorithm. The generator matrix corresponding to this parity-check matrix can be determined and used to encode a block of information bits.
Parallel concatenated convolutional codes (PCCC)	Original turbo convolutional code introduced in 1993.
Reed–Solomon (RS)	A type of block code for FEC. Developed in 1960 by Irving S. Reed and Gustave Solomon. It is the most widely used algorithm for error correcting. The RS code is relatively easy to implement and provides a good tolerance to error bursts. It corrects many bit errors within large message blocks.
RS–Viterbi	Reed–Solomon/Viterbi coding technology. RS is a block-oriented coding system that is generally applied on top of standard Viterbi coding to correct the bulk of the data errors that are not detected by the other coding systems, thereby further reducing the BERs.
Serially concatenated convolutional codes (SCCC)	A type of turbo convolutional code.
Symbol error rate (SER)	The probability of receiving a symbol in error.
Syndrome	In RS ECC chips, typically 32 syndromes make up the syndrome polynomial that contains the information necessary to find and correct errors in a given message. The syndrome is a unique collection of bits that identifies error.
Trellis	A tree diagram where the branches following certain nodes are identical. These nodes (or states), when merged with similar states, form a graph that does not grow beyond $2k-1$ where k is the constraint length [HEN200201].
Trellis coded modulation (TCM)	Method of coding multiple symbols to give improved symbol error rate performance in noisy conditions.
Turbo convolutional codes (TCC)	A coding and decoding scheme based on the use of two concatenated convolutional codes in parallel that achieves near capacity performance on additive white Gaussian noise channel.
Turbo product code (TPC)	An FEC coding scheme used to transmit digital data with the greatest efficiency and reliability of any coding scheme currently available.
Viterbi algorithm (VA)	A technique for searching a decoding trellis to yield a path with the smallest distance. This is also known as *maximum likelihood decoding*.
Viterbi decoder	Decoder that uses the VA for decoding a bitstream that has been encoded using an FEC based on a convolutional code.
Viterbi decoding	A maximum-likelihood decoding algorithm devised by A. J. Viterbi in 1967. The decoder uses a search tree or trellis structure and continually calculates the Hamming (or Euclidean) distance between received and valid code words within the constraint length [HEN200201]. A widely used FEC scheme for satellite and other noisy communication channels. There are two important components of a channel using Viterbi encoding: the Viterbi encoder (at the transmitter) and Viterbi decoder (at the receiver). A Viterbi encoder includes extra information in the transmitted signal to reduce the probability of errors in the received signal that may be corrupted by noise.

5.1.1 Bose–Chaudhuri–Hocquenghem (BCH)

Block FEC codes work as follows. The input to a block FEC encoder is k source symbols and a number n. The encoder generates a total of n encoding symbols. The encoder is systematic if it generates $n - k$ redundant symbols yielding an encoding block of n encoding symbols in total, composed of the k source symbols and the $n - k$ redundant symbols. A block FEC decoder has the property that any k of the n encoding symbols in the encoding block is sufficient to reconstruct the original k source symbols [LUB200201].

The key example of the block code is the Bose–Chaudhuri–Hocquenghem (BCH) code. Just like other block codes, BCH encodes k data bits in n code bits by adding $n - k$ parity checking bits for the purpose of detecting and checking errors. If the length of the codes is $n = 2^m - 1$, for any integer $m \geq 3$, the bound of the error correction is t, where $t < 2^m - 1$; that is, BCH can correct any combination of errors (burst or separate) fewer than t in the n-bit codes. The number of parity checking bits is $n - k \leq mt$ [KOR200701].

An important concept of FEC is the Galois fields (GF). A GF is a finite set of elements on which two binary addition and multiplication can be defined. For any prime number p, one is interested in GF(p); GF(p^m) is called *extended field of GF(p)*—one often uses GF(2^m) in BCH code. A GF can be constructed over a primitive polynomial such as $p(x) = x^4 + x + 1$. Usually, a GF table records all the variables, including expressions for the elements, minimal polynomial, and generator polynomial. By referring to the table, one can locate a proper generator polynomial for the encoder. For example, when $(n, k, t) = (15, 7, 2)$, a possible generator is $g(x) = x^8 + x^7 + x^6 + x^4 + 1$. If one has a data stream $d = d_0, d_1, d_2, \ldots, d_{k-1})$, the code word would be $g(x) \cdot d$ and have the style of $v(x) = v_0 + v_1 x + \ldots v_{n-1} x^{n-1}$.

The decoder of BCH is complex because it has to locate and correct the errors. Suppose one has a received code word $r(x) = r_0 + r_1 x + r_2 x^2 + \ldots r_{n-1} x^{n-1}$; then, $r(x) = v(x) + e(x)$, where $v(x)$ is the correct code word, and $e(x)$ is the error. First, one must compute a syndrome vector $s = (s_1, s_2 \ldots s_{2t})$, which can be achieved by calculating $r \cdot H^T$, where H is parity-check matrix defined as:

$$H = \begin{bmatrix} 1 & \alpha & \alpha^2 & \cdots & \alpha^{n-1} \\ 1 & \alpha^2 & (\alpha^2)^2 & \cdots & (\alpha^2)^{n-1} \\ \vdots & & & & \vdots \\ \vdots & & & & \vdots \\ 1 & \alpha^2 & (\alpha^{2t})^2 & \cdots & (\alpha^{2t})^{n-1} \end{bmatrix}$$

Here, α is the element of the GF field and can be located in the GF table. The error-location polynomial $\sigma(x)$ can be determined. Berkekamp's iterative algorithm is one of the solutions to calculate the error-location polynomial. By finding roots of $\sigma(x)$, the location numbers for the errors can be achieved [KOR200701].

5.1.2 Reed–Solomon (RS)

The RS codec (coder–decoder) is a block-oriented coding system that is applied on top of the standard Viterbi coding. It corrects data errors not detected by the other coding systems, significantly reducing the BERs at nominal signal-to-noise levels (4–8 dB E_b/N_0). Bandwidth expansion is small, with the increase due to code rate in the range of 6 to 12 percent. Typically, RS coding is used in areas where sensitivity to transmission errors is particularly high. The "Reed–Solomon + Viterbi" option is particularly well suited to data communication applications with little or no packet acknowledgment or no packet retransmission, as is the case in satellite communication. Figure 5.3 shows the advantage of combining the technologies [COM199801]. Satellite modems in common use during the past decade have employed Reed–Solomon + Viterbi encoding extensively; newer systems are using either turbo codes or LDPC.

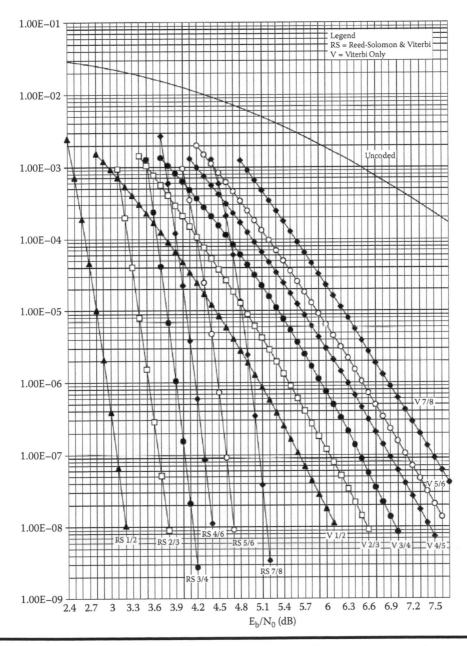

Figure 5.3 Improvement of "Reed–Solomon + Viterbi" over simple Viterbi for typical QPSK system. (Courtesy: ComStream Corporation)

5.1.3 Viterbi Algorithm

As noted, convolutional coding is a bit-level encoding technique rather than block-level techniques such as RS coding. The term *Viterbi* specifically indicates the use of the Viterbi algorithm (VA) for decoding, although the term is often used to describe the entire error-correction process. The encoding method is referred to as convolutional coding or trellis-coded modulation. A state diagram illustrating the sequence of possible codes creates a constrained structure called a *trellis*. The coded data is usually modulated; hence the name trellis-coded modulation. The outputs are generated by convolving a signal with itself, which adds a level of dependence on past values [HEN200201]. Advantages of convolutional codes over block-level codes include [EDW199001]

- With soft-decision data, convolutionally encoded system gain degrades gracefully as the error rate increases. Block-level codes correct errors up to a point, after which the gain drops off rapidly.
- Convolutional codes are decoded after an arbitrary length of data, whereas block-level codes introduce latency by requiring the reception of an entire data block before decoding begins.
- Convolutional codes do not require block synchronization. Although bit-level codes do not allow reconstruction of burst errors block-level codes, interleaving techniques spread out burst errors to make them correctable.

Convolutional codes are decoded by using the trellis to find the most likely sequence of codes. VA simplifies the decoding task by limiting the number of sequences examined. The most likely path to each state is retained for each new symbol. Note that RS is a block-oriented coding system that is generally applied on top of the standard Viterbi coding to correct the bulk of the data errors that are not detected by the other coding systems, thereby further reducing the BERs.

Convolutional encoder error-correction capabilities result from outputs that depend on past data values. Each coded bit is generated by convolving the input bit with previous uncoded bits. Error correction is dependent on the number of past samples that form the code symbols. The number of input bits used in the encoding process is the constraint length, and is calculated as the number of unit delays plus one. The constraint length represents the total span of values used and is determined regardless of the number of taps used to form the code words. The symbol K represents the constraint length. The constraint length indicates the number of possible delay states. The output-bit combination is described by a polynomial. Polynomial selection is important because each polynomial has different error-correcting properties; selecting polynomials that provide the highest degree of orthogonality maximizes the probability of finding the correct sequence. See Figure 5.4 for an example [HEN200201].

Convolutionally encoded data is decoded through the knowledge of the possible state transitions created from the dependence of the current symbol on past data. The allowable state transitions are represented by a trellis diagram. A trellis diagram for a $K = 3$, 1/2-rate encoder is shown in Figure 5.5. The delay states represent the state of the encoder (the actual bits in the encoder shift register), whereas the path states represent the symbols that are output from the encoder. Each column of delay states indicates one symbol interval. The number of delay states is determined by the constraint length. In this example, the constraint length is three, and the number of possible states is $2^{K-1} = 2^2 = 4$. Knowledge of the delay states is very useful in data decoding, but the path states are the actual encoded and transmitted values. The number of bits representing the path states is a function of the coding rate. In this example, two output bits are generated for every input bit,

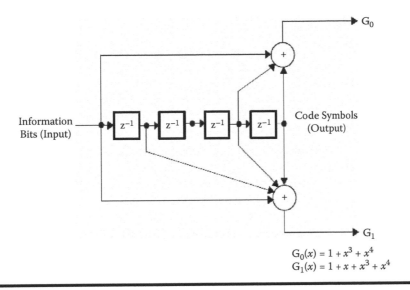

Figure 5.4 Constraint length 5, rate 1/2 convolutional encoder.

resulting in 2-bit path states. Because they represent the actual transmitted values, path states correspond to constellation points, the specific magnitude and phase values used by the modulator. The decoding process estimates the delay state sequence, based on received data symbols, to reconstruct a path through the trellis. The delay states directly represent an encoded data because the states correspond to bits in the encoder shift register. The correspondence between data values and states allows easy data reconstruction once the path through the trellis is determined.

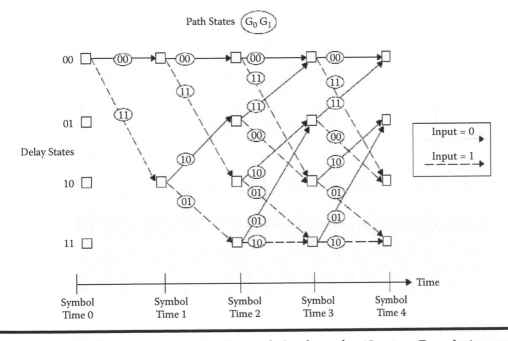

Figure 5.5 Trellis diagram for K = 3, rate 1/2 convolutional encoder. (Courtesy: Texas Instruments)

5.1.4 Turbo Codes

In 1967, Forney introduced a concatenated scheme of inner and outer FEC codes: the inner code is decoded using soft-decision channel information, whereas the outer RS code uses errors and erasures. In 1993, a new coding and decoding scheme, dubbed "turbo codes" by its discoverers, was reported to achieve near-capacity performance on additive white Gaussian noise channel. This technique was based on the use of two concatenated convolutional codes in parallel; it has a floor error near 10^{-5}, but the design of the interleaver is very critical to achieve good results in low BER (it is possible to avoid the floor error with an adequate design of the interleaver). In 1996, a new technique was proposed, based on the same idea of turbo codes, called *serial concatenated convolutional codes* (SCCD); this new codification technique achieves better results for low BER of that turbo code. It avoids both problems of turbo codes: first, the floor error disappears and, second, the design of the interleaver is easier because the input data of the two encoders is different [TOR199801]. Since the introduction of turbo codes, one has seen the emergence of a plethora of codes that exploit a turbo-like structure. The common aspect of all turbo-like codes (TLC) is the concatenation of two or more simple codes separated by an interleaver, combined with an iterative decoding strategy. Members of the TLC family include

- Parallel concatenated convolutional code (PCCC) (this is the classical turbo code)
- Serially concatenated convolutional code (SCCC)
- Low-density parity-check code (LDPC)
- Turbo product code (TPC)

Turbo codes and TLC offer good coding gain compared to traditional FEC approaches, as much as 3 dB in many cases. The error probability performance of TLCs was shown in Chapter 4 (Figure 4.23). The area where TLCs have traditionally excelled is in the waterfall region—there are TLCs that are within a small fraction of a decibel of the Shannon limit in this region; however, many TLCs have an almost flat error floor region, or one that starts at a very high error rate, or both. This means that the coding gain of these TLCs rapidly diminishes as the target error rate is reduced. The performance of a TLC in the error floor region depends on several factors, such as the constituent code design and the interleaver design, as noted, but it typically worsens as the code rate increases or the block size decreases. Many TLC designs only perform well at high error rates, or low code rates, or large block sizes. Furthermore, these designs often only target a single code rate, block size, and modulation scheme, or suffer degraded performance or increased complexity to achieve flexibility in these areas [TRE200401].

5.1.5 Parallel Concatenated Convolutional Code (PCCC)

PCCCs consist of the parallel concatenation of two convolutional codes (see Figure 5.6, top). One encoder is fed the block of information bits directly, whereas the other encoder is fed an interleaved version of the information bits. The encoded outputs of the two encoders must be mapped to the signal set used on the channel. A PCCC encoder is formed by two (or more) constituent systematic encoders joined through one or more interleavers. The input information bits feed the first encoder and, after having been scrambled by the interleaver, they enter the second encoder. A code word of a parallel concatenated code consists of the input bits to the first encoder followed by the parity-check bits of both encoders. PCCC achieves near-Shannon-limit error correction performance. Bit error probabilities as low as 10^{-5} at $E_b/N_0 = 0.6$ dB have been shown. PCCCs yield very

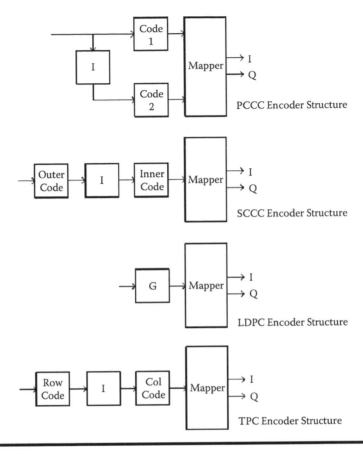

Figure 5.6 Encoder structures.

large coding gains (10 or 11 dB) at the expense of a small data reduction or bandwidth increase [TOR199801]. The characteristics of PCCCs can be summarized as follows [TRE200401]:

- Good threshold performance. Among the best threshold performance of all TLCs.
- Poor floor performance. However, PCCCs also have the worst floor performance of all TLCs. With 8-state constituent codes, BER floors are typically in the range of 10^{-6} to 10^{-8}, but this can be reduced to the 10^{-8} to 10^{-10} range by using 16-state constituent codes. However, achieving floors below 10^{-10} is difficult, particularly for high code rates or small block sizes. There are PCCC variants that have improved floors, such as concatenating more than two constituent codes, but only at the expense of increased complexity.
- High-complexity code constraints. With 8- or 16-state constituent codes, the PCCC decoder complexity is rather high.
- Fast convergence. Among the best convergence of all TLCs. Typically, only 4–8 iterations are required.
- Fair code rate flexibility. Code rate flexibility is easily achieved by puncturing the outputs of the constituent codes. However, for very high code rates, the amount of puncturing required is rather high, which degrades performance and increases decoder complexity (particularly with windowed decoding algorithms).

- Good block-size flexibility. The block size is easily modified by changing the size of the interleaver, and there exist many flexible interleaver algorithms that achieve good performance.
- Good modulation flexibility. The systematic and parity bits from each constituent code must be combined and mapped to the signal set.

5.1.6 Serially Concatenated Convolutional Code (SCCC)

SCCCs consist of the serial concatenation of two convolutional codes (see Figure 5.6, second row). The outer encoder is fed the block of information bits, and its encoded output is interleaved before being input to the inner encoder. The encoded outputs of only the inner encoder must be mapped to the signal set used on the channel. A code word of a serial concatenated code consists of the input bits to the first encoder followed by the parity check bits of both encoders. SCCC achieves near-Shannon-limit error correction performance. Bit error probabilities as low as 10^{-7} at $E_b/N_0 = 1$ dB have been shown. SCCCs yield very large coding gains (10 or 11 dB) at the expense of a small data reduction, or bandwidth increase [TOR199801]. The characteristics of SCCCs can be summarized as follows [TRE200401]:

- Medium threshold performance. The threshold performance of SCCCs is typically 0.3 dB worse than PCCCs.
- Good floor performance. Among the best floor performance of all TLCs. It is possible to have BER floors in the range of 10^{-8} to 10^{-10} with 4-state constituent codes, and below 10^{-10} with 8-state constituent codes. However, floor performance is degraded with high code rates.
- Medium complexity code constraints. Code complexity constraints are typically low but are higher for high code rates. Also, constituent decoder complexity is higher than for the equivalent code in a PCCC because soft decisions of both systematic and parity bits must be formed.
- Fast convergence. Convergence is even faster than for PCCCs, with typically 4–6 iterations required.
- Poor code rate flexibility. As for PCCCs, code rate flexibility is achieved by puncturing the outputs of constituent encoders. However, due to serial concatenation, this puncturing must be higher for SCCCs than for PCCCs for an equivalent overall code rate;
- Good block-size flexibility. As for PCCCs, block-size flexibility is achieved simply by changing the size of the interleaver. However, for equivalent information block sizes, the interleaver of an SCCC is larger then for a PCCC, so there is a complexity penalty in SCCCs for large block sizes. Also, if the code rate is adjusted by puncturing the outer code, then the interleaver size will depend on both the code rate and block size, which complicates reconfigurability.
- Very good modulation flexibility. Because the inner code on an SCCC is connected directly to the channel, it is relatively simple to map the bits into the signal set.

5.1.7 Low-Density Parity Check Code (LDPC)

LDPCs are block codes defined by a sparse parity check matrix. This sparseness allows the use of a low-complexity iterative decoding algorithm. The generator matrix corresponding to this parity check matrix can be determined and used to encode a block of information bits. These encoded bits must then be mapped to the signal set used on the channel (see Figure 5.6, second from bottom). The performance of LDPC codes is within 1 dB of the theoretical maximum performance known as the Shannon limit. The characteristics of LDPCs can be summarized as follows [TRE200401]:

- Good threshold performance. LDPCs have been reported that have threshold performance within a tiny fraction of a decibel of the Shannon limit. However, for practical decoders, their threshold performance is usually comparable to that of PCCCs.
- Medium floor performance. Floors are typically better than PCCCs but worse than SCCCs.
- Low-complexity code constraints. LDPCs have the lowest complexity code constraints of all TLCs. However, high-throughput LDPC decoders require large routing resources or inefficient memory architectures, which may dominate decoder complexity. Also, they are typically a lot more complex than other TLC encoders.
- Slow convergence. LDPC decoders have the slowest convergence of all TLCs. Many published results are for 100 iterations or more. However, practical LDPC decoders will typically use 20 to 30 iterations.
- Good code rate flexibility. LDPCs can achieve good code rate flexibility.
- Poor block-size flexibility. For LDPCs to change block size, they must change their parity check matrix, which is quite difficult in a practical, high-throughput decoder.
- Good modulation flexibility. As with PCCCs, the output bits of an LDPC must be mapped into the signal sets of different modulation schemes.

5.1.8 Turbo Product Code (TPC)

In a TPC system, the information bits are arranged as an array of equal-length rows and columns. The rows are encoded by one block code, and then the columns (including the parity bits generated by the first encoding) are encoded by a second block code (see Figure 5.6, bottom). The encoded bits must then be mapped into the signal set of the channel. The characteristics of TPCs are as follows [TRE200401]:

- Poor threshold performance. TPCs have the worst threshold performance of all TLCs; they can have thresholds that are as much as 1 dB worse than PCCCs. However, for very high code rates (0.95 and above), they will typically outperform other TLCs.
- Medium floor performance. Their floors are typically lower than that of PCCC but not as low as that of SCCC.
- Low-complexity code constraints. TPC decoders have the lowest complexity of all TLCs, and high-throughput parallel decoders can readily be constructed.
- Medium convergence. TPC decoders converge quickly, with 8 to 16 iterations being typically required.
- Poor rate flexibility. The overall rate of a TPC is dictated by the rate of its constituent codes. There is some flexibility available in these rates, but it is difficult to choose an arbitrary rate.
- Poor block-size flexibility. Likewise, the overall block size of a TPC is dictated by the block size of its constituent codes. It is difficult to choose an arbitrary block size, and especially difficult to choose an arbitrary code rate and an arbitrary block size.
- Good modulation flexibility. As with PCCCs and LDPCs, the output bits must be mapped into the signal sets.

5.2 Specific Satellite Applications (DVB-S2)

Because of the nature of the satellite link, 8-PSK is a current technical "sweet spot." This technology is included in the DVB-S2 specification developed by the Digital Video Broadcasting (DVB) Project adopted by European Telecommunications Standards Institute (ETSI) standards now used worldwide. As the name implies, DVB-S2 is the second-generation specification for satellite broadcasting; it was developed in 2003.

The DVB-S2 standard has been specified with the goals of (i) achieving best transmission performance, (ii) embodying flexibility, and (iii) requiring reasonably low receiver complexity (using existing chip technology). The DVB-S2 system has been designed for satellite broadband applications such as broadcast services for MPEG-2/MPEG-4 standard-definition television (SDTV) and high-definition television (HDTV) (including single or multiple Motion Picture Experts Group (MPEG) transport streams, continuous bit streams, IP packets, and even ATM cells), and interactive services, including Internet access. It may be used in "single-carrier-per-transponder" or in "multicarriers-per-transponder" configurations.

Many view the new DVB-S2 standard as a quantum leap in terms of bandwidth efficiency compared to the former DVB-S and DVB-DSNG (Digital Satellite News Gathering) standards. This improvement is due not only to the use of LDPC for FEC but also to new modulation schemes and new modes of operations, specifically Variable Coding and Modulation (VCM) and Adaptive Coding and Modulation (ACM) [BRE200501]. AMC is a link adaptation method (typically used in 3G mobile wireless communication) that provides the flexibility to match the modulation-coding scheme to the average channel conditions for each user. When using ACM, the modulation and coding format is changed dynamically to match the current received signal quality or channel conditions (but the power of the transmitted signal is held constant over a frame interval). A DVB-S2 system can operate at C/N from –2.4 dB (using quadrature phase shift key (QPSK) with FEC at 1/4 rate) to 16 dB (using 32-APSK [amplitude phase shift key] with FEC at 9/10 rate), assuming an additive white Gaussian noise (AWGN) channel and ideal demodulator. The distance from the Shannon limit ranges from 0.7 to 1.2 dB. On AWGN, the result is typically a 20–35 percent capacity increase over DVB-S and DVB-DSNG under the same transmission conditions, and 2–2.5 dB more robust reception for the same spectrum efficiency (a comparison follows later.)

The basic documents for DVB-S2 are the following:

- ETSI: Draft EN 302 307: *Digital Video Broadcasting (DVB); Second generation framing structure, channel coding and modulation systems for Broadcasting, Interactive Services, News Gathering and other broadband satellite applications (DVB-S2).*
- ETSI: EN 300 421: *Digital Video Broadcasting (DVB); Framing structure, channel coding and modulation for 11/12 GHz satellite services*
- ETSI: EN 301 210: *DVB: Framing structure, channel coding and modulation for Digital Satellite News Gathering (DSNG) and other contribution applications by satellite.*

DVB-S is the original DVB specification for satellite-based television distribution. This specification covers FEC and modulation. It was standardized in 1994. The broadcast industry has adopted the format because it established a universal framework for MPEG-2-based digital television services to be broadcast over satellite using Viterbi concatenated with RS (Viterbi + RS) FEC and QPSK modulation. The performance of this code is such that a coding rate of at least 2/3 QPSK is required (to achieve a BER < 10^{-7}); a typical E_b/N_0 of 6.5 dB is needed. The actual (payload) bit

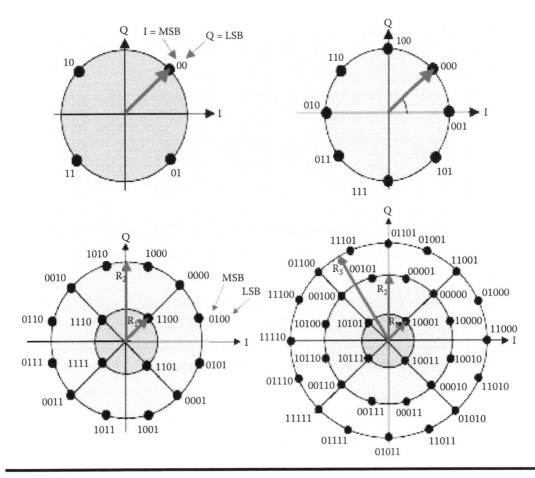

Figure 5.7   DVB-S2 modulation: 4-PSK (QPSK), 3-PSK, 16-APSK, and 32-APSK.

rate for 30 Mbaud with 2/3 QPSK is $1.229 \times 30 = 36.87$ Mbps. In 1999, the DVB-S standard was extended and became the DVB–DSNG standard. This newer standard allowed for more efficient modes of modulation (such as 8-PSK and 16-QAM [quadrature amplitude modulation]) to be utilized. Introducing the higher-order modulation resulted in overall savings in bandwidth but also required an increase of power to achieve similar E_b/N_0 results [RAD200501].

As noted earlier, to achieve the optimal performance complexity trade-off (about 30 percent capacity gain over DVB-S), DVB-S2 makes use of recent advancements in channel coding and modulation. Specifically, it uses an LDPC concatenated with BCH coding. The chosen LDPC FEC codes utilize large blocks (64,800 bits for applications not too critical for delays, and 16,200 bits for application that are more critical). Code rates of 1/4, 1/3, 2/5, 1/2, 3/5, 2/3, 3/4, 4/5, 5/6, 8/9, and 9/10 are available, depending on the selected modulation and the system requirements. Coding rates 1/4, 1/3, and 2/5 have been introduced to operate, in combination with QPSK, under poor link conditions, where the signal level is below the noise level. [MOR200401].

Four modulation modes can be selected for the transmitted payload as shown in Figure 5.7: QPSK, 8-PSK, 16-APSK, and 32-APSK. The QPSK and 8-PSK modes are used for broadcast applications and can be used in nonlinear satellite transponders driven near saturation. To date,

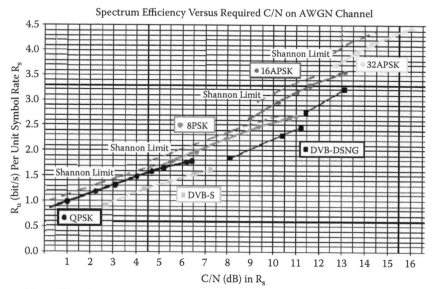

Note: RS(1+ α) corresponds to the theoretical total signal bandwidth after the modulator, with α representing the roll-off factor of the modulation., where RS is the symbol (that corresponds to the −3 dB bandwidth of the modulated signal).

Figure 5.8   Efficiencies for the four modulation schemes supported in DVB-S2.

most commercial product implementations have focused on the 8-PSK scheme. The 16-APSK and 32-APSK modes are targeted at professional applications (they can also be used for broadcasting), but these require a higher level of available C/N and the adoption of advanced predistortion methods in the uplink station to minimize the effect of transponder nonlinearity. Although these modes are not as power-efficient as the other modes, the spectrum efficiency is greater. The 16-APSK and 32-APSK constellations have been optimized to operate over a nonlinear transponder by placing the points on circles. Nevertheless, their performance on a linear channel is comparable with those of 16-QAM and 32-QAM, respectively. By selecting the modulation constellation and code rates, spectrum efficiencies from 0.5 to 4.5 bit per symbol are available, and can be chosen based on the capabilities and restrictions of the satellite transponder used [MOR200401]. Figure 5.8 depicts these efficiencies for the four modulation schemes supported. DVB-S2 has three "roll-off factor" choices to determine spectrum shape: 0.35 as in DVB-S, and 0.25 and 0.20 for tighter bandwidth restrictions. (The use of the narrower roll-off $\alpha = 0.25$, and $\alpha = 0.20$ allows transmission capacity to increase but may also produce larger nonlinear degradations by satellite for single-carrier operation.) (As seen earlier, the DVB-S2 standard allows up to 5 bits per symbol with a 32-APSK constellation; however, 32 APSK receiver chips were not yet available commercially at the time this book went to press.)

Constant coding modulation (CCM) is the simplest mode of DVB-S2; it is similar to the DVB-S, in the sense that all data frames are modulated and coded using the same fixed parameters. The LDPC code has a performance such that a coding rate of 4/5 is sufficient for the same channel conditions. For example, a minimum E_s/N_0 of 5.4 dB supports QPSK 4/5. The useful bit rate is then given by 30 Mbaud × 1.587 = 47.61 Mbps. Compared to DVB-S, this is an efficiency improvement of around 30 percent, and represents a 2–3 dB improvement [BRE200501].

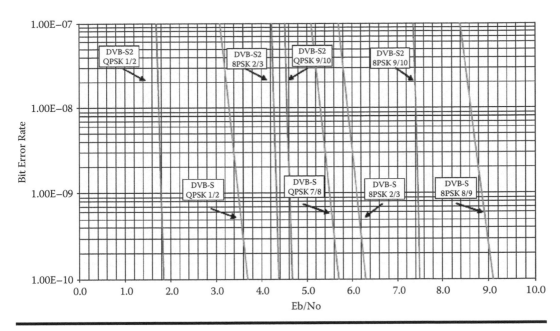

Figure 5.9 DVB versus DBB-S2.

Figure 5.9 compares DVB-S and DVB-S2 performance for QPSK and 8-PSK modulations with a variety of FEC rates. When one compares the 8-PSK 2/3 FEC rate for an 1×10^{-9} BER, the required E_b/N_0 level for DVB-S is 6.10 dB, whereas it is 4.35 dB for DVB-S2, equating to a 1.75 dB reduction in required power for the same E_b/N_0, reducing the required spectral power by 28.7 percent. For a data rate of 6 Mbps using the 8-PSK 2/3 FEC rate, the DVB-S would have a symbol rate of 3.26 Msps, assuming a carrier spacing factor of 1.3 would require 4.23 MHz of bandwidth on the satellite. In comparison, the DVB-S2 would have a symbol rate of 3.03 Msps requiring 3.94 MHz of bandwidth on the satellite, resulting in a bandwidth saving of approximately 6.9 percent [RAD200501].

A feature of the DVB-S2 standard is that different services can be transmitted on the same carrier, each using its own modulation scheme and coding rate (see Figure 5.10). This "multiplexing" at the physical layer is known as VCM. VCM is useful when different services do not need the same protection level, or different services are intended for different stations with different average receiving conditions. Using VCM, a different coding/modulation (CM) mechanisms can be selected for each station. Typically, these CMs range between QPSK 4/5 and 16-APSK 2/3. If the total baud rate (30 Mbaud) is distributed equally over each station, each station will use a $30/20 = 1.5$ Mbaud part of the carrier (assuming there were 20 stations in the system). The corresponding bit rates are varying between 2.38 and 3.96 Mbps according to the selected CM. The total available rate is then 61.09 Mbps. This is an improvement of around 65 percent compared to DVB-S [BRE200501].

When a return channel is available from each receiving site to the transmit site, DVB-S2 offers an even more versatile feature known as ACM. With ACM, it is possible to dynamically modify the coding rate and modulation scheme for every single frame, according to the measured channel conditions in which the frame must be received. The return channel is used to dynamically report the receiving conditions at each receiving site. In this scenario, the CMs range between 8-PSK 3/4 and 16-PSK 5/6. If distributed equally over each station, each station will use a $30/20 = 1.5$ Mbaud

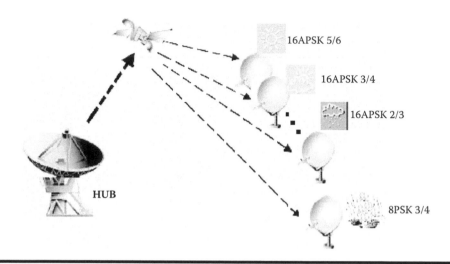

Figure 5.10   DVB-S2 variable coding and modulation (VCM).

part of the carrier (assuming there were 20 stations in the system). The corresponding bit rates are varying between 3.34 and 4.95 Mbps to each site. The total available rate is then 86.3 Mbps. This is more than 130 percent higher than for DVB-S.

The relatively large number of DVB-S receivers already installed in the field implies that there may not be a sudden change of technology in favor of DVB-S2; however, as noted earlier in the chapter, there is always an interest in optimizing bandwidth, driving to an eventual migration. To address this transition period, optional backward-compatible modes have been defined in DVB-S2. These modes allow one to send two transport streams on a single satellite channel. The first (high priority, HP) stream is compatible with DVB-S receivers as well as with DVB-S2 receivers, whereas the second (low priority, LP) stream is compatible with DVB-S2 receivers only.

When DVB-S2 is transmitted by satellite, quasi-constant envelope modulations such as QPSK and 8-PSK are power efficient in the single-carrier-per-transponder configuration because they can operate on transponders driven near saturation. The 16-APSK and 32-APSK modes, which are inherently more sensitive to nonlinear distortions and would require quasi-linear transponders (i.e., with larger output back off [OBO]), may be improved in terms of power efficiency by using

Table 5.2 C_{SAT}/N

Transmission Mode	No predistortion No phase noise	Dynamic predistortion No phase noise	Dynamic predistortion Phase noise
QPSK 1/2	0.6 (OBO = 0.4)	0.5 (IBO = 0; OBO = 0.4)	0.6
8-PSK 2/3	1.0 (OBO = 0.3)	0.6 (IBO = 0; OBO = 0.4)	0.9
16-APSK 3/4	3.2 (OBO = 1.7)	1.5 (IBO = l.0; OBO = 1.1)	1.8
32-APSK 4/5	6.2 (OBO = 3.8)	2.8 (IBO = 36; OBO = 2.0)	3.5

Note 1: C_{SAT} is the unmodulated carrier power at HPA saturation.

Note 2: OBO is the measured power ratio (dB) between the unmodulated carrier at saturation and the modulated carrier.

Table 5.3   Comparison of DVB-S2 and DVB-S broadcasting services

Satellite EIRP (dBW)	*51*		*53.7*	
System	DVB-S	DVB-S2	DVB-S	DVB-S2
Modulation and coding	QPSK 2/3	QPSK 3/4	QPSK 7/8	8-PSK 2/3
Symbol rate (Mbaud)	27.5 ($\alpha = 0.35$)	30.9 ($\alpha = 0.0$)	27.5 ($\alpha = 0.35$)	29.7 ($\alpha = 0.25$)
C/N (in 27.5 MHz) (dB)	5.1	5.1	7.8	7.8
Useful bitrate (Mbit/s)	33.8	46 (gain = 36 percent)	44.4	53.8 (gain = 32 percent)
Number of SDTV programs	7 MPEG-2 15 AVC	10 MPEG-2 21 AVC	10 MPEG-2 20 AVC	13 MPEG-2 26 AVC
Number of HDTV programs	1–2 MPEG-2 3–4 AVC	2 MPEG-2 5 AVC	2 MPEG-2 5 AVC	3 MPEG-2 6 AVC

nonlinear compensation techniques in the uplink station. In frequency-division multiplexing (FDM) configurations, where multiple carriers occupy the same transponder, this power must be kept in the quasi-linear operating region (i.e., with large OBO) to avoid excessive intermodulation interference between signals. In this case, the AWGN performance figures may be adopted for link budget computations. Table 5.2 shows, for the single-carrier-per-transponder configuration, the (simulated) C/N degradation at the optimum operating TWTA point. The figures in the table show the advantage offered by the use of dynamic pre-distortion for 16-APSK and 32-APSK modes [MOR200401]. Note that, for example, a DVB-S2 signal at the symbol rate of 30 Mbaud may be transmitted on a 36 MHz transponder using $\alpha = 0.20$.

Table 5.3 compares DVB-S2 and DVB-S broadcasting services via 36 MHz (at −3 dB) satellite transponders using 60-cm receiving antenna diameters operating at Ku band [MOR200401]. The example of video coding bitrates are 4.4 Mbps (SDTV) and 18 Mbps (HDTV) using traditional MPEG-2 coding, or 2.2 Mbps (SDTV) and 9 Mbps (HDTV) using advanced video coding (AVC) (MPEG-4) systems. The table shows the capacity gain of DVB-S2 versus DVB-S exceeding 30 percent. Furthermore, by combining DVB-S2 and AVC coding, around 25 SDTV channels per transponder are obtained. The combination of DVB-S2 and new AVC coding schemes facilitates the introduction of new HDTV services, with a favorable number of programs per transponder (e.g., five to six), reducing the satellite capacity cost increase with respect to current SDTV services.

References

[AHA200501] Staff, Terms and Definitions, AHA/Comtech Telecommunications Corporation, Moscow, ID 2005.

[ANS200001] ANS T1.523-2001, Telecom Glossary 2000, American National Standard (ANS), an outgrowth of the Federal Standard 1037 series, *Glossary of Telecommunication Terms*, 1996.

[BRE200501] D. Breynaert, M. d'Oreye de Lantremange, "Analysis of the Bandwidth Efficiency of DVB-S2 in a Typical Data Distribution Network," Newtec, CCBN2005, Beijing, March 21–23, 2005.

[COM199801] CM701/DT7000 Reed–Solomon (DVB Version), Option Card, ComStream Corporation, A Spar Company, San Diego, CA, 1998.

[EDW199001] G. Edwards, "Forward Error Correction Encoding and Decoding," Stanford Telecom Application Note 108, 1990. Sunnyvale, CA.

[HEN200201] H. Hendrix, "Viterbi Decoding Techniques for the Tms320c54X DSP Generator," Application Report SPRA071A, Texas Instruments, Dallas, January 2002.

[JOU199901] M. K. Juonolainen, Forward Error Correction in INSTANCE, Cand. Scient. Thesis, 1/2/1999, University of Oslo, Department of Informatics, Oslo, Norway.

[KOR200701] I. Koren and C. M. Krishna, *Fault-Tolerant Systems*, Morgan–Kaufman, San Francisco, CA, 2007, Department of Electrical and Computer Engineering, University of Massachusetts, Amherst, MA.

[LIT200101] L. Litwin, "Error control coding in digital communications systems," RF Design, July 1, 2001.

[LUB200201] M. Luby, L. Vicisano, J. Gemmell, L. Rizzo, M. Handley, and J. Crowcroft, "The Use of Forward Error Correction (FEC) in Reliable Multicast," Request for Comments: 3453, December 2002.

[MOR200401] A. Morello and V. Mignone, "DVB-S2— Ready for lift off," Digital Video Broadcasting, *EBU Technical Review*—October 2004, RAI Radiotelevisione Italiana.

[RAD200501] Radyne ComStream Staff, "DVB-S2 and the Radyne ComStream DM240," White Paper, WP017, Rev. 1.3, January 2005, Radyne ComStream, Inc., Phoenix, AZ.

[TOR199801] A. Torres, and V. Demjanenko, Inclusion of Concatenated Convolutional Codes in the ANSI T1.413 Issue 3, Contribution to Standards Committee T1-Telecommunications, Plano, Texas T1E1.4/98-301R1, November 30–December 4, 1998, VoCAL Technologies Ltd.

[TRE200401] TrellisWare Technologies Staff, "FlexiCodes: A Highly Flexible FEC Solution," April 2004, TrellisWare Technologies, Inc. San Diego, CA.

Chapter 6

Link Budget Analysis

The model of the end-to-end performance of a theoretical (satellite) system is known as a "link budget." This chapter brings together many of the concepts defined in the first five chapters, and it shows how the designer can engineer a satellite link (or links) to meets specific business service goals. As we saw in Chapter 2, *link budget* is a generic term used to describe a series of mathematical calculations designed to model the performance of a communications link; there are many complex trade-offs involved in the design of a satellite link. A link budget is an identification and aggregation of the various system parameters of the satellite link according to pertinent engineering rules and is used to determine

- The link performance from a fixed set of system parameters
- Requisite system parameters given particular link performance criteria

6.1 Overview

In a typical simplex (one-way) satellite link there are two link budget calculations:

- Link from the transmitting ground station to the satellite
- Link from the spacecraft to the receiving ground station

See Figure 6.1 for a reference model.

The satellite link contains transmit stations, receive stations, and the paths that connect them. The satellite is both a transmit station and a receive station. Many factors need to be included in the assessment of the end-to-end performance [USA199801]:

1. Link budget model
   - Uplink and downlink C/kT (carrier-to-receiver noise density)
   - Satellite receive power flux density

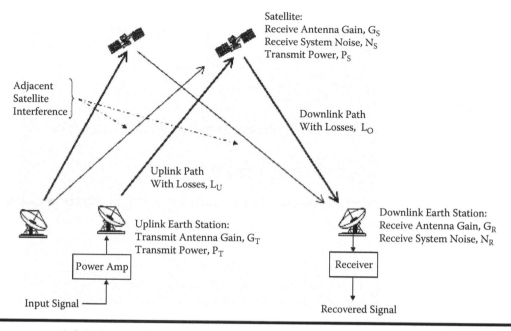

Figure 6.1   Link budget analysis reference model.

2. Positional data model
   - Earth terminal to satellite slant range
   - Earth terminal antenna elevation and azimuth
   - Uplink and downlink Doppler frequency shift
3. Benign atmosphere attenuation
   - Clear air attenuation
   - Rainfall attenuation
   - Atmospheric signal scintillation
4. Modulation and channel encoding
   - Noise equivalent bandwidth
   - Modulation spectral efficiency
   - Demodulator implementation loss
   - Probability of detection of error
   - Convolution channel coding gain
5. Earth terminal model
   - Antenna model
   - Receive system model
   - Transmit system model
   - Satellite model
   - Uplink signal power
   - Downlink signal power
   - Transponder signal and noise power sharing
   - Transponder uplink power flux

The following input items are needed (among others) to produce a link budget calculation for a spacecraft to ground station link:

- Earth station latitude
- Earth station longitude
- Spacecraft longitude
- Downlink frequency
- Antenna gain
- Antenna noise temperature
- Low noise amplifier
- Ortho mode transducer loss
- Effective isotropic radiated power (EIRP)
- Intermediate frequency (IF) receive bandwidth
- Transmit data rate
- Waveguide losses

Table 6.1 defines the major terms; many of these were already defined in previous chapters.

As we saw in earlier chapters, communications measurements and calculations are often done in decibels and units related to decibels. This is done because by using logarithms, multiplication and division are replaced by addition and subtraction, respectively. Link budgets are done largely in dB.

Carriers are modulated by the baseband (intelligent) signal. Each side of a satellite communications link (inbound, outbound) is comprised of an uplink and a downlink. The technical performance of each of these two links determines the quality of the service obtained from the sender to the receiver, measured either as signal-to-noise ratio (*S/N*) for analog communications or bit error rate (*BER*) for digital communications. As implied, several parameters impact the performance of a link, including power, antenna gain (and, so, size), modulation/forward error correction (*FEC*)/filtering choices, rain zone, and location (satellite *EIRP*).

Link performance is described by C/N_0, where C is the received carrier power, and N_0 is the noise power spectral density (C/N_0 is measured in Hz.) Recall that C/N_0 and E_b/N_0 are relatable.

To assess the end-to-end origin-to-destination (overall) performance, one needs to look at the performance of the two sublinks involved in a satellite communications link. These, in turn, are driven by the carrier power budget and the noise contribution budget.

See Figure 6.2 for a diagram of the satellite link.

Table 6.2 depicts some key link budget parameters [FOC200701].

The link analysis is done considering the following:

- The transmit station has an antenna system that radiates the power fed to it by an amplifier; radiated power is measured as *EIRP*.
- The receive station has an antenna system which collects power from an incident *RF* field in the presence of (thermal) *noise*. It has a *gain* (*G*) and a *noise temperature* (*T*). *G/T* is the figure of merit.
- The path between the stations attenuates the *RF* signal in a number of different ways. These are known as *Losses*.

Noise degrades a communications link. As seen in earlier chapters there are various sources of noise in a satellite system, and care must be taken so that these noise components remain within established bounds. Noise produced in communications is commonly specified as a "noise temperature" in

Table 6.1   Glossary of key LBA-related terms

Antenna noise temperature	The temperature of a hypothetical resistor at the input of an ideal noise-free receiver that would generate the same output noise power per unit bandwidth as that at the antenna output at a specified frequency. The antenna noise temperature depends on antenna coupling to all noise sources in its environment, as well as on noise generated within the antenna. The antenna noise temperature is a measure of noise whose value is equal to the actual temperature of a passive device [ANS200001].
Atmospheric signal scintillation	Atmospheric losses for passing radio signals. Scintillation is produced by turbulent air with variations in the refractive index. These losses are dependent on, among other factors, elevation angle and antenna size. Attenuation due to scintillation rapidly increases with increasing frequency (e.g., Ka-band signals that pass through a turbulent region are scattered by the turbulent cells and rapid amplitude changes around the average occurs). In general, scintillation is a random fluctuation of the received field strength caused by irregular changes in the transmission path over time [AER200801].
C/N$_0$	Link performance metric where C is the received carrier power, and N_0 is the noise power spectral density (C/N_0 is measured in Hz.) This term is traditionally used in analog radios where N is the receiver or channel noise equivalent bandwidth.

$N = N_0 \times B$ (linear expression) (B = bandwidth)
$N = N_0 + 10 \log B$ (decibel expression)

Note that $C/N_0 = C/KT_e$ (Noise Density $N_0 = N/B$ or
$N_0 = KT_eB/B = KT_e$)

Equivalent noise temperature:
- $N = KTB$, where K (1.39 × 10^{-23} joules per Kelvin) is Boltzmann's constant, T (Kelvin) is environment temperature, and B (Hz) is the system bandwidth
- $F = 1 + T_e/T$, where F (unitless) is noise factor, and T_e is equivalent noise temperature (Kelvin)
- $T_e = T(F - 1)$

(C/N_0 and Eb/N_0 are relatable.)

Carrier-to-receiver noise density	The ratio of the received carrier power to the receiver noise power density
Clear air attenuation	Attenuation of a signal passing through air (causes of attenuation include rain, scintillation, oxygen and water vapor). The amplitude, phase, and angle-of-arrival of a microwave signal passing through the atmosphere vary due to in-homogeneities in the refractivity of the atmosphere. The clear air attenuation for a satellite system is about 2 dB, and varies to some degree with the location and time.
Convolution channel coding gain	Channel coding is the application of processing algorithms, prior to transmission via a channel (and reverse algorithms at the receiver), used to improve data reliability. Coding gain is the increase in efficiency that a coded signal provides over an uncoded signal. Expressed in decibels, the coding gain indicates a level of power reduction that can be achieved. Specifically, convolution channel

Table 6.1   Glossary of key LBA-related terms (*Continued*)

	coding gain relates to gain achieved with convolutional coding, which is a forward error-correction scheme, where the coded sequence is algorithmically achieved through the use of current data bits plus some of the previous data bits from the incoming stream [AER200801]. Note: The older Reed Solomon Viterbi (RSV) Forward Error Correction (FEC) methodology is inefficient when compared to modem turbo codes (TC) or turbo product codes (TPC); this inefficiency results in higher bandwidth and power required for the same BER as TC or TPC. FEC codes are also referred to as channel codes.
Downlink Doppler frequency shift	A phenomenon that results in a change in the received frequency (here the downlink frequency) when the transmitter and/or receiver are in motion. If the source is moving away (positive velocity) the observed frequency is lower and the observed wavelength is greater; if the source is moving toward (negative velocity) the observed frequency is higher and the wavelength is shorter. The nonrelativistic Doppler shifted frequency of an object moving with speed v with respect to a stationary observer, is (as long as the velocity of the moving object is much less than the speed of light): $$f' = f_0 \left(\frac{1}{1 + V/C_0} \right)$$ and the Doppler shifted wavelength is: $$\lambda' = \lambda_0 \left(1 + V/C_0 \right)$$ where C_0 is the speed of the wave in a stationary medium, and the velocity V is the radial component of the velocity (the part in a straight line from the observer)(as a convention, the velocity is positive if the source is moving away from the observer, and negative if the source is moving toward the observer)
Downlink frequency	Frequency of the downlink for the appropriate band (e.g., C band, Ku band, BSS band, Ka band), also in reference to the specific transponder being used.
Downlink signal power	Power of the signal transmitted by the satellite
Earth station latitude	Latitude of the Earth station
Earth station longitude	Longitude of the Earth station
Earth terminal antenna azimuth	Azimuth is the magnetic compass direction (angle of sighting) at which one points the dish. It is a side-to-side adjustment of the antenna. Azimuth is the angular distance from true north along the horizon to a satellite, measured in degrees [LAU200701].
Earth terminal antenna elevation	Elevation is the angle above the horizon at which one points the dish when seeking to establish satellite reception/transmission; this is an up-and-down adjustment of the antenna.
Earth terminal to satellite slant range	The length of a line drawn from the antenna to the satellite

(*Continued*)

Table 6.1 Glossary of key LBA-related terms (*Continued*)

E_b/N_0	Commonly used figure-of-merit ratio for systems making use of digital modulation. Digital modulation schemes are typically specified in terms of theoretical BER versus E_b/N_0 performance. Both E_b and N_0 are expressed in units of dBm/Hz (or Joules per second), hence the ratio is dimensionless and is expressed in dB.
	$E_b = C/$(data rate)　　　　　(linear expression) $E_b = C - 10 \times \log$(data rate)　　　　　(decibel expression)
	E_b/N_0 is a good measure to compare digital systems using different modulation or encoding schemes and data rates.
	$E_b = P_t \times T_b$ or P_t/f_b, where P_t is transmit power in W (or Joules per second), T_b is bit duration in seconds, and f_b is the bit rate in bits per second (bps)
Effective isotropic radiated power (EIRP)	A measure of the signal strength that a satellite transmits toward the earth, or an earth station toward a satellite, expressed in dBW. Also, the arithmetic product of (a) the power supplied to an antenna and (b) its gain [ANS200001].
	$EIRP = P_T \times G_T/ L_T$, where G_T is the transmit antenna gain and L_T is transmission line losses including back-off loss, feeder loss, etc.
	$EIRP$ (dBW) $= P_T$ (dB) $- L_{bo}$(dB) $- L_{bf}$(dB) $+ G_T$ (dB), where 　P_T = actual power output of the transmitter (dBW) 　L_{bo} = back-off losses of HPA (dB) 　L_{bf} = total branching and feeder loss (dB) 　G_T = transmit antenna gain (dB)
Low noise amplifier	Receive station amplifier
Modulation spectral efficiency	The ratio of the transmitted bit rate R_c to the bandwidth occupied by the carrier. The bandwidth occupied by the carrier depends on the spectrum of the modulated carrier and the filtering it undergoes. Filtering is employed at the transmitter and receiver side to limit the interference to adjacent out-of-band carriers; however, filtering can generate intersymbol interference, which in turn impacts the bit error rate (BER) of the received signal.
Noise equivalent bandwidth (NEB)	Given the amplitude frequency response of an RF device, the noise-equivalent bandwidth is defined as the width of a rectangle whose area is equal to the total area under the response curve and whose height is that of the maximum amplitude of the response. This is the bandwidth used to compute the total noise power passed by a device and is generally not the same as the 3 dB bandwidth [ORT200801].
Ortho mode transfer loss	The ortho mode transducer is used at the antenna for splitting the signal into vertical and horizontal polarization channels. There are losses associated with this device, these being known as ortho mode transfer loss.
Rainfall attenuation	Atmospheric losses for passing radio signals in the presence of (moderate-to-heavy) rain

Table 6.1   Glossary of key LBA-related terms (*Continued*)

Satellite receive power flux density	Satellite uplink power flux density is a calculated parameter. The figure is derived from the power flux density to saturate the transponder (PFD/sat); this figure is available from the uplink coverage map and the satellite transponder specification. A user needs to adjust the uplink earth station transmit power (and possibly uplink antenna size) to obtain the figure one requires at the satellite. For example, the PFD/sat at beam center might be −83 dBW/m²; the PFD/sat on the −4dB contour would then be −79 dBW/m². (Note that the PFD/sat figures can be altered by commanding changes in the gain setting of the satellite transponder; it may be possible to negotiate changes in gain setting, particularly if one leases an entire transponder.) The value just described refers to a whole transponder: if one transmits an uplink signal with sufficient power to produce the PFD/sat just described, one will saturate the transponder [SAT200801].
Satellite transmit power flux density	The power flux density (PFD) is defined as $EIRP/4\pi d^2$
Slant range	The length of a line drawn from the antenna to the satellite
Spacecraft longitude	Orbital location of (geosynchronous) satellite
Transmit data rate	Line rate speed of the digital signal being transmitted (including the FEC overhead)
Uplink Doppler frequency shift	See Downlink Doppler frequency shift
Uplink signal power	Transmit power
Waveguide losses	Losses experienced between the antenna feed and the transmitting or receiving amplifier.

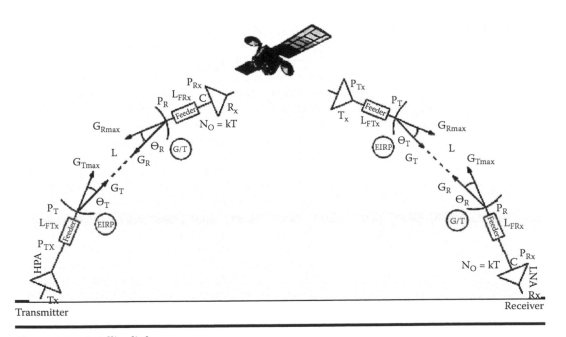

Figure 6.2   Satellite link.

Table 6.2 Overview of key link budget parameters

Earth station parameters	*G/T*	This parameter is used along with the power received from the satellite to determine how much noise impacts a receiving satellite modem. Obviously, if there is too much noise the modems will not be able to demodulate the signals. A larger antenna will increase the G/T; in turn, this will increase the signal level far enough above the noise so that the modems perform correctly.
	HPA power level	For a transmission system one needs to be concerned with the amount of power the high power amplifier (HPA) can produce. One needs to be able to transmit enough power to the satellite to ensure adequate signal level on the downlink into the target earth station. If one does not transmit enough power, the receive station(s) will not be able to demodulate the signals correctly.
	Antenna gain/size	Gain (G) is directly proportional to the size of the antenna's reflecting surface. Hence, the larger the antenna, the larger the gain. This applies to both transmit gain and receive gain.
		A larger antenna has a larger G/T and allows a smaller HPA to be used to get the same power to the satellite (but larger antennas increase the total system cost).
		As a tradeoff, one can look at increasing the HPA power for the transmitting station, and/or increase the power coming from the satellite. Or, there could be cost savings if the satellite resources (power/bandwidth) and HPA power capability are reduced and the antenna is made larger.
Satellite parameters	**Satellite antenna patterns**	Antenna patterns and contour EIRP values need to be taken into consideration. Typically, one is able to receive more power from the satellite in some geographic locations compared to other geographical locations (therefore, at some locations one may be able to reduce antenna size, reduce HPA size, or use less satellite resources.)
	Satellite *saturated* flux density (*SFD*)	This value determines how much power one would need to transmit to the satellite in order for it to produce its maximum output power. If the SFD increases, one needs to get more power to the satellite to have the satellite transmit the power one requires. This can affect the transmit station requirements, for example, the transmit antenna gain and HPA power size.
	Satellite effective isotropic radiated power (*EIRP*)	This value is the amount of power seen at the receiving antenna on the earth. Each satellite channel has a saturated EIRP that can be transmitted, which is the maximum power available from the satellite. When one leases space on a satellite, one is given a maximum EIRP and a maximum frequency bandwidth that can be used (typically only one of these is reached, and it is usually the EIRP.)

Table 6.3   Noise in satellite links

Uplink thermal noise	Noise generated by the satellite's own receiving system. The satellite itself has a receiving antenna and LNA and has a *G/T* value. This *G/T* value, losses, power transmitted from the earth station and other parameters are used to calculate the ratio of the signal (or carrier) power to the thermal noise of the uplink system. This component is referred to as uplink *C/T*.
Transponder intermodulation noise	A transponder is a system of antennas and amplifiers. Typically, an amplifier creates noise when multiple carriers are present because signals tend to combine in frequency and create new signals (in the same way a mixer does), but at lower power levels. Because these signals are usually small they can be viewed as noise, and so they are characterized in the same way as uplink thermal noise. The noise from satellite intermodulation is presented as a ratio of the power of the carrier to the noise temperature, or *C/Tim* (intemodulation). (Note that this term is different from the uplink thermal noise contribution, uplink *C/T*)
Downlink thermal noise	The downlink portion of the satellite link also adds noise; this is dominated by the *G/T* of the satellite earth station. The downlink thermal noise is defined as the ratio of the power in the carrier that is being received to the noise added to the system by the downlink, and (again) is designated *C/T*.
Overall noise performance	The three noise components listed above are combined into an overall noise specification for the entire link; the overall noise performance is used in the link analysis to determine how much noise is present in the received signal.
	If the overall combined noise is too high, the performance of the receiving modems will degrade. For example, if a system has most of the noise in the uplink, one may not be able to get beyond a specific noise level for the entire system. Using a larger antenna will increase the signal level and decrease the noise level only for the downlink portion of the system only.

Kelvin (the noise measurement is referred to as "thermal noise," because it is given in units of temperature). The main noise components in the satellite system are shown in Table 6.3 [FOC200701].

Interference also degrades a communications link. Interference can arise from other communication systems that use the same transmission and/or reception frequencies ("inter-system interference"), or from other users within the same satellite system ("intra-system interference"). Sources of inter-system interference may be other satellite communication systems, and/or terrestrial radio communication systems [FOC200701].

Interference has the effect of adding to the overall noise on the link; in turn, this increases the receiving system noise temperature and degrades the quality of the received signal. If a satellite system is poorly designed, the impact of interference is significant. Earth stations must comply with regulatory-dictated performance criteria to ascertain that the various users of a satellite system can share the satellite resources without negatively impacting each other.

The three principal sources of interference in satellite communication systems are shown in Table 6.4 [FOC200701].

As covered in previous chapters, the transmitter *Tx* (left bottom of Figure 6.2) is connected via an HPA with power P_{Tx} and via a feeder to the antenna (the feeder could be a long waveguide or a fiberoptic-based system that makes use of a fairly short waveguide); the power at the end of the feeder is P_T. The antenna has a gain G_T in the direction of the satellite (receiver). The power radiated in the direction of the satellite receiver is the *EIRP* with

$$EIRP = P_T\, G_T \text{(expressed in } W)$$

Table 6.4   Principal sources of interference in satellite communication systems

Adjacent satellite interference (ASI)	A transmit earth station can inadvertently impose some of its radiated power onto satellites that are operating at orbital positions adjacent to that of the wanted satellite. This could occur if the transmit antenna is poorly pointed toward the wanted satellite, or if the earth station antenna beam is not sufficiently concentrated in the direction of the wanted satellite (this may arise because the antenna is too small). The unintended radiation can interfere with services that use the same or similar frequencies on the adjacent satellites.
	Interference into adjacent satellite systems is minimized by ensuring that the transmit earth station antenna is accurately pointed toward the satellite and that its radiation pattern is sufficient tight to suppress radiation toward the adjacent satellites. In general, a larger uplink antenna has less potential for causing adjacent satellite interference (but it will typically be more expensive).
	A receive earth station could inadvertently receive transmissions from adjacent satellite systems, thus interfering with the wanted signal. As is the case for a transmit earth station, it is also critical to accurately point the antenna toward the wanted satellite to minimize adjacent satellite interference effects.
Adjacent channel interference (ACI)	The design goal is to keep the power of a carrier transmitted by an earth station entirely contained within the assigned transponder bandwidth. This facilitates the use of frequency division multiplexing (FDM), that is, it permits carriers from different earth stations to be placed side by side in frequency with no interference between them, so long as their bandwidths do not overlap.
	In reality, some carrier power is radiated outside of the nominal bandwidth of the carrier, for example, in digital modulation environments. This sideband power can interfere with the carriers that are adjacent in frequency to the wanted carrier. The key factor that drives the amount of interference is amplifier nonlinearity. One way to control this is to use output backoff of the earth station's HPA (the larger the backoff, the lower the potential for ACI).
Cross-channel interference (CCI)	Satellite systems typically employ polarization to achieve frequency reuse. This implies that two signals may share the same frequency within a satellite system, so long as they employ opposite polarization states (e.g., horizontal linear polarization and vertical linear polarization). Fortunately, the frequency bands for the two signals are usually shifted by 20 MHz, so that the potential for interference is mitigated. In principle, each signal can be received without interference from the cofrequency signal on the opposite polarization; pragmatically, the earth station feeds and satellite antennas' feeds are not able to perfectly separate the two polarization states. This results in a (small) proportion of the unwanted "cross-polar" signal being transmitted or received along with the wanted signal, causing cross-channel interference.
	CCI is maintained at an acceptable level by ensuring that the earth station antenna has adequate cross-polar performance (cross-polar discrimination). For linear polarization, it is important to ascertain that the antenna feed is properly aligned with the linear polarization plane of the satellite antennas (by rotating the feed appropriately).

On its way to the satellite receiver the power has a path loss L.

The (satellite) receiving equipment consists of a receive antenna with gain G_R in the direction of the transmit equipment, connected with a (short) feeder to the receiver Rx. At the receiver's input the power of the modulated carrier is C. Various sources contribute to the system noise temperature T; the noise temperature drives the noise power spectral density N_0 (since $N_0 = kT$) and in turn the link performance C/N_0. The receiving equipment performance is measured using G/T, with G the overall G of the receiving equipment.

C/N_0 is the overall parameter of interest.

As covered in Chapter 3, the gain is the ratio of the power radiated (or received) per unit solid angle by an isotropic antenna fed with the same power. As we have seen, the gain is maximum in the direction of maximum radiation (the electromagnetic axis of the antenna, the boresight).

$$G_{max} = (4\pi/\lambda)^2 A_{eff}$$

A_{eff} is the effective aperture area of the antenna. For a circular dish (with $A = \pi D^2/4$), we saw that $A_{eff} = \eta A$, where η is the efficiency of the antenna (usually between 55 and 75 percent). Then

$$G_{max} = \eta \, (\pi D/\lambda)^2 = \eta \, (\pi Df/c)^2$$

In dBi, relative to an isotropic antenna, the maximum antenna gain is

$$G_{max, \, dBi} = 10\log \eta(\pi D/\lambda)^2 = 10\log \eta(\pi Df/c)^2$$

Figure 6.3 shows a plot of $G_{max, \, dBi}$ as a function of the antenna size and operating frequency.

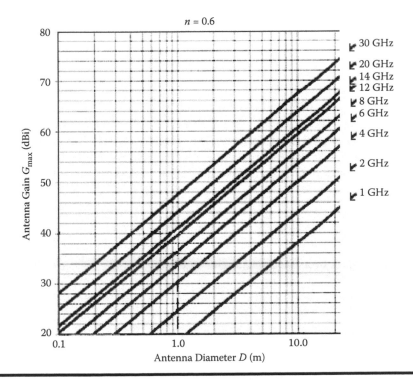

Figure 6.3   **Plot of $G_{max, \, dBi}$ as a function of the antenna size.**

6.2 Parameters Required to Analyze Link

6.2.1 Losses

Losses* are degradations in the RF power level of the transmitted signal. These are due to a number of factors, including:

- **Free space loss**—Attenuation due to the spreading of the wave front in free space (vacuum).
- **Atmospheric loss**—Attenuation due to absorption and scattering caused by the composition of the atmosphere, including precipitation.
- **Pointing loss**—Lost signal power due to antenna mispointing at the earth station and the spacecraft.
- **Multipath**—Variations in received signal power as the signal finds its way to the receiver over more than one path. In satellite applications this occurs at low look angles.

6.2.2 Free Space Loss (FSL)

While the other losses are not insignificant, the Free Space Loss (FSL) is much larger than the other losses combined. The FSL is the attenuation of the RF signal along the path from transmitter to receiver. This is dependent on the RF wavelength (or the frequency) and the distance along the path. To define this loss, we need to consider the concept of the isotropic transmit antenna, that is, one that radiates energy equally in all directions, as covered in Chapter 3. The energy from the isotropic antenna radiates outward as in the shape of a sphere and the energy is spread equally over that surface. The power density, P_{av}, available at any point on the sphere is equal to the transmitted power, P_T, divided by the area of the surface of the sphere. If we let R be the radius of the sphere, then,

$$P_{av} = P_T/(4\pi R^2)$$

This is the power density available to the isotropic receive antenna. The isotropic antenna absorbs power from the incident RF field that has power density P_{av}. As we saw in Chapter 3, the amount of power any antenna absorbs is related to its *effective aperture*. The received power for an isotropic antenna is related to the wavelength of the incident RF field, λ, and is equal to $(\lambda^2/4\pi)$. The power P_R received by the antenna is equal to the available RF field density times the effective aperture. Then,

$$P_R = P_{av} \times (\lambda^2/4\pi) = P_T \times (\lambda/4\pi R)^2$$

This is the power received by an isotropic receive antenna.

The transmission loss between a transmit and receive antenna pair is the ratio of the power transmitted to the power received, that is,

$$L = P_T/P_R$$

---

* Some key portions of the rest of this chapter were developed by Joe Schiavino, SES Engineering, Princeton, New Jersey. His valuable contributions are herewith acknowledged and emphasized.

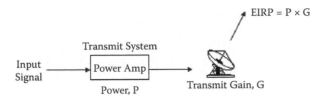

Effective Isotropic Radiated Power (EIRP) – At the transmitter, the product
of the transmit antenna gain and the amplifier power, in a direction of interest.

Figure 6.4   Transmission chain.

As noted in earlier chapters, in link analysis it is common to work in decibels (dB); after substitution one gets

$$L_{dB} = 10 \times \log_{10} [(\lambda /4\pi R)^2]$$

Letting $R = d$, for the distance between the antenna, the expression for free space loss becomes

$$L_{dB} = 21.98 + 20 \times \log_{10} [d/\lambda]$$

6.2.3   Effective Isotropic Radiated Power (EIRP)

Now that we have an expression available for the loss encountered over the isotropic transmission path, it is convenient to speak of the power radiated by an antenna system isotropically. This power level is known as the effective isotropic radiated power. An actual antenna radiates the power fed to it directionally according to its particular physical characteristics. However, one is typically only interested in the transmission in the direction of the receive antenna. Therefore, antenna gains are given as effective isotropic figures, meaning in the direction of interest. The *EIRP* at the output of the antenna system is given as the product of the power fed into the antenna, *P*, multiplied by the isotropic antenna gain, *G*, or *EIRP* = *P* × *G*, as shown in Figure 6.4. Expressed in decibels this becomes

$$(EIRP)_{dBW} = P_{dBW} + G_{dBi}$$

(the subscripts indicate the units.) Power levels in watts become dBW and the isotropic gain (unitless) of the antenna becomes dBi. These are decibels related to 1 watt and decibels-isotropic, respectively.

6.2.4   Receive System Noise

At the receive station, the antenna collects the incident RF energy and feeds it to a receiver. The receive system, consisting of the antenna and receiver, adds noise to the received signal. There are a number of various sources and types of noise. However, for link analysis the noise that is of interest is known as *thermal noise*. The amount of thermal noise power available per unit bandwidth or *noise power density* (N_0), is given by

$$N_0 = kT \qquad \text{(Watts/Hz)}$$

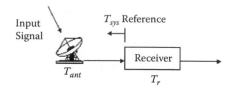

Figure 6.5   Receive chain.

where k is the Boltzman constant (1.38×10^{-23} *J/K*) and T is the absolute temperature in Kelvins (K). *Total noise power*, N, in the bandwidth of interest can thus be calculated by multiplying by the bandwidth,

$$N = kTB \qquad \text{(Watts)}$$

where B is the bandwidth of interest (in Hz).

6.2.5 Receive System Total System Noise Temperature

The noise contributed by the receive system (consisting of the antenna and receiver) is given by the *system temperature*, which is the factor T in the above expression. The system temperature is measured in units of Kelvin and for the receive system is the sum of the antenna temperature and the receiver temperature, or

$$T_{sys} = T_{ant} + T_r$$

The system temperature is referenced at the input of the receiver. See Figure 6.5.

6.2.6 Receive System Figure of Merit—(G/T)

Clearly the higher the system temperature, the more noise the receive system adds to the received signal (recall $N_0 = kT$). A way to measure how well the receive system works is to look at the ratio of antenna gain to system temperature ratio, or G/T. This is calculated at the reference as

$$(G/T)_{dB} = G_{dB} - 10 \times \log_{10} (T_{sys}) \text{ (dB/K)}$$

(see Figure 6.6).

The concepts discussed thus far will be used to analyze the RF link.

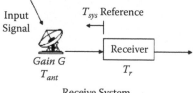

Receive System

Receive Figure of Merit (G/T) – At the receive site, the ratio of the receive antenna gain and the receive system noise temperature.

Figure 6.6   G/T concept.

6.3 Basic Link Analysis Approach

The basic link budget procedure is as follows:

- For each path, calculate the carrier power-to-noise power ratio (*C/N*).
- Combine the ratios to obtain the total, or system *C/N*.
- Compare system *C/N* to proper performance equations to determine system link performance.
- Iterations may be necessary if attempting to design a suitable system, as opposed to performance verification.

6.3.1 System Equations

The system equations are used to calculate the overall system *C/N* of the complete satellite link. The *C/N* parameter is used to determine performance of the overall link.

There are three key equations:

- The uplink equation
- The downlink equation
- The system equation

Some notation follows.

- $(C/N)_U$, $(C/N)_D$, $(C/N)_T$: Carrier power-to-noise power ratio, subscripts indicating uplink, downlink, and system (total), respectively, in dB.
- $(EIRP)_U$, $(EIRP)_D$: Uplink (from the earth station) and downlink (from the satellite) EIRP, in units of dBW.
- $(G/T)_{ES}$, $(G/T)_{sat}$: Earth station and satellite receive system "figure of merit," respectively, in units of dB/K.
- $(FSL)_U$, $(FSL)_D$: Free space loss, subscripts indicating uplink and downlink, respectively, in units of dB.
- $(B)_{dB}$: Carrier noise bandwidth, in dB-Hz.
- (k): Boltzman's constant, which is equal to −228.6 dBJ/K.
- $(losses)_U$, $(losses)_D$: All link losses other than free space, for the uplink and downlink, respectively, in dB. For example, losses due to rain, pointing errors, and atmospheric absorption.
- (C/I): Carrier power-to-interference power ratio, in dB. Lumped parameter accounting for interference from a number of sources.

With this notation, the system equations are

- Uplink equation (in dB)
  $(C/N)_U = (EIRP)_U - (FSL)_U + (G/T)_{sat} - (k) - (B)_{dB} - (losses)_U$

- Downlink equation (in dB)
  $(C/N)_D = (EIRP)_D - (FSL)_D + (G/T)_{ES} - (k) - (B)_{dB} - (losses)_D$

- System (*C/N*)
  $(C/N)_T = (C/N)_U \oplus (C/N)_D \oplus (C/I)$

  Note: This equation is not in dB.

"Power" addition (⊕) is defined as follows:

- This is a "product over sum" function:
  $Z = X \oplus Y \Rightarrow Z = XY/(X + Y)$

- Or "parallel resistors":
  $Z = X \oplus Y \Rightarrow (1/Z) = (1/X) + (1/Y)$

These two expressions are equivalent.

6.3.2 Example Link Budget

Table 6.5 depicts a link budget example.

6.4 Auxiliary Equations

The system equations discussed above assume that values such as *EIRP* and *G/T* are known, or are required to be calculated. Often, these are not directly given, nor are they the desired result. Often, these values must be related to other earth station parameters (e.g., antenna size), or satellite parameters (e.g., transponder operating point).

6.4.1 Antenna Gain

As we saw in Chapter 3, the power gain, *G*, of an antenna with an aperture area, *A*, relative to the isotropic antenna can be expressed as

$$G = 4\pi\eta A/\lambda^2$$

where η is the efficiency of the aperture and λ is the wavelength of RF energy. For a parabolic dish antenna the area can be taken to be a circle with diameter *D*. Substituting for the area, transforming the wavelength, λ, to frequency in GHz, f_{GHz}, and finally converting to decibels yields the expression

$$G_{dB} = 20.4 + 10 \times \log_{10} [\eta] + 20 \times \log_{10} [D] + 20 \times \log_{10} [f_{GHz}]$$

This is used for both transmit and receive gain calculations of a parabolic antenna.

6.4.2 "Back-Off"

As we saw in earlier chapters, the concept of Back-Off is simply the ratio of the maximum power level, P_{max}, of a device to the operating level, P_0, of the device expressed in decibels. This concept is also used for power flux density ratios. As an equation (in decibels) this is

$$(BO) = (P_{max}) - (P_0)$$

Table 6.5 Link budget example

Inputs:
Uplink location: St. Paul, MN (establish size/power)
Downlink locations: State of Minnesota, where all remote antennas are 3.8 m.
2 SCPC carriers of 19.3 Mbps each (to nominally carry 2 MPEG-2 HD video signals)
Availability: 99.99%, DVB-S (QPSK), FEC as required (used 3/4)
AMC6 C-Band. 8dB FCA
Outputs:

sat	AM06					
xpdr	D8C					
Data rate	19300 kbps			Uplink site	St. Paul, MN	St. Paul
Number of phases	4			Uplink SFD		−97.61 dBW/m²
FEC	3	4		Attenuator		8.0 dB
Reed–Solomon	188	204		Uplink G/T		3.03 dB/K
Required Eb/No	5.50 dB			Uplink frequency		6085.00 MHz
If bandwidth	13961.7 kHz			Downlink frequency		3660.00 MHz
C/N required	6.9 dB			Uplink EIRP		58.3 dBW
Faded system margin	1.0 dB			Uncompensated U/L fade		1.75 dB
Antenna efficiency	65.00%			Uplink C/N		16.9 dB
				Assumed C/I		16.0 dB
Spacing factor	1.35			Minimum C/N down		9.3 dB
Required bandwidth	18.848 MHz					
Allocated bandwidth	18.850 MHz			Transponder bandwidth		36.0 MHz
Transponder SCPC IBO	7.0 dB			Required bandwidth		52.36 %
Transponder SCPC OBO	4.0 dB			Allocated bandwidth		52.36 %
Carrier IBO	14.60 dB			Carrier allocated power		17.38 %
Carrier OBO	11.60 dB					

(Continued)

Table 6.5   Link budget example (Continued)

Receive site	ANT DIA (m)	ES Tsys (K)	ANT G/T (dB/K)	XPDR SAT EIRP (dBW)	CXR EIRP (dBW)	Clear-sky Downlink C/N (dB)	Margin to Min C/Ndn (dB)	Availability (%)	Outage (hr/year)	Fails Criteria 0	Required Availability (%)
St. Paul, MN	3.80	90.0	22.3	41.02	27.67	11.3	1.9	99.996%	0.3		99.990%
Karlstad, MN	3.80	90.0	22.3	40.74	27.39	10.9	1.6	99.992%	0.7		99.990%
Roseau, MN	3.80	90.0	22.3	40.69	27.34	10.9	1.5	99.991%	0.8		99.990%
International Falls, MN	3.80	90.0	22.3	40.65	27.30	10.8	1.5	99.992%	0.7		99.990%
Grand Portage, MN	3.80	90.0	22.3	40.59	27.24	10.8	1.5	99.992%	0.7		99.990%
Crookston, MN	3.80	90.0	22.3	40.85	27.50	11.0	1.7	99.994%	0.6		99.990%
Cass Lake, MN	3.80	90.0	22.3	40.84	27.49	11.0	1.7	99.994%	0.5		99.990%
Duluth, MN	3.80	90.0	22.3	40.81	27.46	11.0	1.7	99.995%	0.5		99.990%
Moorhead, MN	3.80	90.0	22.3	40.97	27.62	11.2	1.8	99.995%	0.4		99.990%
Wadena, MN	3.80	90.0	22.3	40.96	27.61	11.2	1.8	99.995%	0.4		99.990%
Brainerd, MN	3.80	90.0	22.3	40.93	27.58	11.2	1.8	99.995%	0.4		99.990%
Ortonville, MN	3.80	90.0	22.3	41.13	27.78	11.4	2.0	99.996%	0.3		99.990%
Pipestone, MN	3.80	90.0	22.3	41.25	27.90	11.5	2.2	99.997%	0.2		99.990%
Mankato, MN	3.80	90.0	22.3	41.13	27.78	11.4	2.1	99.997%	0.3		99.990%
Rochester, MN	3.80	90.0	22.3	41.06	27.71	11.3	2.0	99.997%	0.3		99.990%
Harmony, MN	3.80	90.0	22.3	41.08	27.73	11.4	2.0	99.997%	0.3		99.990%

Using an alpha = 1.35, a 19.3 Mbps link requires 18.85 MHz (At alpha = 1.25, a 19.3 Mbps link requires 17.50 MHz.)
At the transmit site, the following combinations could be used: 3.8 m/51 W HPA; 4.5 m/36 W HPA; 5 m/30 W HPA; 6.1 m/20 W HPA.

This concept is used to specify:

■ Earth station high-power amplifier levels
■ Satellite transponder carrier output power levels
■ Satellite transponder carrier input flux density levels

Some notation follows.

■ (HPA): Total rated power of the earth station high power amplifier, in units of dBW.
■ (BO): Back-Off of earth station HPA, which is the ratio of the HPA output power to its rated power, in units of dB.
■ $(losses)_{ES}$: Miscellaneous ohmic losses between the HPA and the antenna flange, in units of dB.
■ $(G)_{Ant}$: Isotropic gain of earth station uplink antenna, in units of dBi.
■ (D): Antenna reflector diameter, given in meters (m).
■ (f_{GHz}): Frequency at which calculating antenna gain, in GHz.
■ (η) : Antenna system efficiency, unitless.

With this notation, the auxiliary equations for the earth station are

■ High power amplifier

$$(EIRP)_U = (HPA) - (BO) - (losses)_{ES} + (G)_{Ant}$$

■ Antenna gain

$$(G)_{Ant} = 20.4 + 20 \times \log_{10}(D) + 20 \times \log_{10}(f_{GHz}) + 10 \times \log_{10}(\eta)$$

6.4.3 Satellite Link Budget Model

For RF link analysis purposes the satellite is reduced to several parameters. In the system equations the satellite parameters are the satellite G/T, reference as $(G/T)_{sat}$, and the satellite downlink EIRP, referenced as $(EIRP)_D$.

The satellite G/T is a *fixed value* which is provided by the manufacturer.

The downlink EIRP in the equation is a function of two other fixed value parameters. These are the *saturated EIRP* and the *saturated flux density (SFD)* of the satellite transponder.

6.4.4 Satellite Power Transfer and Back-Off

The relationship in the satellite transponder between the input flux density of the uplink receive RF and the transmit downlink EIRP is the satellite power transfer. When the input flux density is equal to the specified SFD for a transponder, then the output EIRP is equal to the saturated EIRP for that transponder. At flux density levels lower than the SFD, the downlink

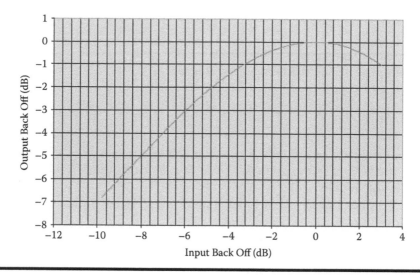

Figure 6.7   Typical power transfer curve.

EIRP will be lower than the saturated *EIRP*. The power transfer curve is indicated in terms of back-off. That is, the level of input flux density and output *EIRP* are referenced to the saturated levels being 0.0 dB. All other levels are referenced as some level of input back-off and output back-off. See Figure 6.7.

For single carrier operation, the full power of the transponder is often used. On the transfer curve this is (*IBO, OBO*) = (0, 0) on the graph.

For multicarrier operation, the transponder is operated in the "linear" region, that is, somewhere less than (*IBO, OBO*) = (−6, −3) on this particular curve.

6.4.5   Satellite Back-Off

The auxiliary equations specify a number of different back-off types:

- Transponder back-off: Transponder back-off is the aggregate back-off of a transponder referenced to the transponder's saturated value.
- Carrier back-off: Carrier back-off is the back-off of an RF carrier in a transponder referenced to the transponder's saturated value.
- Input back-off: Input back-off is a back-off referenced to the SFD of a transponder.
- Output back-off: Output back-off is a back-off reference to the saturated EIRP of a transponder.

This yields four types of back-off that are relevant for the satellite auxiliary equations:

- Transponder input back-off—(*XIBO*)
- Transponder output back-off—(*XOBO*)
- Carrier input back-off—(*CIBO*)
- Carrier output back-off—(*COBO*)

6.4.6 Flux Density Attenuation

To increase noise immunity and uplink *C/N*, the capability exists to adjust the effective *SFD*. That is, while the *SFD* for a particular transponder remains fixed, an adjustment can be made so the effective *SFD* is increased. To do this, each satellite transponder is equipped with a flux control attenuator (*FCA*), or a similar device (this is sometimes simply referred to as an attenuator). From the link budget standpoint the level of attenuation is given in units of dB. Though from the view of the spacecraft, the specification may be more involved.

The *SFD* plus the attenuator setting can be referred to as the effective *SFD* of the transponder, that is,

$$(\textit{Effective SFD}) = (SFD) + (\textit{Attenuator})$$

6.4.7 Unity Antenna–Flux Density Conversion

Since saturated levels at the input of the satellite transponder are given as power density, specifically as dB referenced to watts/square meter (dBW/m²), it is convenient to have antenna gain referenced to a 1 square meter aperture size antenna. This gain of this type of antenna is equal to $4\pi/\lambda^2$. It is also convenient have the equation in terms of frequency in GHz. The resulting equation is then

$$(\textit{Gain of } 1 \ m^2) = 20*\log_{10}[f_{GHz}] + 21.44 \ (\text{dB})$$

Some notation for spacecraft auxiliary equations follows.

- $(EIRP)_{\text{sat}}$: The maximum EIRP available from a transponder, in units of dBW.
- (SFD): Saturated flux density, the power density of carrier at the transponder input required to saturate the transponder, in units of dBW/m². The transponder output RF will have *EIRP* equal to the saturated *EIRP* of the transponder. That is, $(EIRP)_{\text{D}} = (EIRP)_{\text{sat}}$
- $(ATTN)$: A variable transponder setting which allows the (SFD) to be increased, in units of dB.
- $(G)_{1m}$: Gain of 1 square meter of antenna surface at a given frequency, in units of dB.
- $(CIBO)$: Carrier input back-off, the ratio of an incoming carrier flux density to the effective saturated flux density, in units of dB.
- $(COBO)$: Carrier output back-off, the ratio of an outgoing carrier *EIRP* to the saturated *EIRP*, in units of dB.
- $(XIBO)$: Transponder input back-off, the ratio of the total flux density of all incoming carriers to the effective saturated flux density, in dB.
- $(XOBO)$: Transponder output back-off, the ratio of the total outgoing carrier *EIRP* to the saturated *EIRP*, in dB.

With this notation, the auxiliary equations for the satellite are

- Uplink *EIRP*

$$(EIRP)_U = (SFD) + (ATTN) - (CIBO) - (G)_{1m} + (FSL)_U$$

- Downlink *EIRP*

$$(EIRP)_D = (EIRP)_{\text{sat}} - (COBO)$$

- Satellite Power Transfer
- $(COBO) = (CIBO) + (XOBO) - (XIBO)$

6.5 Performance Calculations

The performance equations are used to determine the actual performance of the satellite link. These use the calculated system C/N of the link and convert that to some performance parameter.

- Analog transmission uses the concept of signal-to-noise ratio (S/N). The S/N definition is discussed below. The equation varies slightly depending on the baseband signal type (e.g., video or audio).
- For digital transmission, performance is specified as a *bit error rate* (BER). The BER is the rate at which the receiver makes an incorrect decision, or an error, on the status of a demodulated bit, that is, a "0" for a "1," or vice versa. For link budget use, the BER is converted to the energy per bit-to-noise power density ratio, or E_b/N_0. This ratio is related to the C/N as a function of modulation.

Subcarrier transmissions—that is, those which use an already-modulated signal to in turn modulate a higher frequency carrier—require an additional calculation to arrive at a subcarrier C/N, and then calculate performance as an analog or digital transmission as appropriate.

6.5.1 Margin

In general, the link is said to "close" if the total calculated $(C/N)_T$ is greater than the required (C/N). This difference is known as the *margin*:

$$(Margin) = (C/N)_T - (C/N)_{Req}$$

Simply put, for the link to work in the desired manner then the result of this equation must be greater than zero.

6.5.2 Performance Equations–FM

Analog systems use the baseband signal to frequency modulate the carrier. The baseband signal could be of any type signal. In recent practice, only TV is still encountered as the baseband signal. The calculations below apply to FM in general, and TV transmission in particular (see next section).

The S/N for FM referenced to the calculated C/N for some link is

$$(S/N) = (C/N) + 10 \times \log_{10}[3 \times (\Delta f/f_m)^2] + 10 \times \log_{10}[B_{if}/(2 \times B)] + E$$

where
  Δf: FM peak deviation of baseband signal
  f_m: highest frequency of baseband signal
  B_{if}: IF noise bandwidth
  B: noise bandwidth of baseband signal
  E: emphasis, weighting, and conversion factors

(The importance of analog systems is decreasing over time, as digital systems become ubiquitous.)

6.5.3 TV Performance

When television is the baseband signal in the FM signal in a 36 MHz transponder, this equation for *S/N* can be written with the following parameter settings:

Δf: FM peak deviation of video signal = 10.75 MHz
f_m: highest frequency of video signal = 4.2 MHz
B_{if}: IF filter bandwidth = 36 MHz
B: noise bandwidth of video signal = 4.2 MHz
E: Pre/de-emphasis, weighting, and peak to RMS conversion, respectively

$$2.9 \text{ dB} + 9.9 \text{ dB} + 6 \text{ dB} = 18.8 \text{ dB}$$

Finally,

$$(S/N) = (C/N) + 38.1 \text{ dB}$$

The constant in the equation is referred to as the *video transfer function*.

(The importance of analog systems is decreasing over time, as digital systems become ubiquitous.)

6.5.4 FM Performance Threshold

Wideband FM systems experience a threshold effect below which the signal quality deteriorates rapidly and the *S/N* equations are not valid. The actual threshold depends on the modulation index, which is $(\Delta f/f_m)$, and of course the *C/N*. For this case, the common "FM threshold" of 10 dB is chosen as a minimum,

$$(C/N)_{\text{Req}} = 10 \text{ dB}$$

Wideband FM is typically defined with the modulation index to be "large," and generally that means greater than 0.6 approximately.

6.5.5 Digital Modulation Performance

For a particular digital modulation scheme, the required BER will dictate the required E_b/N_0. However, the bit rate at the input to the modulator and the noise bandwidth at the modulator output must also be taken into account. The performance equation is

$$(C/N)_{\text{Req}} = (E_b/N_0)_{\text{Req}} + 10 \times log_{10}(R/B)$$

The *R/B* term is the *spectral efficiency* of the modulator, where *R* is the input bit rate (in bits/second) and *B* the bandwidth of the modulator output (in Hertz).

(More detail on digital modulation is covered later.)

An example of a margin calculation follows.

A digital transmission system requires and E_b/N_0 of 5.5 dB and has a spectral efficiency of 3 bps/Hz. The uplink C/N is 25 dB and the system C/N is 16 dB. (1) What is the system margin? (2) What degradation in downlink C/N will cause the system margin to be zero?

1. $(C/N)_{\text{Req}} = (E_b/N_0)_{\text{Req}} + 10*\log_{10}(R/B) = 5.5 + 10*\log_{10}(3) = 5.5 + 4.8 = 10.3$ dB

$$(Margin) = (C/N)_T - (C/N)_{\text{Req}} = 16 - 10.3 = 5.7 \text{ dB}$$

2. $(C/N)_T = (C/N)_U \oplus (C/N)_D \Rightarrow 1/39.81 = 1/316.23 + 1/(C/N)_D$

$$\Rightarrow (C/N)_D = 45.54 \Rightarrow 16.6 \text{ dB}$$

No margin $(C/N)_T = 10.3$ dB $\Rightarrow 10.72$; $(C/N)_U = 25$ dB $\Rightarrow 316.23$

$$(C/N)_T = (C/N)_U \oplus (C/N)_D \Rightarrow 1/10.72 = 1/316.23 + 1/(C/N)_D$$

$$\Rightarrow (C/N)_D = 11.1 \Rightarrow 10.5 \text{ dB}$$

$$\Rightarrow Margin \text{ on } (C/N)_D = 16.6 - 10.5 = 6.1 \text{ dB}$$

This is known as the *downlink margin*.

6.5.6 Rain Fade

Propagation through the atmosphere encounter a number of other sources of loss in addition to the free space loss described earlier. While free space loss remains fixed, at least for normal geostationary operation, these other losses may exhibit some variability. These other losses are caused by rain, atmospheric gaseous absorption, tropospheric scintillation, depolarization, and multipath, as well as effects from the ionosphere (these topics were already discussed in previous chapters). Generally, most of these will have little effect on the satellite link and can be ignored for most purposes. Rain, however, will place large losses into the link and make the link unusable from time to time. The impact of rain needs to be accounted for in any link design.

6.5.6.1 Fade Characterization–Path Geometry

Fade levels are characterized as a loss that will not be exceeded for a certain percentage of the time, known as the *availability*, at a particular ground location in the direction of a certain orbital location. The levels are determined using a particular rain model employing local rain statistics and the RF link path geometry.

Example of typical fade levels:
Earth Station: Trenton, NJ
Frequency: 11.94 GHz
Orbital Slot 1: 105 degrees west longitude
Orbital Slot 2: 72 degrees west longitude

The availability of the service (link) is as follows:

Availability (%)	Fade level in dB	
	Slot 1	Slot 2
99.999	23.5	20.4
99.998	20.0	17.2
99.995	14.9	12.8
99.990	11.4	9.8
99.980	8.0	6.8
99.950	4.8	4.1
99.900	3.2	2.7
99.800	1.9	1.6
99.500	1.0	0.8
99.000	0.5	0.4
98.000	0.3	0.2

6.5.6.2 *Fade Characterization–Transmission Frequency*

The previous data listing attempted to illustrate that a particular earth station will have different fade levels for different orbital positions. Transmission frequency also plays a role.

Some typical fade levels:
Earth station: Trenton, NJ
Orbital slot: 105 degrees west longitude
Frequency 1: 14.24 GHz
Frequency 2: 11.94 GHz

The availability of the service (link) is as follows:

Availability (%)	Fade level in dB	
	Frequency 1	Frequency 2
99.999%	32.4	23.5
99.998%	27.7	20.0
99.995%	20.8	14.9
99.990%	16.0	11.4
99.980%	11.3	8.0
99.950%	6.9	4.8
99.900%	4.6	3.2
99.800%	2.8	1.9
99.500%	1.4	1.0
99.000%	0.8	0.5
98.000%	0.4	0.3

6.5.6.3 *Use of Rain Fade Margins*

Rain fade margins enable the use of availability as a performance criterion. That is, the link is said to be available when it is performing above its established criterion, such as BER for digital links, or desired *S/N* level for FM links.

Rain fade margins can be used in essentially two ways:

- In design, power levels can be adjusted to obtain enough margin to achieve a desired availability level.
- In analysis, clear-sky margins are calculated to determine predicted availability.

An example of availability calculation follows. A previous example calculated the downlink margin to be equal to 6.1 dB. Calculate the downlink availability using the rain fades given on the previous slide for 11.94 GHz. We can use linear interpolation to determine the availability. The downlink margin is bracketed by 8.0 dB at 99.98% availability and 4.8 dB at 99.95% availability. Then

$$\text{Downlink availability} = 99.95\% + (99.98\% - 99.95\%) \times [(6.1 - 4.8)/(8.0 - 4.8)]$$
$$= 99.95\% + 0.05\% \times [1.3/3.2] = 99.970\%$$

6.5.7 Digital Modulation

Recall the digital performance equation discussed earlier,

$$(C/N)_{\text{Req}} = (E_b/N_0) + 10^*\log10(R/B)$$

where

- (E_b/N_0) is the bit energy-to-noise density ratio. This is a figure of merit used to compare the error rate performance of a one error coding and modulation scheme to another.
- (R/B) is the spectral efficiency, where R is the data rate and B is the noise bandwidth of the carrier.

The noise bandwidth is a function of the error correction and modulation schemes, and of the data rate. *FEC* techniques are used to improve demodulator performance. *FEC* is an integral part of almost all digital communication systems to guarantee a reliable bit transmission despite the presence of noise and other disturbances in the communication channel. *FEC* requires two additional signal processing elements to the communication system: (1) an encoder at the transmitter and (2) a channel decoder at the receiver. The encoder adds redundancy to the bit sequence in a well-defined manner. This redundancy can then be exploited by the channel decoder to detect and correct transmission errors. *FEC* not only reduces the number of transmission errors but also facilitates an extended operating range and/or a reduced transmit power [MOR200701]. As we saw in Chapter 5, this involves placing additional bits into the data stream prior to the modulator input. The *coding rate*, or *code rate*, *RC*, refers to how often the additional bits are placed into the data stream. The rate is given as a number between zero and one, often as ratio of integers, such as ½ or ¾ or *k/n*. For every *k* bits that are input to the *encoder*, *n* bits are output. For that reason, higher code rates are associated with smaller fractions. The bit rate at the output of the FEC encoder is equal to the ratio R/R_C, where R is the data rate at the encoder input.

As we saw in Chapter 5, many different code schemes exist, with varying degrees of performance. The schemes are sometimes *concatenated*, to construct a super code. The overall code rate is merely the product of the rates of the constructing codes. Channel codes can be grouped into the class of block and convolutional codes depending on whether the information bits are encoded one block at a time or in a more continuous fashion. A special case are tailbiting convolutional codes which can also be treated as block codes. In 1966 Forney demonstrated that powerful codes can be generated based on a concatenation of simple component codes. The key advantage of this approach is that sequential processing of the component codes in the receiver has only a moderate complexity compared to processing the overall code in one step. Serially concatenated codes found their first application in the area of deep space communications. A major breakthrough was achieved with the parallel concatenation of recursive systematic convolutional codes by Berrou et al. in 1993. This class of codes was termed *turbo codes*. The discovery of turbo codes raised a lot of attention in the research community and Spielman et al. and MacKay et al. rediscovered LDPC codes. LDPC codes were invented by Gallager in the early 1960s. However, soon afterwards they were largely forgotten because the excellent error correcting performance of these codes could not be demonstrated on early computers with very limited processing power. LDPC codes are essentially block codes, but can also be interpreted as a serial concatenation of very simple block codes [MOR200701].

6.5.8 Digital Modulator Output Bandwidth

The noise bandwidth of the carrier is determined by the modulation scheme being used. In general, the bandwidth, B, is some function of the symbol rate, S_R, which is the modulator output rate. The symbol rate is a function of the modulation type, the modulation order, and the input bit rate to the modulator. For PSK, QAM, and ASK signaling schemes the relationship is straightforward, with the bandwidth being equal to the symbol rate, that is, $B = S_R$. These schemes are commonly encountered in satellite transmissions.

6.5.9 Modulation Order

The process of modulation takes an integral number of input bits and maps them into a single output symbol. Current satellite modem technology uses an integer value from 1 to 5, inclusive. This number is determined by the total of the phase, amplitude, and frequency combinations in the carrier signal available to the modulator. Each combination is referred to as a symbol. The total number of bits mapped into a symbol is the *order* of the modulation. An example of some of the schemes, the associated bit/symbol ratio, and number of symbols are:

BPSK—1 bit/symbol, two total symbols
QPSK—2 bits/symbol, four total symbols
8-PSK—3 bits/symbol, eight total symbols
16-QAM—4 bits/symbol, sixteen total symbols
General *M*-ary Modulation—*m* bits/symbol, *M* total symbols

A simple relation is seen here between the number of bits per symbol, m, and the number of symbols available, M, or modulation order:

$$m = \log_2 (M) \text{ or } M = 2^m$$

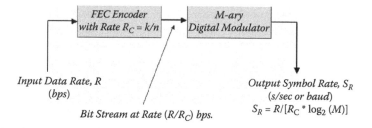

Input Data Rate, R
(bps)

Bit Stream at Rate (R/R_C) bps.

Output Symbol Rate, S_R
(s/sec or baud)
$S_R = R/[R_C * \log_2 (M)]$

Figure 6.8   FEC encoder and digital modulator block diagram.

6.5.10 Bandwidth Calculation and Spectral Efficiency

The FEC encoder output is used as the input to the modulator (see Figure 6.8). Calculation of the symbol rate, and thus bandwidth at the modulator output, can be calculated using the input data rate. For the M-PSK, M-ASK, and M-QAM schemes this is:

$$B = S_R = (R/R_C) / \log_2 (M) = R / [R_C \times \log_2 (M)]$$

The spectral efficiency, (R/B), can be calculated as we now have an expression for the bandwidth:

$$(R/B) = R_C \times \log_2 (M) = R_C \times m$$

The spectral efficiency can be calculated using the modulation order and FEC rate.

6.5.11 Example Bandwidth Calculation

In this example one wants to determine the noise bandwidth associated with the following parameters:

Input data rate: 16.4 Mbps
Modulation type: 8-PSK
FEC rates: Viterbi 5/6
Reed–Solomon 188/204

It follows that for PSK modulation:

$$B = S_R = R/[R_C \times \log_2 (M)]$$

Code rate:

$$R_C = R_V \times R_{RS} = (5/6) \times (188/204) = 0.7697$$

Then,

$$B = 16.4/[0.7697 \times \log_2 (8)] = 7.118 \text{ MHz}$$

Occupied bandwidth on the transponder is larger due to spectral roll-off, typically 1.2 to 1.4 times the noise bandwidth.

6.5.12 Example Spectral Efficiency Calculation

In this example one wants to determine the spectral efficiency required to transmit 52 Mbps in a noise bandwidth of 30 MHz. Recommend a possible modulation and coding, using DVB-S.

It follows that

$$(R/B) = R_C \times \log_2(M) = R_C \times m$$

and then

$$(R/B) = (52/30) = 1.733 \text{ bps/Hz}$$

Hence, one has:

$$1.733 = R_C \times m$$

Finally, one varies the modulation, $m = 2, 3$, etc. (QPSK, 8-PSK, etc.) and solve for code rates.

QPSK:
$R_C = 1.733/2 = 0.866 \Rightarrow$ Most efficient is 7/8 at 0.8064; try again.

8-PSK:
$R_C = 1.733/3 = 0.577 \Rightarrow$ Efficiency of 2/3 is 0.6144; Use 8-PSK 2/3.
Check: $B = R/[R_C \times \log_2(M)] = 52/[(2/3) \times (188/204) \times 3] = 28.212 \text{ MHz} \Rightarrow$ OK

6.5.13 Error Performance of Digital Modulation

Again, for digital modulation,

$$(C/N)_{\text{Req}} = (E_b/N_0) + 10 \times \log_{10}(R/B)$$

E_b/N_0 is the bit energy-to-noise density ratio.

E_b/N_0 is actually a theoretical construct used to compare the *probability of error* (P_e or $P(e)$) or the *bit error rate* (BER) of different signaling schemes. BER is used in actual systems while $P(e)$ is used in analysis and design of different schemes. See Figure 6.9.

6.5.14 Coding Gain

The E_b/N_0 for any modulation scheme can be lowered for a particular BER by using forward error correction. The difference between uncoded modulation and coded modulation for a particular code is known as the *coding gain*. Virtually all satellite modems available support some type of forward error correction. Use of this feature significantly reduces power requirements and requirements for related parameters—for example, antenna size—LNA noise figure, and transponder power utilization.

The required (E_b/N_0) versus spectral efficiency, (R/B), plot for some common modulation and/or coding schemes is shown in Figure 6.10.

Some summary observations related to digital modulation follow:

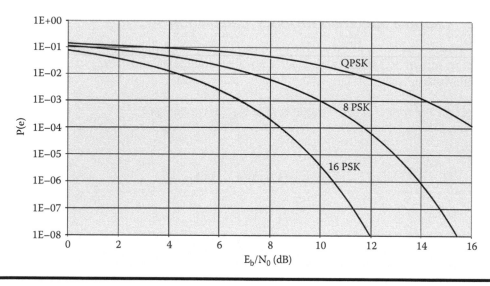

Figure 6.9 *P(e)* curves for *uncoded* **M-PSK modulation.**

- Higher-order modulation schemes will require less bandwidth for a given input data rate.
- However, these schemes require more power to achieve comparable bit error rate performance.
- Forward error correction will result in a significant decrease in power requirements for any modulation type over the same type uncoded. Bandwidth is slightly increased.
- Generally speaking, higher-order modulation schemes and lower rate codes result in less required bandwidth, but at the expense of using more power.

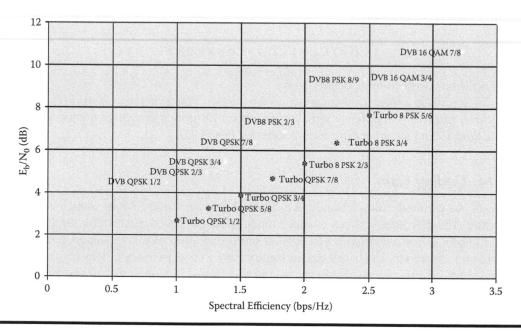

Figure 6.10   Required E_b/N_0.

6.5.15   G/T Degradation

Rain conditions will cause a degradation in earth station G/T which needs to be accounted for as a downlink loss. This is characterized as an increase in antenna system noise temperature because the antenna no longer looks into the clear sky. The additional loss can be converted to a noise temperature in the usual manner:

$$\Delta T = T_0 \times [1 - (1/L_{\text{rain}})]$$

where T_0 is the reference temperature and L_{rain} are the losses due to the rain.

This becomes a degradation in the G/T for the earth station:

$$(G/T) = 10 \times \log_{10}[\, 1 + (T_0 \,/\, TS\,) \times (1 - (1/L_{\text{rain}})]$$

In the link budget this may appear as an individual line item, but often is grouped with the earth station G/T.

6.5.16   Rain Models

Rain impacts the performance of an RF link, as noted several times in the text. Generally, the higher the frequency, the higher the attenuation caused by rainfall. Moisture can degrade the link such that the overall signal to noise ratio drops. In heavy rain there could be portions of time when the link is unusable (outage); heavy rain can also cause depolarization, where signals from one polarization appear in the opposite polarization. Rain models are used to calculate the amount of attenuation seen by the RF link. There are a number of models available but all models ultimately use the following equation to calculate the rain loss:

$$A = a \times R^b \times L \text{ (dB)}$$

where A is the attenuation in dB, R is the rain rate seen along the susceptible portion of the RF path, and L is the length of the transmission path which is susceptible to rain. (See Figure 6.11.) The coefficients a and b are dependent on RF frequency and polarization. Specifically [RA0200401]

$$a = 4.21 \times 10^{-5} f^{2.42} \;(2.9 < f < 52 \text{ GHz})$$
$$b = 1.41 f^{-0.0776} \;(8.5 < f < 25 \text{ GHz})$$

The rain rate is a constant which depends on the site location of the RF terminal.

There are a number of rain models in existence in varying degrees of revision. Practically, there are two rain models that are used extensively: the global crane model, and the ITU rain model. The models differ in the techniques and data used in calculating the various parameters in the attenuation

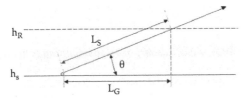

Θ – Elevation Angle h_R – Rain Height h_S – Height Above Mean Sea Level

L_S – Rain Slant-path Length L_G – Horizontal Projection of L_S

Figure 6.11   Rain model geometry.

Table 6.6 Example of waveguide losses

System Component	DBS band		Ku-band		C-band	
	Tallguide[a] TG87 17 GHz 90 ft loss	Standard Waveguide WR62 17 GHz 90 ft loss	Tallguide[a] TG115 14 GHz 120 ft loss	Standard Waveguide WR75 14 GHz 120 ft loss	Tallguide[a] TG215 5.9 GHz 400 ft loss	Standard Waveguide WR137 5.9 GHz 400 ft. loss
Transitions	2*0.07 = 0.14 dB		2*0.07 = 0.14 dB		2*0.07 = 0.14 dB	
H-plane Bends	3*0.02 = 0.06 dB	3*0.04 = 0.12 dB	3*0.02 = 0.06 dB	3*0.04 = 0.12 dB	3*0.02 = 0.06 dB	3*0.04 = 0.12 dB
Straight Sections	0.87*90/100 = 0.78 dB	4.98*90/100 = 4.5 dB	0.635*120/100 = 0.76 dB	3.73*120/100 = 4.5 dB	0.281* 400/100 = 1.12 dB	1.88*400/100 = 7.52 dB
Mode Suppressor	0.12 dB		0.12 dB		0.12 dB	
Flanges	0.01*10 = 0.1 dB	0.02*8 = 0.16 dB	0.01*13 = 0.13 dB	0.02*10 = 0.2 dB	0.01*23 = 0.23 dB	0.02*20 = 0.4 dB
Total Loss	**1.2 dB**	**4.8 dB**	**1.2 dB**	**4.8 dB**	**1.7 dB**	**8.0 dB**

Note: Tallguide is a Registered Trademark of Antennas for Communications.

equation. The ITU model is gaining wider acceptance of late, and is updated periodically. The global crane model is still in widespread use but has not been updated in a number of years.

6.5.17 Waveguide Losses

Waveguide losses at an earthstation (particularly at a teleport where there is a need for many dispersed antennas) can be non trivial. Table 6.6 depicts the comparison of a traditional waveguide to a better quality waveguide [ANT200801]. The loss of 8 dB right at the antenna is significant. This is why many teleports now use a fiberoptic interfacility link (IFL) to transmit the information to a shelter in close proximity to the antenna and only use a short waveguide run.

References

[AER200801] The Aerospace Corporation, Satellite Communications Glossary, In Crosslink (ISSN 1527-5264 [print], ISSN 1527-5272 [Web]). Corporate Communications, Los Angeles. The Aerospace Press, Los Angeles.

[ANS200001] ANS T1.523-2001, Telecom Glossary 2000, American National Standard (ANS), an outgrowth of the Federal Standard 1037 series, *Glossary of Telecommunication Terms*, 1996.

[ANT200801] Antennas for Communications, Inc., Ocala, FL, http://www.tallguide.com.

[FOC200701] Staff, An Introduction to Earth Stations, Focalpoint Consultiong, St Gaudens, France, 2007.

[ITE200401] Staff, White Paper, Iterative Connections Pty Ltd, Adelaide, South Australia, May 2004.

[KAS200201] K. Kastamonitis, B. Grémont, and M. Filip, "Online Extraction of Scintillations for Satellite Links with Up–Link Power Control," 1st International Workshop: COST Action 280 (Propagation Impairment Mitigation for Millimetre Wave Radio Systems), July 2002.

[LAU200701] J. E. Laube, HughesNet, "Introduction to the Satellite Mobility Support Network, HughesNet User Guide" 2005–2007.

[MOR200701] M. Morz, Analog Signal Processing in Forward Error Correction (FEC) Decoders, Die Dissertation wurde am 18.01.2007 bei der Technischen Universität München eingereicht und durch die Fakultät für Elektrotechnik und Informationstechnik am 21.06.2007 angenommen (Institute for Communications Engineering [LNT] at the Munich University of Technology [TUM]).

[ORT200801] ORTEL/EMCORE, "RF and Microwave Fiber-Optic Design Guide," Application Note, March 7, 2003, Albuquerque, NM.

[RA0200401] K. N. Raja Rao, *Fundamentals of Satellite Communication,* Prentice Hall of India, 2004.

[SAT 200801] http://www.satsig.net

[USA199801] U.S. Army Information Systems Engineering Command Fort Huachuca, Arizona, "Automated Information Systems, Design Guidance, Commercial Satellite Transmission," August 1998.

Appendix 6A: Formulas Generally Used in Link Budget Analysis

Antenna, area aperture	For a circular antenna, $A = \pi D^2/4$
Antenna, effective aperture area	$A_{eff} = \eta A$, where η is the efficiency of the antenna (usually between 55% and 75%)
Antenna, maximum gain	Maximum in the direction of maximum radiation (the electromagnetic axis of the antenna, the boresight) $$G_{max} = (4\pi/\lambda)^2 A_{eff} = \eta (\pi D/\lambda)^2 = \eta (\pi Df/c)^2$$ In dBi, relative to an isotropic antenna, the maximum antenna gain is $$G_{max, dBi} = 10\log \eta(\pi D/\lambda)^2 = 10\log \eta(\pi Df/c)^2$$
Carrier-to-noise ratio	$C/N = EIRP + G/T - (L_p + k + B_{IF})$ $EIRP$ = effective radiated power of satellite (dBW) L_p = path loss k = Boltzman's constant (−228.6 dBW/K/Hz) B_{IF} = 10 log (bandwidth of IF in Hz)
Decibel	Defined as being equal to ten times the common (or base ten) logarithm of a ratio of power measurements, that is: $X = 10 \times \log_{10} [P_1/P_0]$ To recover the original ratio the equation can be inverted: $[P_1/P_0] = 10^{(X/10)}$
Figure of merit	G/T = antenna gain in dB/(10 log (antenna + LNA noise) LNA noise is in K
IF bandwidth	B_{IF} = 10 log (Bandwidth of IF), bandwidth is in Hz
Noise power spectral density N_0	$N_0 = kT$ (watts/Hz) where k is the Boltzman constant (1.38×10^{-23} J/K) and T is the absolute temperature in Kelvins (K).
Total noise power, N	(in the bandwidth of interest) is calculated by multiplying the noise power spectral density N_0 by the bandwidth: $N = k T B$ (watts) where B is the bandwidth of interest (in Hz)
Path loss	$L_p = 37 + 20 \log F + 20 \log D$ F = frequency in MHz D = distance in miles

(Continued)

Appendix 6A: Formulas Generally Used in Link Budget Analysis (*Continued*)

Probability of a transmission error on the AWGN channel	$P_b = \dfrac{1}{2}\text{erfc}\left(\dfrac{m_{\tilde{y}}}{\sqrt{2\sigma_{\tilde{y}}^2}}\right) = \dfrac{1}{2}\text{erfc}\left(\sqrt{\dfrac{E_S}{N_0}}\right),$ With $\text{erfc}(x) = \dfrac{2}{\sqrt{\pi}}\int_x^\infty e^{-t^2}\,dt$
Converting noise figure (NF) (dB) to noise temperature (NT) (K)	$NF = 10 \log\left((NT/290) + 1\right)$
Converting noise temperature (NT) (K) to noise figure (NF) (dB)	$NT = 290\,(10^{(NF/10)} - 1)$
Free space loss becomes	$L_{dB} = 21.98 + 20 \times \log_{10}[d/\lambda]$ $d = distance$
Satellite link budget at the receiver (summarized)	C/N_0 [dB] $= E_b/N_0$ [dB] + 10logR where C/N_0 is carrier to noise density ratio, E_b/N_0 is energy per information bit to noise density ratio, and R is the information bit rate. For a data rate R of 2 Mbps (= 63 dB) a satellite receiver using QPSK modulation and rate 1/2 convolutional/Reed–Solomon coding, requires $E_b/N_0 = 5$ dB to achieve a target bit error rate (BER) less than 10^{-10} over a satellite channel. Therefore, the required C/N_0 is more than 68dB [ITE200401].
Clear–air attenuation known as gaseous attenuation, $A_g(t)$	$A_g(t) = 10\log\left[\dfrac{\hat{T}_m(t) - T_{sky}^{cs}(t)}{\hat{T}_m - \hat{T}_{bg}(t)}\right]$ [DB], Where $\hat{T}_m(t)$ is the mean radiating temperature, and $\hat{T}_{bg}(t)$ is the cosmic background temperature, and $T_{sky}^{cs}(t)$ [K] is the sky noise temperature
Earth terminal to satellite slant range	The range calculation for a geostationary satellite in terms of the satellite orbital slot, and the earth station longitude and latitude is shown following.

Appendix 6A: Formulas Generally Used in Link Budget Analysis (*Continued*)

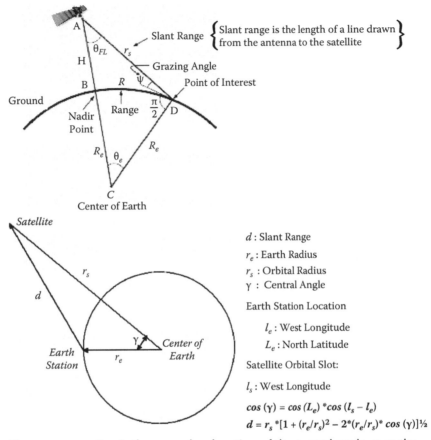

d : Slant Range

r_e : Earth Radius

r_s : Orbital Radius

γ : Central Angle

Earth Station Location

    l_e : West Longitude

    L_e : North Latitude

Satellite Orbital Slot:

l_s : West Longitude

$cos (γ) = cos (L_e) {}^*cos (l_s - l_e)$

$d = r_s {}^*[1 + (r_e/r_s)^2 - 2{}^*(r_e/r_s)^* cos (γ)]^{1/2}$

The range equation is then merely a function of the central angle, γ, as the other values are known. Specifically,

$r_s = 42,242$ *km for geostationary*

$r_e = 6,370$ *km for average Earth radius*

The equation can then be written

$d = 42,242 * [1.02274 - 0.301596* cos(γ)]^{1/2}$

with the result in kilometers. The result is used to calculate free space loss in the appropriate equation.

Chapter 7

IPv6 Overview

Previous chapters discussed satellite links from the perspective of providing a physical transmission path, either in a point-to-point mode or in a point-to-multipoint (broadcast) mode. As is the case with other transmission links, users of satellite links typically utilize a data link layer protocol to define framing mechanisms to support communications over a single such link, as well as a network layer protocol to route packets (datagrams) over a number of such links (some of which could also be terrestrial hops). Modern-day communication is characterized by the use of the Internet Protocol (IP) at the network layer. At this time, most of the systems deployed in the field or just about being ready for deployment make use of IP version 4 (IPv4), which has been around for about two decades. However, the expectation is that by 2010 and beyond there will be increased use of IP version 6 (IPv6). Hence, this chapter shifts the focus to the topic of IPv6; the chapter that follows will cover the topic of Transmission Control Protocol (TCP).

IPv6 offers the potential of achieving the scalability, reachability, end-to-end interworking, quality of service (QoS), and commercial-grade robustness for data communication, as well as for VoIP/triple-pay networks. IPv6 is now gaining momentum globally, with major interest and activity in Europe and Asia, and there also is some traction in the United States. For example, the U.S. Department of Defense (DoD) announced in 2003 that from October 1, 2003, all new developments and procurements need to be IPv6-capable. The DoD's goal is to complete the transition to IPv6 for all intra- and internetworking across the agency by June of 2008. In 2005 the U.S. Government Accountability Office (GAO) recommended that all agencies become proactive in planning a coherent transition to IPv6*. Corporations

---

* OMB Memorandum M-05-22 directs that agencies must transition from IPv4 agency infrastructures to IPv6 agency infrastructures (network backbones) by June 2008. When specific agency task orders require connectivity and compliance with IPv6 networks, service providers need to ensure that services delivered support federal agencies as required to comply with OMB IPv6 directives. All agency infrastructures must be using IPv6 by June 30, 2008 (meaning that the network backbone is either operating a dual-stack network core or it is operating in a pure IPv6 mode, that is, IPv6-compliant and configured to carry operational IPv6 traffic), and agency networks must interface with this infrastructure.

and institutions need to start planning at this time how to kick off the transition planning process and determining how coexistence can best be maintained during the 3 to 6-year window that will likely be required to achieve the global transition. This book addresses the migration and macro-level scalability requirements for this transition.

7.1 Opportunities Offered by IPv6

IPv6 is considered to be the next-generation Internet Protocol [HUI199701], [HAG200201], [MUR200501], [SOL200401], [ITO200401], [MIL199701], [MIL200001], [GRA200001], [DAV200201], [LOS200301], [LEE200501], [GON199801], [DEM200301], [GOS200301], [MIN200601], and [WEG199901]. The current version of the Internet Protocol, IPv4, has been in use for almost 30 years and exhibits some challenges in supporting emerging demands for address space cardinality, high-density mobility, multimedia, and strong security. This is particularly true in developing domestic and defense department applications that utilize peer-to-peer networking.

IPv6 is an improved version of the Internet Protocol that is designed to coexist with IPv4 and eventually provide better internetworking capabilities than IPv4 [IPV200401]. IPv6 offers the potential of achieving the scalability, reachability, end-to-end interworking, QoS, and commercial-grade robustness for data as well as for Voice-Over-IP (VoIP), IP-based TV (IPTV*) distribution, and triple-play networks; these capabilities are mandatory mileposts of the technology if it is to replace the time division multiplexing (TDM) infrastructure around the world. Every device connected to the internet has at least one IP address, and usually more than one. When the current version of the Internet Protocol (IPv4) was conceived in the 1970s, it provided just over 4 billion addresses; today, that is not enough to provide each person on the planet with one address. Additionally, 74 percent of IPv4 addresses have been assigned to North American organizations. The goal of developers is to be able to assign IP addresses to a new class of Internet-capable devices: mobile phones, car navigation systems, home appliances, industrial equipment, and other devices (such as sensors and Body Area Network medical devices). All of these devices can then be linked together, constantly communicating wirelessly. The current generation of the Internet will "run out of space" in the near future (2010/2011) if IPv6 is not adopted around the world. IPv6 is an essential technology for ambient intelligence and will be a key driver for a multitude of new, innovative mobile/wireless applications and services [DIR200801]. Satellite systems will also need to support IPv6 in the near future.

IPv6 was initially developed in the early 1990s because of the anticipated need for more end system addresses based on anticipated Internet growth, encompassing mobile phone deployment, smart home appliances and billions of new users in developing countries (e.g., in China and India). New technologies and applications such as VoIP, "always-on access" (e.g., Digital Subscriber Line and cable), Ethernet-to-the-home, converged networks, and evolving ubiquitous computing applications will be driving this need even more in the next few years [IPV200501]. Figures 7.1 thru 7.3 depict some examples of IPv6 networks. Figure 7.1 shows an example of a converged network utilizing IPv6, with both local and wide area components as well as private and carrier-provided

---

* IPTV is the delivery of (entertainment-quality) video programming over an IP-based network. Traditional telecom carriers are looking to compete with Cable TV companies by deploying IP video services over their networks.

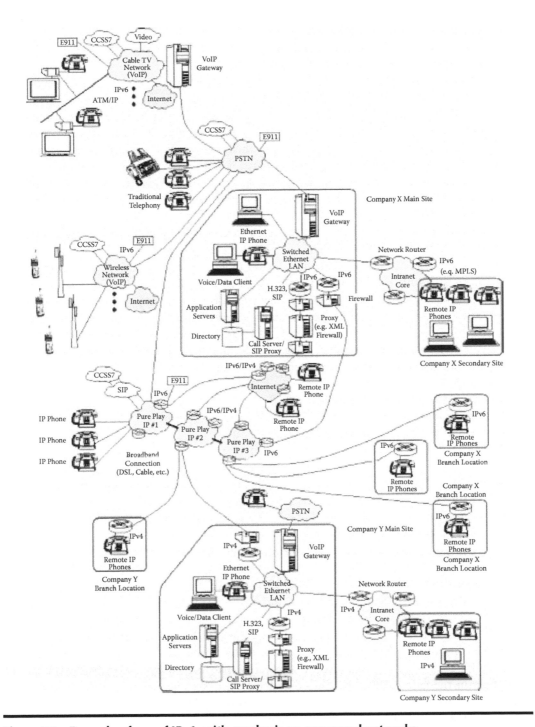

Figure 7.1   Example of use of IPv6, with emphasis on converged networks.

communications domains; the IPv6/IPv4 network shown in this figure supports video delivery, VoIP, Internet, intranet, and wireless services. Figure 7.2 and 7.3 depict examples of IPTV networks that employ both satellite technology and IPv6.

Basic Network Address Translation (NAT) is a method by which IP addresses (specifically IPv4 addresses) are transparently mapped from one group to another. Specifically, private "non-registered" addresses are mapped to a small set (as small as one) of public registered addresses; this impacts the general addressability, accessibility, and "individuality" of the device. Network address port translation (NAPT) is a method by which many network addresses and their TCP/UDP (User Datagram Protocol) ports are translated into a single network address and its TCP/UDP ports. Together, these two methods, referred to as traditional NAT, provide a mechanism to connect a realm with private addresses to an external realm with globally unique registered addresses [SRI200101].

NAT is only a short-term solution for the anticipated Internet growth phenomenon, and a better solution is needed for address exhaustion. There is a recognition that NAT techniques make the Internet, the applications, and even the devices more complex and this means, in the end, a cost overhead [IPV200501]. The expectation is that IPv6 can make IP devices less expensive, more powerful, and even consume less power; the power issue is not only important for environmental reasons, but also improves operability (e.g., longer battery life in portable devices such as mobile phones).

Corporations and government agencies will be able to achieve a number of improvements with IPv6. IPv6 can improve a firm's intranet, with benefits such as:

- Expanded addressing capabilities
- Serverless autoconfiguration ("plug-n-play") and reconfiguration
- Streamlined header format and flow identification
- End-to-end security, with built-in, strong IP-layer encryption and authentication (embedded security support with mandatory Internet Protocol Security (IPSec) implementation)
- Enhanced support for multicast and QoS (more refined support for Flow Control and QoS for the near-real-time delivery of data)
- More efficient and robust mobility mechanisms (enhanced support for mobile IP and mobile computing devices)
- Extensibility: improved support for feature options/extensions

While the basic function of the Internet Protocol is to move information across networks, IPv6 has more capabilities built into its foundation than IPv4. A key capability is the significant increase in address space. For example, all devices could have a public IP address, so that they can be uniquely tracked. Today, inventory management of dispersed IT assets cannot be achieved with IP mechanisms; during the inventory cycle someone has to manually verify the location of each device. With IPv6 one can use the network to verify that such equipment is there; even non-IT equipment in the field can also be tracked by having an IP address permanently assigned to it. IPv6 also has extensive automatic configuration (autoconfiguration) mechanisms and reduces the IT burden, making configuration essentially plug-and-play.

Corporations and institutions need to start planning the migration process and how coexistence can be maintained during the two to five year window that will likely be required to achieve the global transition.

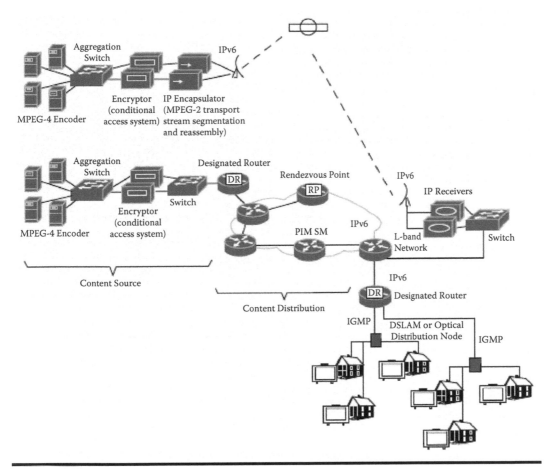

Figure 7.2   Example of use of IPv6, with emphasis on IPTV and satellite communication.

At a minimum, satellite-based services will have to support the following capabilities to comply with GSA/OMB IPv6 Compliance Requirements for SATCOM providers (also see Chapters 8 and 9):

1. *Physical layer transport.* The satellite service must support any native IPv6 application that uses the satellite link as physical layer transport with its peered-router exchanging IPv6 datagrams. The satellite service provider may need to support IPv6 network transport on its backbone.
2. *Dual-stack devices.* The satellite service must support any dual-stack devices that use IPv6 when communicating "internal" to their cloud and IPv4 when communicating "external" to their cloud.
3. *Platforms.* The satellite service provider may use VSAT platforms using commercial products such as iDirect. The upgrade to IPv6 for these platforms is the responsibility of the respective hardware vendor. If these vendors do upgrade their platforms and/or add blades to support some IPv6 coexistent island, then the service provider may be able to support native applications using this hardware.

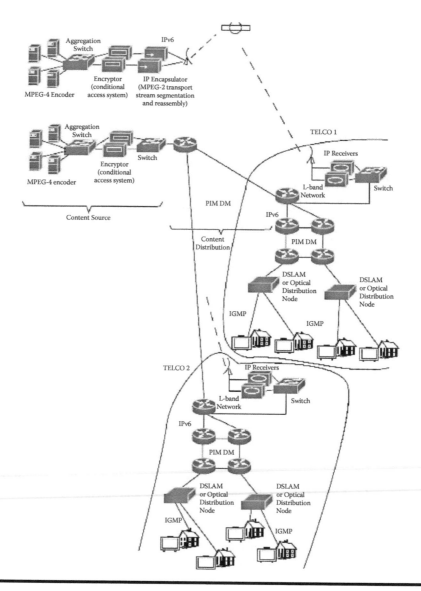

Figure 7.3 Another example of use of IPv6.

4. *IPv6 ISP.* The satellite service provider may peer with any IPv6-based ISP. If or when peering with an IPv6-based ISP becomes a task requirement, then the satellite service provider will need to deploy IPv6 and/or dual-stack peering routers to comply with this requirement.

7.2 Introductory Overview of IPv6

IP was designed in the 1970s for the purpose of connecting computers that were in separate geographic locations. Computers in a campus were connected by means of local networks, but these local networks were separated into essentially stand-alone islands. "Internet," as a name to designate the protocol and, more recently, the worldwide information network, simply means "internetwork," that is, a connection between networks. In the beginning, the protocol had only military use, but computers from universities and enterprises were quickly added. The Internet as a worldwide information network is the result of the practical application of the IP protocol, that is, the result of the interconnection of a large set of information networks [IPV200501]. Starting in the early 1990s developers realized that the communication needs of the 21st century needed a protocol with some new features and capabilities, while at the same time retaining the useful features of the existing protocol.

Although link-level communication does not generally require a node identifier (address) since the device is intrinsically identified with the link-level address, communication over a group of links (a network) does require unique node identifiers (addresses). The IP address is an identifier that is applied to each device connected to an IP network. In this setup, different elements taking part in the network (servers, routers, user computers, etc.) communicate with one another using their IP address as an entity identifier. In version 4 of the IP protocol, addresses consist of four octets. For ease of human conversation, IP protocol addresses are represented as separated by periods, for example, 166.74.110.83, where the decimal numbers are a short-hand and correspond to the binary code described by the byte in question (an 8 bit number takes a value in the 0-255 range). Since the IPv4 address has 32 bits, there are nominally 2^{32} different IP addresses (approximately 4 billion nodes, if all combinations are used).

IPv6 is the Internet's next-generation protocol, which was at first called IPng ("Internet Next Generation"). The Internet Engineering Task Force (IETF) developed the basic specifications during the 1990s to support a migration to a new environment. IPv6 is defined in RFC 2460, "Internet Protocol, Version 6 (IPv6) Specification" (S. Deering, R. Hinden [December 1998]), which obsoletes RFC1883. (The "Version 5" nomenclature was employed for another use (an experimental real-time streaming protocol), and to avoid any confusion, it was decided not to use this nomenclature).

7.2.1 IPv6 Benefits

IPv4 has proved, by means of its long life, to be a flexible and powerful networking mechanism. However, IPv4 is starting to exhibit limitations, not only with respect to the need for an increase of the IP address space, driven, for example, by new populations of users in countries such as China and India, and by new technologies with "always-connected devices" (DSL, cable, networked PDAs, 2.5G/3G mobile telephones, etc.), but also in reference to a potential global rollout of VoIP. IPv6 creates a new IP address format, so that the number of IP addresses will not be exhausted for several decades or longer, even though an entire new crop of devices are expected to connect to the Internet.

IPv6 also adds improvements in areas such as routing and network autoconfiguration. Specifically, new devices that connect to the Internet will be "plug-and-play" devices. With IPv6 one is not required to configure dynamic nonpublished local IP addresses, the gateway address, the subnetwork mask or any other parameters. The equipment, when plugged into the network, automatically obtains all requisite configuration data [IPV200501].

The advantages of IPv6 can be summarized as follows:

■ *Scalability:* IPv6 has 128-bit addresses versus 32-bit IPv4 addresses. With IPv4 the theoretical number of available IP addresses is $2^{32} \sim 10^{10}$. IPv6 offers a 2^{128} space. Hence, the number of available unique node addressees are $2^{128} \sim 10^{38}$.

- *Security:* IPv6 includes security features such as payload encryption and authentication of the source of the communication in its specifications.
- *Real-time applications:* To provide better support for real time traffic (e.g., VoIP), IPv6 includes "labeled flows" in its specifications. By means of this mechanism, routers can recognize the end-to-end flow to which transmitted packets belong. This is similar to the service offered by MultiProtocol Label Switching (MPLS), but it is intrinsic to the IP mechanism rather than an add-on. Also, it preceded this MPLS feature by a number of years.
- *"Plug-and-play":* IPv6 includes a "plug-and-play" mechanism that facilitates the connection of equipment to the network. The requisite configuration is automatic.
- *Mobility:* IPv6 includes more efficient and enhanced mobility mechanisms, particularly important for mobile networks.
- *Optimized protocol:* IPv6 embodies IPv4 best practices but removes unused or obsolete IPv4 characteristics. This results in a better-optimized Internet protocol.
- *Addressing and routing:* IPv6 improves the addressing and routing hierarchy.
- *Extensibility:* IPv6 has been designed to be extensible and offers support for new options and extensions.

7.2.2 *Traditional Addressing Classes for IPv4*

With IPv4, the 32-bit address can be represented as AdrClass|netID|host ID. The network portion can contain either a network ID or a network ID and a subnet. Every network and every host or device has a unique address, by definition. Figure 7.4 depicts the traditional address classes.

- *Traditional Class A Address.* Class A uses the first bit of the 32-bit space (bit 0) to identify it as a Class A address; this bit is set to 0. Bits 1 to 7 represent the network ID and bits 8 through 31 identify the PC, terminal device, or host/server on the network. This address space supports $2^7 - 2 = 126$ networks and approximately 16 million devices (2^{24}) on each network. By convention, the use of an "all 1's" or "all 0's" address for both the Network ID and the Host ID is prohibited (which is the reason for subtracting the 2 above.)

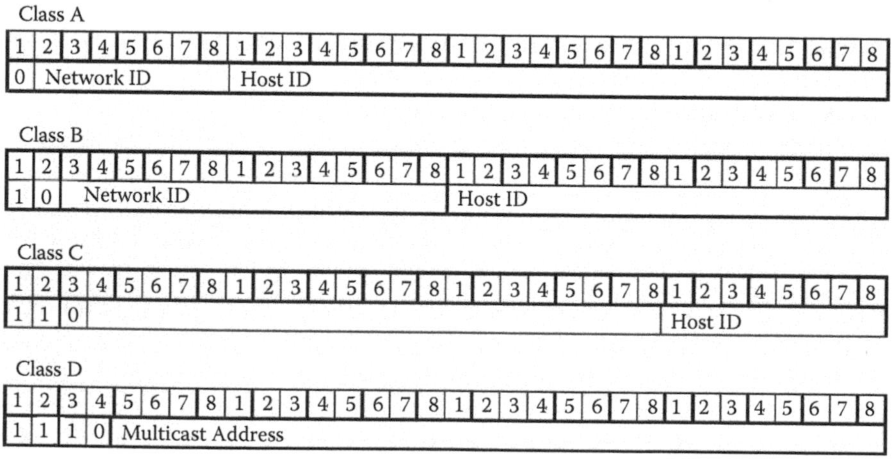

Figure 7.4   Traditional address classes for IP address.

- *Traditional Class B Address.* Class B uses the first 2 bits (bit 0 and bit 1) of the 32-bit space to identify it as a Class B address; these bits are set to 10. Bits 2 to 15 represent the network ID and bits 16 through 31 identify the PC, terminal device, or host/server on the network. This address space supports $2^{14} - 2 = 16,382$ networks and $2^{16} - 2 = 65,134$ devices on each network.
- *Traditional Class C Address.* Class C uses the first 3 bits (bit 0, bit 1, and bit 2) of the 32-bit space to identify it as a Class C address; these bits are set to 110. Bits 3 to 23 represent the network ID and bits 24 through 31 identify the PC, terminal device, or host/server on the network. This address space supports about 2 million networks ($2^{21} - 2$) and $2^8 - 2 = 254$ devices on each network.
- *Traditional Class D Address.* This class is used for broadcasting: all devices on the network receive the same packet. Class D uses the first 4 bits (bit 0, bit 1, bit 2, and bit 3) of the 32-bit space to identify it as a Class D address; these bits are set to 1110.

Classless Interdomain Routing (CIDR), described in RFC 1518, RFC1519, and RFC 2050, is a mechanism that was developed to help alleviate the problem of exhaustion of IP addresses and growth of routing tables. The concept behind CIDR is that blocks of multiple addresses (e.g., blocks of Class C addresses) can be combined, or aggregated, to create a larger classless set of IP addresses, with more hosts allowed. Blocks of Class C network numbers are allocated to each network service provider; organizations using the network service provider for Internet connectivity are allocated subsets of the service provider's address space as required. These multiple Class C addresses can then be summarized in routing tables, resulting in fewer route advertisements. CIDR mechanism can also be applied to blocks of Class A and B addresses [TEA200401]. All of this assumes, however, that the institution in question already has an assigned set of public, registered IP addresses; it does not address the issue of how to get additional public, registered globally unique IP addresses. NAT is the principal mechanism used today to deal with IPv4 address exhaustion.

7.2.3 Network Address Translation Issues in IPv4

IPv4 addresses can be from an officially assigned public range or from an internal intranet private (but not globally unique) block. Internal intranet addresses may be in the ranges 10.0.0.0/8, 172.16.0.0/12, and 192.168.0.0/16. In the internal intranet private address case, a NAT function is employed to map the internal addresses to an external public address when the private-to-public network boundary is crossed. This, however, imposes a number of limitations, particularly since the number of registered public addresses available to a company is almost invariably much smaller (as small as 1) than the number of internal devices requiring an address.

As noted, IPv4 theoretically allows up to 2^{32} addresses, based on a four-octet address space. Public, globally unique addresses are assigned by the Internet Assigned Numbers Authority (IANA). IP addresses are addresses of network nodes at layer 3; each device on a network (whether the Internet or an intranet) must have a unique address. In IPv4 it is a 32-bit (4-byte) binary address used to identify the device. It is represented by the nomenclature a.b.c.d (each of a, b, c, and d being from 1 to 255 (0 has a special meaning). Examples are 167.168.169.170, 232.233.229.209, and 200.100.200.100.

The problem is that during the 1980s many public, registered addresses were allocated to firms and organizations without any consistent control. As a result, some organizations have more addresses that they actually need, giving rise to the present dearth of available "registerable" Layer 3 addresses. Furthermore, not all IP addresses can be used due to the fragmentation described earlier.

One approach to the issue would be a renumbering and a reallocation of the IPv4 addressing space. However, this is not as simple as it appears since it requires worldwide coordination efforts. Moreover, it would still be limited for the human population and the number of devices that will be connected to the Internet in the medium-term future [IPV200501]. At this juncture, and as a temporary and pragmatic approach to alleviate the dearth of addresses, NAT mechanisms are employed by organizations and even home users. This mechanism consists of using only a small set of public IPv4 addresses for an entire network to access to Internet. The myriad of internal devices are assigned IP addresses from a specifically designated range of Class A or Class C addresses that are locally unique but are duplicatively used and reused within various organizations. In some cases (e.g., residential Internet access use via DSL or cable), the legal IP address is only provided to a user on a time-lease basis, rather than permanently.

A number of protocols cannot easily travel through a NAT device, and hence the use of NAT implies that many applications (e.g., VoIP) cannot be used effectively in all instances. As a consequence, these applications can only be used in intranets. Examples include [IPV200501]:

■ Multimedia applications such as videoconferencing, VoIP, or video-on-demand/IPTV do not work smoothly through NAT devices. Multimedia applications make use of Real-Time Transport Protocol (RTP) and Real-Time Control Protocol (RTCP). These in turn use UDP with dynamic allocation of ports, and NAT does not directly support this environment.
■ IPSec is used extensively for data authentication, integrity, and confidentiality. However, when NAT is used, IPSec operation is impacted, since NAT changes the address in the IP header.
■ Multicast, although possible in theory, requires complex configuration in a NAT environment and, hence, in practice, is not utilized as often as could be the case.

The need for obligatory use of NAT disappears with IPv6.

7.2.4 IPv6 Address Space

The format of IPv6 addressing is described in RFC 2373. As noted, an IPv6 address consists of 128 bits, rather than 32 bits as with IPv4 addresses. The number of bits correlates to the address space, as follows:

IP Version	Size of Address Space
IPv6	128 bits, which allows for 2^{128} or 340,282,366,920,938,463,463,374,607,431,768,211,456 (3.4×10^{38}) possible addresses.
IPv4	32 bits, which allows for 2^{32} or 4,294,967,296 possible addresses.

The relatively large size of the IPv6 address is designed to be subdivided into hierarchical routing domains that reflect the topology of the modern-day Internet. The use of 128 bits provides multiple levels of hierarchy and flexibility in designing hierarchical addressing and routing. The IPv4-based Internet currently lacks this flexibility [MSD200401].

The IPv6 address is represented as 8 groups of 16 bits each, separated by the ":" character. Each 16-bit group is represented by four hexadecimal digits, that is, each digit has a value between 0 and F (0,1, 2, ... A, B, C, D, E, F with A = 10, B = 11, etc., to F = 15). What follows is an IPv6 address example

3223:0BA0:01E0:D001:0000:0000:D0F0:0010

If one or more four-digit groups is 0000, the zeros may be omitted and replaced with two colons (::). For example

3223:0BA0::

is the abbreviated form of the following address:

3223:0BA0:0000:0000:0000:0000:0000:0000

Similarly, only one 0 is written, removing 0's in the left side, and four 0's in the middle of the address. For example, the address

3223:BA0:0:0:0:0::1234

is the abbreviated form of the following address:

3223:0BA0:0000:0000:0000:0000:0000:1234

There is also a method to designate groups of IP addresses or subnetworks that is based on specifying the number of bits that designate the subnetwork, beginning from left to right, using remaining bits to designate single devices inside the network. For example, the notation

3223:0BA0:01A0::/48

indicates that the part of the IP address used to represent the subnetwork has 48 bits. Since each hexadecimal digit has 4 bits, this points out that the part used to represent the subnetwork is formed by 12 digits, that is, "3223:0BA0:01A0". The remaining digits of the IP address would be used to represent nodes inside the network.

There are a number of special IPv6 addresses, as follows:

- Auto-return or loopback virtual address. This address is specified in IPv4 as the 127.0.0.1 address. In IPv6 this address is represented as ::1.
- Unspecified address (::). This address is not allocated to any node since it is used to indicate the absence of an address.
- IPv6 over IPv4 dynamic/automatic tunnel addresses. These addresses are designated as IPv4-compatible IPv6 addresses and allow the sending of IPv6 traffic over IPv4 networks in a transparent manner. They are represented as, for example, ::156.55.23.5.
- IPv4 over IPv6 addresses automatic representation. These addresses allow for IPv4-only-nodes to still work in IPv6 networks. They are designated as "IPv4-mapped IPv6 addresses" and are represented as ::FFFF:, for example ::FFFF:156.55.43.3.

7.2.5 *Basic Protocol Constructs*

Table 7.1 lists basic IPv6 terminology, while Table 7.2 shows the core protocols that comprise IPv6.

Like IPv4, IPv6 is a connectionless, unreliable datagram protocol used primarily for addressing and routing packets between hosts. Connectionless means that a session is not established before exchanging data. Unreliable means that delivery is not guaranteed. IPv6 always makes a best-effort attempt to deliver a packet. An IPv6 packet might be lost, delivered out of sequence,

Table 7.1   Basic IPv6 terminology

Address	An IP layer identifier for an interface or a set of interfaces.
Host	Any node that is not a router.
Interface	A node's attachment to a link.
Link	A communication facility or medium over which nodes can communicate at the link layer, that is, the layer immediately below IP. Examples are Ethernet (simple or bridged); PPP links, X.25, Frame Relay, ATM networks, and Internet (or higher) layer "tunnels," such as tunnels over IPv4 or IPv6 itself.
Link-local address	An IPv6 address having a link-only scope, indicated by having the prefix (FE80::/10), that can be used to reach neighboring nodes attached to the same link. Every interface has a link-local address.
Multicast address	An identifier for a set of interfaces typically belonging to different nodes. A packet sent to a multicast address is delivered to all interfaces identified by that address.
Neighbor	A node attached to the same link.
Node	A device that implements IP.
Packet	An IP header plus payload.
Prefix	The initial bits of an address, or a set of IP addresses that share the same initial bits.
Prefix length	The number of bits in a prefix.
Router	A node that forwards IP packets not explicitly addressed to itself.
Unicast address	An identifier for a single interface. A packet sent to a unicast address is delivered to the interface identified by that address.

duplicated, or delayed. IPv6 per se does not attempt to recover from these types of errors. The acknowledgment of packets delivered and the recovery of lost packets is done by a higher-layer protocol, such as TCP [MSD200401]. From a packet-forwarding perspective, IPv6 operates just like IPv4. An IPv6 packet, also known as an IPv6 datagram, consists of an IPv6 header and an IPv6 payload, as shown Figure 7.5.

The IPv6 header consists of two parts, the IPv6 base header, and optional extension headers. Functionally, the optional extension headers and upper-layer protocols—for example, TCP—are considered part of the IPv6 payload. Table 7.3 shows the fields in the IPv6 base header. IPv4 headers and IPv6 headers are not directly interoperable: hosts and/or routers must use an implementation of both IPv4 and IPv6 to recognize and process both header formats. This gives rise to a number of complexities in the migration process between the IPv4 and the IPv6 environments.

7.2.6 IPv6 Autoconfiguration

Autoconfiguration is a new characteristic of the IPv6 protocol that facilitates network management and system set-up tasks by users. This characteristic is often called "plug-and-play" or "connect-and-work." Autoconfiguration facilitates initialization of user devices: after connecting

Table 7.2 Key IPv6 protocols

Protocol	*Description*
Internet Protocol version 6 (IPv6): RFC 2460	IPv6 is a connectionless datagram protocol used for routing packets between hosts.
Internet Control Message Protocol for IPv6 (ICMPv6): RFC 2463	ICMPv6 is a mechanism that enables hosts and routers that use IPv6 communication to report errors and send simple status messages.
Multicast Listener Discovery (MLD): RFC 2710, RFC 3590, RFC 3810	MLD is a mechanism that enables one to manage subnet multicast membership for IPv6. MLD uses a series of three ICMPv6 messages. MLD replaces the Internet Group Management Protocol (IGMP) v3 that is employed for IPv4.
Neighbor Discovery (ND): RFC 2461	ND is a mechanism that is used to manage node-to-node communication on a link. ND uses a series of five ICMPv6 messages. ND replaces Address Resolution Protocol (ARP), ICMPv4 Router Discovery, and the ICMPv4 Redirect message; it also provides additional functions.

a device to an IPv6 network, one or several IPv6 globally unique addresses are automatically allocated.

The autoconfiguration process is flexible, but it is also somewhat complex. The complexity arises from the fact that various policies are defined and implemented by the network administrator. Specifically, the administrator determines the parameters that will be assigned automatically. At a minimum (and/or when there is no network administrator), the allocation of a "link" local address is often included. The link local address allows communication with other nodes placed in the same physical network. Note that "link" has somewhat of a special meaning in IPv6, as follows: a communication facility or medium over which nodes can communicate at the link layer, that is, the layer immediately below IPv6. Examples are Ethernets (simple or bridged); PPP links; an X.25 packet-switched network; a Frame Relay network; a Cell Relay/Asynchronous Transfer Mode (ATM) network, and Internet (working) layer (or higher layer) tunnels, such as tunnels over IPv4 or IPv6 itself [DEE199801].

Two autoconfiguration basic mechanisms exist: (1) "stateful" and (2) "stateless." Both mechanisms can be used in a complementary manner and/or simultaneously to define parameter

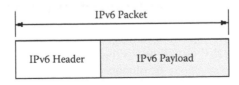

Figure 7.5 IPv6 packet.

Table 7.3 IPv6 base header

IPv6 header field	Length (bits)	Function
Version	4	Identifies the version of the protocol. For IPv6, the version is 6.
Traffic Class	8	Intended for originating nodes and forwarding routers to identify and distinguish between different classes or priorities of IPv6 packets.
Flow Label	20	(Sometimes referred to as Flow ID.) Defines how traffic is handled and identified. A flow is a sequence of packets sent either to a unicast or a multicast destination. This field identifies packets that require special handling by the IPv6 node. The following list shows the ways the field is handled if a host or router does not support flow label field functions: If the packet is being sent, the field is set to zero. If the packet is being received, the field is ignored.
Payload Length	16	Identifies the length, in octets, of the payload. This field is a 16-bit unsigned integer. The payload includes the optional extension headers, as well as the upper-layer protocols, for example, TCP.
Next Header	8	Identifies the header immediately following the IPv6 header. The following shows examples of the next header: 00 = Hop-by-Hop Options 01 = ICMPv4 04 = IP in IP (encapsulation) 06 = TCP 17 = UDP 43 = Routing 44 = Fragment 50 = Encapsulating security payload 51 = Authentication 58 = ICMPv6
Hop Limit	8	Identifies the number of network segments, also known as links or subnets, on which the packet is allowed to travel before being discarded by a router. The Hop Limit is set by the sending host and is used to prevent packets from endlessly circulating on an IPv6 internetwork. When forwarding an IPv6 packet, IPv6 routers must decrease the Hop Limit by 1, and must discard the IPv6 packet when the Hop Limit is 0.
Source Address	128	Identifies the IPv6 address of the original source of the IPv6 packet.
Destination Address	128	Identifies the IPv6 address of the intermediate or final destination of the IPv6 packet.

configurations. Stateful autoconfiguration is often employed when there is a need for rigorous control in reference to the address allocated to hosts; in stateless autoconfiguration, the only concern is that the address be unique [IPV200501].

Stateless autoconfiguration is also described as *serverless*. Here, the presence of configuration servers to supply profile information is not required. The host generates its own address using a combination of the information that it possesses (in its interface or network card) and the information that is periodically supplied by the routers. Routers determine the prefix that identifies networks associated to the link under discussion. The "interface identifier" identifies an interface within a subnetwork and is often, and by default, generated from the media access control (MAC) address of the network card. The IPv6 address is built combining the 64 bits of the interface identifier with the prefixes that routers determine as belonging to the subnetwork. If there is no router, the interface identifier is self-sufficient to allow the PC to generate an address.

"Stateful" configuration requires a server to send the information and parameters of network connectivity to nodes and hosts. Servers maintain a database with all addresses allocated and a mapping of the hosts to which these addresses have been allocated, along with any information related to all requisite parameters. In general, this mechanism is based on the use of DHCPv6.

IPv6 addresses are "leased" to an interface for a fixed established time (including an infinite time.) When this "lifetime" expires, the link between the interface and the address is invalidated and the address can be reallocated to other interfaces. For the suitable management of addresses' expiration time, an address goes through two states (stages) while it is affiliated to an interface [IPV200501]:

1. At first, an address is in a "preferred" state, so its use in any communication is not restricted.
2. After that, an address becomes "deprecated," indicating that its affiliation with the current interface will (soon) be invalidated.

While it is in a deprecated state, the use of the address is discouraged, although it is not forbidden. However, when possible, any new communication (e.g., the opening of a new TCP connection) must use a preferred address. A deprecated address should only be used by applications that already used it before and in cases where it is difficult to change this address to another address without causing a service interruption.

To insure that allocated addresses (granted either by manual mechanisms or by autoconfiguration) are unique in a specific link the *link duplicated addresses detection algorithm* is used. The address to which the duplicated address detection algorithm is being applied to is designated (until the end of this algorithmic session) as an "attempt address." In this case, it does not matter that such address has been allocated to an interface and received packets are discarded.

Next, we describe how an IPv6 address is formed. The lowest 64 bits of the address identify a specific interface, and these bits are designated as "interface identifier." The highest 64 bits of the address identify the "path" or the "prefix" of the network or router in one of the links to which such interface is connected. The IPv6 address is formed by combining the prefix with the interface identifier.

It is possible for a host or device to have IPv6 and IPv4 addresses simultaneously. Most of the systems that currently support IPv6 allow the simultaneous use of both protocols. In this way it is possible to support communication with IPv4-only-networks as well as IPv6-only-networks and the use of the applications developed for both protocols [IPV200501].

One can transmit IPv6 traffic over IPv4 networks using tunneling methods. This approach consists of "wrapping" the IPv6 traffic as IPv4 payload data: IPv6 traffic is sent "encapsulated" into IPv4 traffic and at the receiving end this traffic is parsed as IPv6 traffic. Transition mechanisms are methods used for the coexistence of IPv4 and/or IPv6 devices and networks. For example, an "IPv6-in-IPv4 tunnel" is a transition mechanism that allows IPv6 devices to communicate through an IPv4 network. The mechanism consists of creating the IPv6 packets in a normal way and encapsulating them in an IPv4 packet. The reverse process is undertaken in the destination machine, which de-encapsulates the IPv6 packet.

There is a significant difference between the procedures to allocate IPv4 addresses that focus on the parsimonious use of addresses (since addresses are a scarce resource and should be managed with caution), and the procedures to allocate IPv6 addresses that focus on flexibility. Internet service providers (ISPs) deploying IPv6 systems follow the Regional Internet Registries' (RIRs) policies relating to how to assign IPv6 addressing space among their clients. RIRs are recommending ISPs and operators allocate to each IPv6 client a/48 subnetwork; this allows clients to manage their own subnetworks without using NAT. (The implication is that the need for NAT disappears in IPv6).

To allow its maximum scalability, the IPv6 protocol uses an approach based on a basic header, with minimum information. This differentiates it from IPv4 where different options are included in addition to the basic header. IPv6 uses a header "concatenation" mechanism to support supplementary capabilities. The advantages of this approach include the following:

- The size of the basic header is always the same, and is well known. The basic header has been simplified compared with IPv4, since only 8 fields are used instead of 12. The basic IPv6 header has a fixed size; hence, its processing by nodes and routers is more straightforward. Also, the header's structure aligns to 64 bits, so that new and future processors (64 bits minimum) can process it in a more efficient way.
- Routers placed between a source point and a destination point (i.e., the route that a specific packet has to pass through), do not need to process or understand any "following headers." In other words, in general, interior (core) points of the network (routers) only have to process the basic header, while in IPv4 all headers must be processed. This flow mechanism is similar to the operation in MPLS, yet precedes it by several years.
- There is no limit to the number of options that the headers can support (the IPv6 basic header is 40 octets in length, while IPv4 one varies from 20 to 60 octets, depending on the options used).

In IPv6, interior/core routers do not perform packets fragmentation, but the fragmentation is performed end-to-end. That is, source and destination nodes perform, by means of the IPv6 stack, the fragmentation of a packet and the reassembly, respectively. The fragmentation process consists of dividing the source packet into smaller packets or fragments. [IPV200501].

A jumbogram is an option that allows an IPv6 packet to have a payload greater than 65,535 bytes. Jumbograms are identified with a 0 value in the payload length in the IPv6 header field and include a Jumbo Payload Option in the Hop-by-Hop Option header. It is anticipated that such packets will be used in particular for multimedia traffic.

This preliminary overview of IPv6 highlights the advantages of the new protocol and its applicability to a whole range of applications. The sections that follow provide more details on these topics.

7.3 Migration and Coexistence

Migration is expected to be fairly complex. Initially, internetworking between the two environments will be critical. Existing IPv4-endpoints and/or nodes will need to run dual-stack nodes or convert to IPv6 systems. Fortunately the new protocol supports IPv4-compatible IPv6 addresses, which is an IPv6 address format that employs embedded IPv4 addresses. Tunneling, which we already described in passing, will play a major role in the beginning.

There are a number of requirements that are typically applicable to an organization wishing to introduce an IPv6 service [6NE200501]:

- The existing IPv4 service should not be adversely disrupted (e.g., as it might be by router loading of encapsulating IPv6 in IPv4 for tunnels)
- The IPv6 service should perform as well as the IPv4 service (e.g., at the IPv4 line rate and with similar network characteristics)
- The service must be manageable and be able to be monitored (thus tools should be available for IPv6 as they are for IPv4)
- The security of the network should not be compromised, due to the additional protocol itself or as a weakness of any transition mechanism used
- An IPv6 address allocation plan must be drawn up

Well-known interworking mechanisms include the following [GIL200001].

- Dual IP layer (also known as dual stack): A technique for providing complete support for both Internet protocols—IPv4 and IPv6—in hosts and routers.
- Configured tunneling of IPv6 over IPv4: Point-to-point tunnels made by encapsulating IPv6 packets within IPv4 headers to carry them over IPv4 routing infrastructures.
- Automatic tunneling of IPv6 over IPv4: A mechanism for using IPv4-compatible addresses to automatically tunnel IPv6 packets over IPv4 networks.

Tunneling techniques include the following [GIL200001]:

- IPv6-over-IPv4 tunneling: The technique of encapsulating IPv6 packets within IPv4 so that they can be carried across IPv4 routing infrastructures.
- Configured tunneling: IPv6-over-IPv4 tunneling where the IPv4 tunnel endpoint address is determined by configuration information on the encapsulating node. The tunnels can be either unidirectional or bidirectional. Bidirectional configured tunnels behave as virtual point-to-point links.
- Automatic tunneling: IPv6-over-IPv4 tunneling where the IPv4 tunnel endpoint address is determined from the IPv4 address embedded in the IPv4-compatible destination address of the IPv6 packet being tunneled.
- IPv4 multicast tunneling: IPv6-over-IPv4 tunneling where the IPv4 tunnel endpoint address is determined using Neighbor Discovery. Unlike configured tunneling this does not require any address configuration and unlike automatic tunneling it does not require the use of IPv4-compatible addresses. However, the mechanism assumes that the IPv4 infrastructure supports IPv4 multicast.

Applications (and the lower-layer protocol stack) need to be properly equipped. There are four cases [SHI200501]:

Case 1: IPv4-only applications in a dual-stack node. IPv6 protocol is introduced in a node, but applications are not yet ported to support IPv6. The protocol stack is as follows:

```
+--------------------+
|       appv4        |   (appv4—IPv4-only applications)
+--------------------+
|TCP / UDP / others|   (transport protocols—TCP, UDP, etc.)
+--------------------+
|    IPv4 | IPv6     |   (IP protocols supported/enabled in the OS)
+--------------------+
```

Case 2: IPv4-only applications and IPv6-only applications in a dual-stack node. Applications are ported for IPv6-only. Therefore, there are two similar applications, one for each protocol version (e.g., ping and ping6). The protocol stack is as follows:

```
+--------------------+
|    appv4 | appv6   |   (appv4—IPv4-only applications)
+--------------------+   (appv6—IPv6-only applications)
|TCP / UDP / others|   (transport protocols—TCP, UDP, etc.)
+--------------------+
|    IPv4 | IPv6     |   (IP protocols supported/enabled in the OS)
+--------------------+
```

Case 3: Applications supporting both IPv4 and IPv6 in a dual-stack node. Applications are ported for both IPv4 and IPv6 support. Therefore, the existing IPv4 applications can be removed. The protocol stack is as follows:

```
+--------------------+
|      appv4/v6      |   (appv4/v6—applications supporting both
+--------------------+   IPv4 and IPv6)
|TCP / UDP / others|   (transport protocols—TCP, UDP, etc.)
+--------------------+
|    IPv4 | IPv6     |   (IP protocols supported/enabled in the OS)
+--------------------+
```

Case 4: Applications supporting both IPv4 and IPv6 in an IPv4-only node. Applications are ported for both IPv4 and IPv6 support, but the same applications may also have to work when IPv6 is not being used (e.g., disabled from the OS). The protocol stack is as follows:

```
+--------------------+
|      appv4/v6      |   (appv4/v6—applications supporting both
+--------------------+   IPv4 and IPv6)
|TCP / UDP / others|   (transport protocols—TCP, UDP, etc.)
+--------------------+
|       IPv4         |   (IP protocols supported/enabled in the OS)
+--------------------+
```

The first two cases are not interesting in the longer term; only a few applications are inherently IPv4- or IPv6-specific and should work with both protocols without having to care about which one is being used.

A (near) press-time summary of the state of affairs in reference to IPv6 is as follows [LAD200601]:

1. IPv6 is a mature technology with significant deployment experience worldwide. The majority of deployment is in academic networks, but commercial deployment is now growing, particularly in the Far East.
2. IPv6 has clear technical advantages but these need to be translated to business advantages for various sectors, with detailed but clear business models. This is a task for economists rather than standards developers and implementers.
3. IPv6 is supported fully by Microsoft; they have deployed it in their own worldwide enterprise network, and Windows Vista will ship preferring use of IPv6 by default.
4. A number of companies have decided to support IPv6 as a core strategy, building products and services in advance of demand (e.g., Microsoft, NTT, KDDI).
5. A wide range of new IPv6 application scenarios is available to be exploited; many of these are green field scenarios (e.g., supply chain, sensor networks or transport networks) that can use IPv6 from the outset.
6. IPv6 networks can enrich educational experiences, with the right support and vision.
7. IPv6 can facilitate convergence both between delivery platforms and between business sectors. This has the potential for streamlining services.
8. Commodity IPv6 devices are required for consumer (SOHO) deployment, in particular there are no IPv6 DSL routers available to the market; this hinders ISP deployment.
9. For IPv6 to be widely deployed in all commercial sectors, the immediate and realistic market needs need to be addressed—in particular, site multi-homing and ISP independence, but also IPv6 capability in operations support systems (OSS) and management tools.
10. Training and education capacity needs to be increased. Best practice, roadmap, and guidance documents are still required (e.g., defining "IPv6 capable" for those making public sector IT procurements).

Typical press time IPv6 carrier services included the following [ATT200801]:

1. IPv6 Internet connectivity service:
   - Provide connectivity to the IPv6 Internet for activities such as Web surfing and database searches
   - Support multiple access methods (PPP, MLPPP, Frame Relay, and ATM) for customer access, typically from large-user locations
2. Remote Access Service to IPv6 Internet:
   - Support IPv6 for small-user locations and individual remote users
   - Establish dynamically configurable IPv6 Tunnel Gateway through IPv4 ISPs through fractional T1, DSL, or dial-up access
   - The Tunnel Setup Protocol (TSP) can be used to create tunnels to transport IPv6 traffic over an IPv4 network to the gateway
3. IPv6 Virtual Private Network (VPN) Service:
   - Use Multi-Protocol Label Switching (MPLS) to create a VPN interconnecting a set of agency locations using the IPv6 protocol for access

7.4 IPv6 Addressing Mechanisms

Sections 7.4 through Section 7.6 cover the IPv6 addressing scheme is some detail. The previous subsections introduced some basic concepts on addressing, and these concepts are expanded in this chapter.

The IPv6 addressing scheme is defined in The IPv6 Addressing Architecture specification, IETF RFC 3513, April 2003 [HIN200401] (RFC 3513 obsoletes RFC 2373). The IPv6 Addressing Architecture specification defines the address scope that can be used in an IPv6 implementation and the various configuration architecture guidelines for network designers of the IPv6 address space. Two advantages of IPv6 are that support for multicast is intrinsic (it is required by the specification), and nodes can create link-local addresses during initialization [DRO200301]. Some portions of this discussion are based on [MSD200401].

7.4.1 Addressing Conventions

As we saw in Chapter 1, the IPv6 128-bit address is divided along 16-bit boundaries; each 16-bit block is then converted to a 4-digit hexadecimal number, separated by colons. The resulting representation is called colon-hexadecimal. This is in contrast to the 32-bit IPv4 address represented in dotted-decimal format, divided along 8-bit boundaries, and then converted to its decimal equivalent, separated by periods. The following examples show 128-bit IPv6 addresses in binary form:

Address 1: 0010000111011010000000000110100110000000000000000000010111100111011 0000000101010101000000000011111111111111110001010001001110001011010

Address 2: 0010000111011010000000000110100110000000000000000000010111100111011 0000000101010101000000000011111111000000000000000001001110001011010

Address 3: 0010000111011010000000000110100110000000000000000001001110001011010 0000000101010101000000000011111111000000000000000001001110001011010

Address 4: 0010000111011010000000000110100110000000000000000001001110001011010 0000000101010101000000000011111111000000000000000010111100111011

The following example shows these same addresses divided along 16-bit boundaries:

Address 1: 0010000111011010:0000000001010011:0000000000000000:0010111100111011: 0000000101010101010:0000000011111111:1111111000101000:1001110001011010:

Address 2: 0010000111011010:0000000001010011:0000000000000000:0010111100111011: 0000000101010101010:0000000011111111:0000000000000000:1001110001011010:

Address 3: 0010000111011010:0000000001010011:0000000000000000:1001110001011010: 0000000101010101010:0000000011111111:0000000000000000:1001110001011010:

Address 4: 0010000111011010:0000000001010011:0000000000000000:1001110001011010: 0000000101010101010:0000000011111111:0000000000000000:0010111100111011:

The following shows each 16-bit block in the address converted to hexadecimal and delimited with colons.

Address 1: 21DA:00D3:0000:2F3B:02AA:00FF:FE28:9C5A

Address 2: 21DA:00D3:0000:2F3B:02AA:00FF:0000:9C5A

Address 3: 21DA:00D3:0000: 9C5A:02AA:00FF:0000:9C5A

Address 4: 21DA:00D3:0000: 9C5A:02AA:00FF:0000:2F3B

IPv6 representations can be further simplified by removing the *leading* zeros (trailing zeros are not removed) within each 16-bit block. However, each block must have, in the abbreviated nomenclature, at least a single digit. The following example shows the addresses without the *leading* zeros:

Address 1: 21DA:D3:0:2F3B:2AA:FF:FE28:9C5A

Address 2: 21DA:D3:0:2F3B:2AA:FF:0:9C5A

Address 3: 21DA:D3:0: 9C5A:2AA:FF:0:9C5A

Address 4: 21DA:D3:0: 9C5A:2AA:FF:0:2F3B

Some types of addresses contain long sequences of zeros. In IPv6 addressing, a contiguous sequence of 16-bit blocks set to 0 in the colon-hexadecimal format can be compressed to :: (known as *double-colon*).
The following list shows examples of compressing zeros:

■ The address 21DA:0:0:0:2AA:FF:9C5A:2F3B can be compressed to 21DA::2AA:FF:9C5A: 2F3B.
■ The multicast address of FF02:0:0:0:0:0:0:2 can be compressed to FF02::2.

Note: Zero-compression can only be used to compress a single contiguous series of 16-bit blocks expressed in colon-hexadecimal notation; one cannot use zero-compression to include part of a 16-bit block—for example, one *cannot* abbreviate FF01:30:0:0:0:0:0:8 as FF01:3::8.). Also, zero-compression can be used only once in an address, which enables one to determine the number of 0 bits represented by each instance of a double-colon (::). To determine how many 0 bits are represented by the ::, one can count the number of blocks in the compressed address, subtract this number from 8, and then multiply the result by 16. For example, in the address FF02::2, there are two blocks (the FF02 block and the 2 block); the number of bits expressed by the :: is 96 (= (8 − 2) × 16) [MSD200401].

7.4.2 *Addressing Issues/Reachability*

Every IPv6 address has a defined reachability scope. Table 7.4 shows the address and associated reachability scopes. The reachability of *node-local addresses* is "the same node"; the reachability of *link-local addresses* is "the local link"; and, the reachability of *global addresses* is "the IPv6-enabled Internet." IPv6 interfaces can have multiple addresses that have different reachability scopes. For example, a node may have a link-local address and a global address.
   Note: Site-Local addresses have recently been obsoleted by the IETF. We only mention them in passing.
   Similarly to the IPv4 address space, the IPv6 address space is partitioned according to the value of high order bits (known as a *format prefix*) in the address. Table 7.5 depicts the IPv6 address

Table 7.4   IPv6 Address and associated reachability scopes

Address scope/reachability	Description
Node-local addresses to reach same node	Used to send Protocol Data Units (PDUs) to the same node: • Loopback address (PDUs addressed to the loopback address are never sent on a link or forwarded by an IPv6 router—this is equivalent to the IPv4 loopback address) • Node-local multicast address
Link-local addresses to reach local link (*)	Used to communicate between hosts devices (e.g., servers, VoIP devices, etc.) on the link; these addresses are always configured automatically: • Unspecified address. It indicates the absence of an address, and is typically used as a source address for PDUs that are attempting to verify the uniqueness of a tentative address (it is equivalent to the IPv4 unspecified address.) The unspecified address is never assigned to an interface or used as a destination address. • Link-local Unicast address • Link-local Multicast address
Site-local addresses to reach the private intranet (internetwork) (*)	Used between nodes that communicate with other nodes in the same site; site-local addresses are configured by router advertisement: Note: Site-local addresses were deprecated in September 2004 by RFC 3879 ("Deprecating Site Local Addresses"); see additional details in the text
Global addresses to reach the Internet (IPv6-enabled); also known as aggregatable global unicast addresses	Globally routable and reachable addresses on the IPv6 portion of the Internet (they are equivalent to public IPv4 addresses); global addresses are configured by router advertisement: • Global unicast address • Other scope multicast address Global addresses are designed to be aggregated or summarized to produce an efficient, hierarchical addressing and routing structure.

(*) *When one specifies a link-local address, one needs to also specify a scope ID, which further defines the reachability scope for these (nonglobal) addresses.*

space allocation by format prefixes. The (current) set of unicast addresses that can be employed by IPv6 nodes consists of aggregatable global unicast addresses, link-local unicast addresses, and site-local unicast addresses (now deprecated; these addresses represent about 12.6 percent of the entire IPv6 address space, but it is still ~3.4×10^{38}.) The prefix is the portion of the address that indicates the bits that have fixed values or are the bits of the network identifier. Prefixes for IPv6 routes and subnet identifiers are expressed in the same way as Classless Inter-Domain Routing

Table 7.5   IPv6 address space allocation

Address space allocation	Format prefix	Percentage of the address space	Hex notation	Fraction of the address space
Reserved	0000 0000	0.391%	0×00	1/256
Reserved for NSAP allocation	0000 001	0.781%	0×0 001	1/128
Aggregatable global unicast addresses	001	12.500%	001	1/8
Link-local unicast addresses	1111 1110 10	0.098%	0×FE 10	1/1024
Site-local unicast addresses (now deprecated)	1111 1110 11	0.098%	0×FE 11	1/1024
Multicast addresses	1111 1111	0.391%	0×FF	1/256
The remainder of the IPv6 address	Unassigned	85.742%		

Note: *0×Y is the hexadecimal notation for digit "Y." Site-local addresses are deprecated in RFC 3879, September 2004*

notation for IPv4. An IPv6 prefix is written in address/prefix-length notation (IPv4 environments use a dotted decimal representation known as the subnet mask to establish the network prefix of a given IP address; the subnet mask approach is *not used* in IPv6, rather, only the prefix-length notation is used).

As noted earlier, the prefix is the part of the address that indicates the bits that have fixed values or are the bits of the network identifier. For example, 21DA:D3::/48 is a 48-bit route prefix

21DA	00D3	0000	16 bits	16 bits	16 bits	16 bits	16 bits
<- route prefix ->							

and

21DA:D3:0:2F3B::/64 is a 64-bit subnet prefix (a 48-bit route prefix plus a site topology identifier for the next 16 bits)

21DA	00D3	0000	2F3B	16 bits	16 bits	16 bits	16 bits
<- route prefix ->			<- subnet prefix ->				

Note: RFC 3879, September 2004, formally deprecated the IPv6 site-local unicast prefix defined in RFC3513, that is, 1111111011 binary or FEC0::/10. The special behavior of this prefix is no longer be supported in new implementations. The prefix has not been reassigned for other use except by a future IETF standards action. A brief discussion follows below, based directly on the RFC [HUI200401].

Studies in IETF outlined several defects of the site local addressing scope originally included in the IPv6 specification. These defects fall in two broad categories: ambiguity of addresses and fuzzy definition of sites. As originally defined, site local addresses are ambiguous: an address such as FEC0::1 can be present in multiple sites, and the address itself does not contain any indication of the site to which it belongs. This creates pain for developers of applications, for the designers of routers and for the network managers. This issue is compounded by the fuzzy nature of the site concept.

Early feedback from developers indicates that site local addresses were hard to use correctly in an application. This is particularly true for multihomed hosts, which can be simultaneously connected to multiple sites, and for mobile hosts, which can be successively connected to multiple sites. Applications would learn or remember that the address of some correspondent was "FEC0::1234:5678:9ABC"; they would try to feed the address in a socket address structure and issue a connect, but the call will fail because they did not fill up the "site identifier" variable, as in "FEC0::1234:5678:9ABC%1." (The % character is used as a delimiter for zone identifiers.) The problem is compounded by the fact that the site identifier varies with the host instantiation, for example, sometimes %1 and sometimes %2, and thus the host identifier cannot be remembered in memory, or learned from a name server.

The issue is caused by the ambiguity of site-local addresses. Since site-local addresses are ambiguous, application developers have to manage the "site identifiers" that qualify the addresses of the hosts. This management of identifiers has proved hard to understand by developers, and also hard to execute by those developers who do understand the concept.

The management of IPv6 site local addresses is in many ways similar to the management of RFC 1918 addresses in some IPv4 networks. In theory, the private addresses defined in RFC 1918 should only be used locally, and should never appear in the Internet. In practice, these addresses "leak." The conjunction of leaks and ambiguity ends up causing management problems. Names and literal addresses of "private" hosts leak in mail messages, Web pages, or files. Private addresses end up being used as source or destination of TCP requests or UDP messages, for example in DNS or trace-route requests, causing the request to fail, or the response to arrive at unsuspecting hosts.

Having a nonambiguous address solves a large part of the developers' pain, as it removes the need to manage site identifiers. The application can use the addresses as if they were regular global addresses, and the stack will be able to use standard techniques to discover which interface should be used. Some level of pain will remain, as these addresses will not always be reachable; however, applications can deal with the unreachability issues by trying connections at a different time, or with a different address. Having nonambiguous addresses will not eliminate the leaks that cause management pain. However, since the addresses are not ambiguous, debugging these leaks will be much simpler.

Having nonambiguous addresses will solve a large part of the router issues: since addresses are not ambiguous, routers will be able to use standard routing techniques, and will not need different routing tables for each interface. Some of the pain will remain at border routers, which will need to filter packets from some ranges of source addresses; this is, however, a fairly common function.

Avoiding the explicit declaration of scope will remove the issues linked to the ambiguity of the site concept. Nonreachability can be obtained by using "firewalls" where appropriate. The firewall rules can explicitly accommodate various network configurations, by accepting or refusing traffic to and from ranges of the new nonambiguous addresses.

7.5 Address Types

This section looks at some more detailed information related to address types. We discuss a number of unicast addresses, multicast addresses, and anycast addresses.

7.5.1 Unicast IPv6 Addresses

A unicast address identifies a single interface within the scope of the unicast address type. This could be a VoIP handset in a VoIPv6 environment, a PC on a LAN, a terminal in a VSAT network and so on. Utilizing an up-to-date unicast routing topology, Protocol Data Units (PDUs) addressed to a unicast address are delivered to a single interface. Unicast addresses fall into the following categories:

- Aggregatable global unicast addresses (e.g., used to reach an Internet-connected VoIP phone)
- Link-local addresses (e.g., used to reach a VoIP phone on the same LAN segment)
- Special addresses, including unspecified and loopback addresses
- Compatibility addresses, including 6to4 addresses

These are discussed next.

7.5.1.1 Aggregatable Global Unicast Addresses

The IPv6-based Internet has been designed to support efficient, hierarchical addressing and routing (this is in contrast IPv4-based Internet, which has a mixture of both flat and hierarchical routing). Aggregatable global unicast addresses are globally routable and globally reachable on the IPv6 portion of the (IPv6) Internet. The region of the Internet over which the aggregatable global unicast address is unique (the scope) is the entire IPv6 Internet. As we saw earlier, aggregatable global unicast addresses (aka global addresses), are identified by the format prefix of 001. This type of addressing can be used, for example, to reach an Internet-connected VoIP Session Initiation Protocol (SIP)-based phone (say, the author's phone given to him by his company and utilized by him while traveling on business and using the Internet for connectivity) from any origination point, be such origination point on the firm's intranet or on any other company's intranet, or even at another Internet point.

Figure 7.6 shows how the fields within the aggregatable global unicast address create a three-level topological structure with globally unique addresses. The first 48 bits comprise the 3-bit format prefix; the Top Level Aggregator* (TLA) id comprises the next 13 bits; the next 8 bits are reserved; and, the next 24 bits represent the Next Level Aggregator (NLA) id. This combination gives the first two levels. The next 16 bits represents the site topology, namely, the Site Level Aggregator (SLA) id. The SLA is used by a firm or organization to identify subnets within its site (intranet); the organization can use the 16 bits within its site to create 65,536 subnets or multiple levels of addressing hierarchy, which, can also facilitate the routing process. (Note that with a 2-octet of address space for subnetting, an aggregatable global unicast prefix assigned to a firm is equivalent to that firm being granted an IPv4 Class A network ID. Also, the structure of the customer's network is not visible to the ISP.) Finally, the Interface ID point to the interface of a node on a specific subnet.

Addresses of this type can, by design, be aggregated (summarized) to produce an efficient routing infrastructure.

---

* The TLA/NLA/SLA field structure has been deprecated by the IETF in RFC 3587, but is covered herewith for historical purposes.

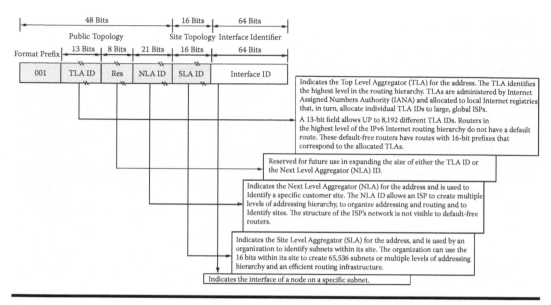

Figure 7.6   Aggregatable global unicast address.

7.5.1.2   Link-Local (Unicast) Addresses

Link-local addresses are utilized by nodes when communicating with neighboring nodes on the same link. For example, link-local addresses are used to communicate between hosts on the link on a single link IPv6 network without the intervention/utilization of a router (e.g., in a LAN segment, a VLAN, etc.). This type of addressing can be used, for example, to reach a company colleague on a LAN-connected VoIP phone (say, for colleagues working in the same building—assuming that both are on the same LAN).

The scope of a link-local address is the local link. An IPv6 router does not forward link-local traffic beyond the link. A link-local address is required for Neighbor Discovery processes and is always automatically configured, even in the absence of all other unicast addresses. As seen earlier, link-local addresses are identified by the Format Prefix of 1111 1110 10. The address starts with FE (e.g., 1111 1110 1000 is 0×FE8; 1111 1110 1001 is 0×FE9; 1111 1110 1010 is 0×FEA; and, 1111 1110 1011 is 0×FEB.) With the 64-bit interface identifier, the prefix for link-local addresses is, by convention, always FE80::/64.

7.5.1.3   Unspecified (Unicast) Address

The unspecified address, 0:0:0:0:0:0:0:0 (i.e., ::) indicates the absence of an address, and is typically used as a source address for PDUs that are attempting to verify the uniqueness of a tentative address. It is equivalent to the IPv4 unspecified address of 0.0.0.0. The unspecified address is never assigned to an interface or used as a destination address.

7.5.1.4   Loopback (Unicast) Address

The loopback address, 0:0:0:0:0:0:0:1 or ::1, identifies a loopback interface, enabling a node to send PDUs to itself. It is equivalent to the IPv4 loopback address of 127.0.0.1. PDUs addressed to the loopback address are never sent on a link or forwarded by an IPv6 router.

7.5.1.5 Compatibility (Unicast) Addresses

IPv6 provides what are called *6to4 addresses* to facilitate the coexistence of IPv4-to-IPv6 environments and the migration from the IPv4 to the IPv6 environment. The 6to4 address is used for communicating between two nodes operating both IPv4 stacks and IPv6 stacks over an IPv4 routing infrastructure. The 6to4 address is formed by combining the prefix 2002::/16 with the 32 bits of the public IPv4 address of the node, forming a 48-bit prefix.

7.5.2 Multicast IPv6 Addresses

A useful feature supported in IPv6 is multicasting. The use of multicasting in IP networks is defined in RFC 1112, which describes addresses and host extensions for the way IP hosts support multicasting—the concepts originally developed for IPv4 also apply to IPv6. Besides a variety of protocol-level functionally supported by multicasting (e.g., MLD and ND), one also can use this mechanism to support VoIP/IPTV functionality (e.g., audioconferencing/bridging and program distribution). Multicast traffic is promulgated by utilizing a single-destination address in the IPv6 header, but the IPv6 datagram is received and processed by multiple hosts. Hosts and devices listening on a specific multicast address comprise a multicast group; these devices receive and process traffic sent to the group address. As seen earlier, IPv6 multicast addresses have the format prefix of 1111 1111; namely, the multicast address always begins with 0xFF.

Group membership in multicast is dynamic, allowing hosts to join and leave the group at any time. Groups can be from multiple network segments (links or subnets) if the connecting routers support forwarding of multicast traffic and group membership information [MSD200401]. A host (e.g., a VoIP SIP proxy or a H.323 gatekeeper) can send traffic to a group address without belonging to the group. In fact, to join a group, a host sends a group membership message. Each multicast group is identified by one IPv6 multicast address. All group members who listen and receive IPv6 messages sent to the group address share the group address. Multicast routers periodically poll membership status.

Some of the reserved IPv6 multicast addresses (RFC 2375) are shown in Table 7.6.

A multicast address is an addressing mechanism that identifies multiple interfaces; it is used for one-to-many communication. With the appropriate multicast routing topology, PDUs addressed

Table 7.6　Reserved multicast IPv6 addresses

IPv6 multicast address	Description
FF02::1	The all-nodes address used to reach all nodes on the same link.
FF02::2	The all-routers address used to reach all routers on the same link.
FF02::4	The address used to reach all Distance Vector Multicast Routing Protocol (DVMRP) multicast routers on the same link.
FF02::5	The address used to reach all Open Shortest Path First (OSPF) routers on the same link.
FF02::1:FF*XX:XXXX*	The solicited-node address used in the address resolution process to resolve the IPv6 address of a link-local node to its link-layer address. The rightmost 24 bits (*XX:XXXX*) of the solicited-node address are the rightmost 24 bits of an IPv6 unicast address.

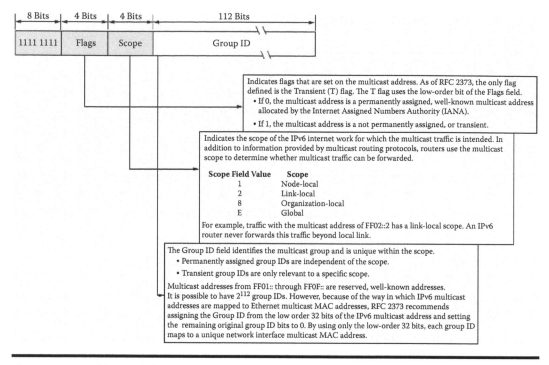

Figure 7.7   Multicast address.

to a multicast address are delivered to all interfaces that are identified by the address. Multicast addresses cannot be utilized as source addresses. Multicast address Flags, Scope, and Group, as shown in Figure 7.7.

To identify all nodes for the node-local and link-local scopes, the following multicast addresses are defined:

- FF01::1 (node-local scope all-nodes address)
- FF02::1 (link-local scope all-nodes address)

To identify all routers for the node-local, and link-local, scopes, the following multicast addresses are defined:

- FF01::2 (node-local scope all-routers address)
- FF02::2 (link-local scope all-routers address)

Next, we briefly look at solicited-node addresses. The solicited-node address supports efficient querying of network nodes for the purpose of address resolution. IPv6 uses the Neighbor Solicitation message to perform address resolution. This multicast address consists of the prefix FF02::1:FF00:0/104 along with the last 24-bits of the IPv6 address that is being resolved. In contrast to IPv4 where the ARP Request frame is sent via a MAC-level broadcast, and in doing so imposing on all nodes on the network segment, in IPv6 the solicited-node multicast address is used as the Neighbor Solicitation message destination. This avoids imposing on all IPv6 nodes on the local link by using the local-link scope all-nodes address.

7.5.3 *Anycast IPv6 Addresses*

An anycast address identifies multiple interfaces, but not an entire broadcast universe. This could be used, for example, to support VoIP voice mail group distribution. With the appropriate routing topology, PDUs addressed to an anycast address are delivered to a single interface for further appropriate handling (a PDU addressed to an anycast address is delivered to the nearest interface identified by the address). To make possible the delivery to the nearest anycast group member, the routing infrastructure must be aware of the interfaces that are assigned anycast addresses and must know their distances in terms of routing metrics. At present, anycast addresses are used only as destination addresses and are assigned only to routers.

7.6 Addresses for Hosts and Routers

In contrast to IPv4 where a host with a single network adapter has a single IPv4 address assigned to that adapter, an IPv6 host (e.g., a SIP proxy, a VSAT station) typically has multiple IPv6 addresses (even in the case of a single interface). (When a computer is configured with more than one IP address, it is referred to as a *multihomed* system.) IPv6 host and router address usage is as follows [MSD200401]:

Host: Typical IPv6 hosts are logically multihomed because they have at least two addresses with which they can receive PDUs. Each host is assigned the following unicast addresses:

- A link-local address for each interface. This address is used for local traffic.
- An address for each interface. This could be one or more global addresses.
- The loopback address (::1) for the loopback interface.

Additionally, each host is listening for traffic on the following multicast addresses:

- The node-local scope all-nodes address (FF01::1)
- The link-local scope all-nodes address (FF02::1)
- The solicited-node address for each unicast address on each interface
- The multicast addresses of joined groups on each interface

Router: An IPv6 router is assigned the following unicast addresses:

- A link-local address for each interface. This address is used for local traffic.
- An address for each interface. This could be one or more global addresses.
- The loopback address (::1) for the loopback interface

An IPv6 router is assigned the following anycast addresses:

- A subnet-router anycast address for each subnet
- Additional anycast addresses (optional)

Each router is listening for traffic on the following multicast addresses:

- The node-local scope all-nodes address (FF01::1)
- The node-local scope all-routers address (FF01::2)
- The link-local scope all-nodes address (FF02::1)
- The link-local scope all-routers address (FF02::2)
- The solicited-node address for each unicast address on each interface
- The addresses of joined groups on each interface

7.6.1 Interface Determination

The last 64 bits of an IPv6 address are the interface identifier that is unique to the 64-bit prefix of the IPv6 address. There are two ways for interface identifier determination: (1) derived from the Electrical and Electronic Engineers (IEEE) Extended Unique Identifier (EUI)-64 address; and, (2) randomly generated and randomly changed over time. IETF RFC 2373 stipulates that unicast addresses that use format prefixes 001 through 111 must use a 64-bit interface identifier that is derived from the EUI-64 address. Related to the second approach, RFC 3041 states that to provide a level of anonymity, the identifier can be randomly generated, and changed over time.

EUI-64 addresses are either *assigned* to a network adapter or *derived* from IEEE 802 addresses. LAN network interface cards (NICs) that (at this point in the development of hardware) typically comprise the physical interface (network adapters) of hosts and devices identifiers use the 48-bit IEEE 802 address. This address (also called the physical, hardware, or media access control (MAC) address) consists of two parts: (1) company ID, and (2) extension ID. The company ID is 24-bit ID uniquely assigned to each manufacturer of network adapters; this is also known as the manufacturer ID. The extension ID (also known as the board ID) is a 24-bit uniquely assigned to each network adapter at the time of assembly. The IEEE 802 address is thus a globally unique 48-bit address. The IEEE EUI-64 address is a newly-defined standard for network interface addressing. The company ID is 24-bits in length but the extension ID is 40 bits, supporting a larger address space for a network adapter manufacturer. See Figure 7.8.

To generate an EUI-64 address from an IEEE 802 address, 16 bits of 11111111 11111110 (0xFFFE) are inserted into the IEEE 802 address between the company ID and the extension ID (see Figure 7.9).

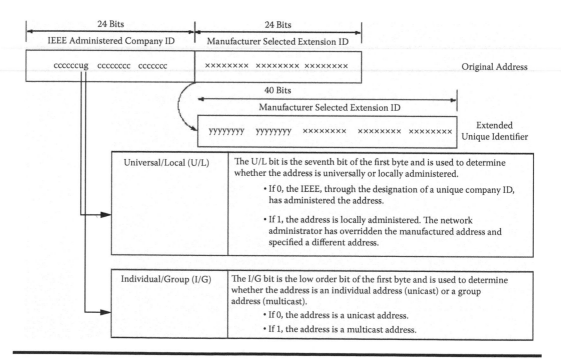

Figure 7.8   IEEE address along with the extended unique identifier.

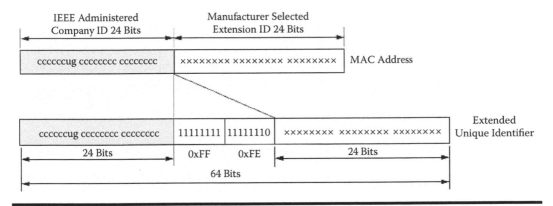

Figure 7.9   Extended unique identifier generated from MAC address.

7.6.2   Mapping EUI-64 Addresses to IPv6 Interface Identifiers

An IPv6 unicast address utilizes a 64-bit interface identifier. To obtain this identifier from a EUI-64 address, the U/L bit in the EUI-64 address is complemented (if it is a 1, it is set to 0; if it is a 0, it is set to 1). The resulting bitstream is used as a universally administered unicast EUI-64 address.

7.6.3   Mapping IEEE 802 Addresses to IPv6 Interface Identifiers

To obtain an IPv6 interface identifier from an IEEE 802 address, one must first map the IEEE 802 address to an EUI-64 address, as discussed previously; then, one must complement (flip) the U/L bit. The resulting bitstream is used as a universally administered unicast IEEE 802 address.

7.6.4   Randomly Generated Interface Identifiers

IPv6 interface identifiers remain static over time, hence, for security reasons, a capability is needed to generate temporary addresses. (Because of NAT/DHCP, in an IPv4 environment it is difficult to track a user's traffic on the basis of IP address.) (However, it should be noted that many—if not most—hacking techniques do not rely on knowing the IP address of a *specific* device on a network; instead, such techniques simply look for *any* available entry point. After that, a deposited Trojan Horse may do the job of perpetrating a full infraction.)

In IPv6, after the connection is made through router discovery and stateless address auto-configuration, the end-user device is assigned a 64-bit prefix. If the interface identifier is based on a EUI-64 address (which, as we saw earlier, is derived from the static IEEE 802 address), the traffic of a specific node can be identified which opens up the possibility of tracking a specific user, should that be of interest to an intruder. To address this issue, an alternative IPv6 interface identifier can be randomly generated and changed over time, as described in RFC 3041. For IPv6 systems that have storage capabilities, a history value is stored. When the IPv6 protocol is initialized, a new interface identifier is created through the following process (the IPv6 address based on this random interface identifier is known as a temporary address):

1. Retrieve the history value from storage and append the interface identifier based on the EUI-64 address of the adapter.
2. Compute the Message Digest-5 (MD5) one-way encryption hash over the quantity in step 1.

3. Save the last 64 bits of the MD5 hash computed in step 2 as the history value for the next interface identifier computation.
4. Take the first 64 bits of the MD5 hash computed in Step 2 and set the seventh bit to zero. The seventh bit corresponds to the U/L bit which, when set to 0, indicates a locally administered interface identifier. The result is the interface identifier.

Temporary addresses are generated for public address prefixes that use stateless address auto-configuration.

7.7 IPv6 Infrastructure

The IPv6 Specification (RFC 2460) and the IPv6 Addressing Architecture (RFC 2373) provide the base architecture and design of IPv6; we covered some of these key concepts in earlier sections. Sections 7.7 and 7.8 look at basic IPv6 network constructs, specifically routing processes. Because there are differences on some of the details of how these IPv6 processes operate compared with IPv4, it is worth looking at some of these issues. Related work in IPv6 that needs to be mastered by implementers and network designers includes the IPv6 Stateless Address Autoconfiguration (RFC 2462); the IPv6 Neighbor Discovery Processing (RFC 2461); the Dynamic Host Configuration Protocol for IPv6 (DHCPv6) (RFC 3315); and, the Dynamic Updates to DNS (RFC 2136). Some portions of this discussion are based on [MSD200401].

7.7.1 Protocol Mechanisms

As we discussed earlier, an IPv6 Protocol Data Unit (PDU) consists of an IPv6 header and an IPv6 payload, as depicted in Figure 7.10. The IPv6 header consists of two parts: the IPv6 base header and optional extension headers. The optional extension headers are considered part of the IPv6 payload, as are the TCP/UDP/RTP PDUs. Obviously, IPv4 headers and IPv6 headers are not automatically interoperable; hence, a router operating in a mixed environment must support an implementation of both IPv4 and IPv6 to deal with both header formats. Figure 7.11 shows for illustration purposes the flows of IPv6 PDUs in a VoIP environment.

As we noted earlier in the chapter, the large size of the IPv6 address allows it to be subdivided into hierarchical routing domains that are supportive of the topology of today's ubiquitous Internet (IPv4-based Internet lacks this flexibility). Conveniently, the use of 128 bits provides multiple levels of hierarchy and flexibility in designing hierarchical addressing and routing.

7.7.2 Protocol-Support Mechanisms

Two support mechanisms are of interest: (1) a mechanism to deal with communication transmission issues and (2) a mechanism to support multicast.

Internet Control Message Protocol for IPv6 (ICMPv6) (defined in RFC 2463) is designed to enable hosts and routers that use IPv6 protocols to report errors and forward along other basic status messages. For example, ICMPv6 messages are sent by Network Elements when an IPv6 PDU cannot be forwarded further along to reach its intended destination. ICMPv6 messages are carried as the payload of IPv6 PDUs (see Figure 7.12), hence, there is *no guarantee* on their delivery.

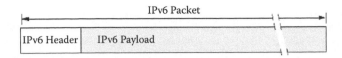

IPv6 Header Field	Function	Length
Version	Identifies the version of the protocol. For IPv6, the field is 6 (1010)	4 Bits
Class	Intended for originating nodes and forwarding routers to identify and distinguish between different classes or priorities of IPv6 packets.	8 Bits
Flow Label	Defines how traffic is handled and identified. A flow is a sequence of packets sent either to a unicast or a multicast destination. Field identifies packets that require special handling by the IPv6 node.	20 Bits
Payload Length	Identifies the length, octets, of the payload. This field is a 16-bit unsigned integer. The payload includes the optional extension headers, as well as the upper-layer protocols.	16 Bits
Next Header	Identifies the header immediately following the IPv6 header.	8 Bits
Hop Limit	Identifies the number of network segments (links or subnets), on which the packet is allowed to travel before being discarded by a router. This parameter is set by the sending host and is used to prevent packets from endlessly circulating on an IPv6 internetwork. When forwarding an IPv6 packet, IPv6 routers must decrease the Hop Limit by 1, and must discard the IPv6 packet when the Hop Limit is 0.	8 Bits
Source Address	Identifies the IPv6 address of the original source of the IPv6 packet.	128 Bits
Destination Address	Identifies the IPv6 address of the intermediate or final destination of the IPv6 packet.	128 Bits

Figure 7.10   IPv6 PDU.

The following list identifies the functionality supported by the basic ICMPv6 mechanisms:

- Destination Unreachable: An error message that informs the sending host that a PDU cannot be delivered.
- Packet Too Big: An error message that informs the sending host that the PDU is too large to forward.
- Time Exceeded: An error message that informs the sending host that the Hop Limit of an IPv6 PDU has expired.
- Parameter Problem: An error message that informs the sending host that an error was encountered in processing the IPv6 header or an IPv6 extension header.
- Echo Request: An informational message that is used to determine whether an IPv6 node is available on the network.
- Echo Reply: An informational message that is used to reply to the ICMPv6 Echo Request message.

The ping command is basically an ICMPv6 Echo Request messages along with the receipt of ICMPv6 Echo Reply messages. Just as is the case with IPv4, one can use pings to detect network or host communication failures and troubleshoot connectivity problems.

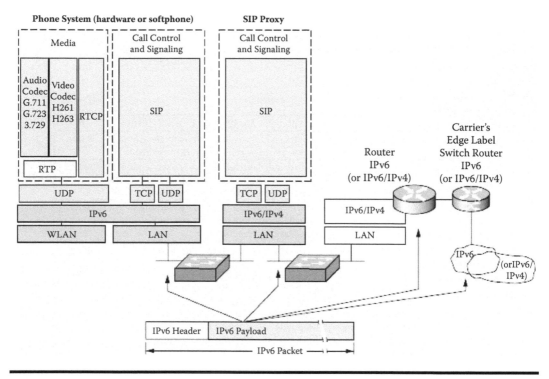

Figure 7.11   Flows of IPv6 packets in a VoIPv6 environment.

ICMPv6 also supports Multicast Listener Discovery (MLD). MLD (RFC 2710, RFC 3590, and RFC 3810) enables one to manage subnet multicast membership for IPv6. MLD is a collection of three ICMPv6 messages that replace the Internet Group Management Protocol (IGMP) version 3 that is employed in IPv4. MLD messages are used to determine group membership on a network segment, also known as a link or subnet. As implied, MLD messages are sent as ICMPv6 messages. They are used in the context of multicast communications (see below):

- Multicast Listener Query: Message issued by a multicast router to poll a network segment for group members. Queries can be general, requesting group membership for all groups, or can request group membership for a specific group.
- Multicast Listener Report: Message issued by a host when it joins a multicast group, or in response to an MLD Multicast Listener Query sent by a router.
- Multicast Listener Done: Message issued by a host when it leaves a host group and is the last member of that group on the network segment.

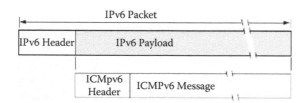

Figure 7.12   ICMPv6 message.

Table 7.7   Key ND processes

Process	Description
Address Autoconfiguration	The process for configuring IP addresses for interfaces in the absence of a stateful address configuration server, for example, via Dynamic Host Configuration Protocol version 6 (DHCPv6).
Address Resolution	The process by which a node resolves a neighboring node's IPv6 address to its link-layer address. The resolved link-layer address becomes an entry in a neighbor cache in the node. The link-layer address is equivalent to ARP in IPv4, and the neighbor cache is equivalent to the ARP cache. The neighbor cache displays the interface identifier for the neighbor cache entry, the neighboring node IPv6 address, the corresponding link-layer address, and the state of the neighbor cache entry.
Duplicate Address Detection	The process by which a node determines that an address considered for use is not already in use by a neighboring node. This is equivalent to the use of ARP frames in IPv4.
Dynamic Updates to DNS (RFC 2136)	A process that supports the dynamic update of DNS records for both IPv4 and IPv6. DHCP can use the dynamic updates to DNS to integrate addresses and name space to not only support autoconfiguration but also auto registration in IPv6 [DRO200301].
IPv6 Neighbor Discovery (RFC 2461)	The node discovery process/protocol in IPv6 which replaces and enhances functions of ARP. To understand IPv6 and stateless address autoconfiguration, implementers and network designers need to understand IPv6 Neighbor Discovery [DRO200301].
Neighbor Unreachability Detection	The process by which a node determines that neighboring hosts or routers are no longer available on the local network segment. After the link-layer address for a neighbor has been determined, the state of the entry in the neighbor cache is tracked. If the neighbor is no longer receiving and sending back PDUs, the neighbor cache entry is eventually removed.
Next-Hop Determination	The process by which a node determines the IPv6 address of the neighbor to which a PDU is being forwarded. The determination is made based on the destination address. The forwarding or next-hop address is either the destination address of the PDU being sent or the address of a neighboring router. The resolved next-hop address for a destination becomes an entry in a node's destination cache, also known as a route cache. The route cache displays the destination address, the interface identifier and next-hop address, the interface identifier and address used as a source address when sending to the destination, and the path maximum transmission unit (MTU) for the destination.
Parameter Discovery	The process by which a host discovers additional operating parameters, including the link MTU and the default hop limit for outbound PDUs.
Prefix Discovery	The process by which a host discovers the network prefixes for local destinations.
Redirect Function	The process by which a router informs a host of a better first-hop IPv6 address to reach a destination. This is equivalent to the function of the IPv4 ICMP Redirect message.
Router Discovery	The process by which a host discovers the local routers on an attached link and automatically configures a default router. In IPv4, this is equivalent to using ICMPv4 router discovery to configure a default gateway.

ICMPv6 also supports Neighbor Discovery (ND). ND (RFC 2461) is a collection of five ICMPv6 messages that manage node-to-node communication on a link. Nodes on the same link are also called neighboring nodes. ND replaces Address Resolution Protocol (ARP), ICMPv4 Router Discovery, and the ICMPv4 Redirect message. Table 7.7 identifies key ND processes [MSD200401]. Hosts (e.g., servers, SIP proxies, H.323 gatekeepers, VSAT nodes, etc.) make use of ND to discover neighboring routers, addresses, address prefixes, and other configuration parameters. Routers make use of ND to advertise their presence, host configuration parameters, and on-link prefixes. Routers also use ND to inform hosts of a better next-hop address to forward PDUs for a specific destination. Nodes make use of ND to resolve the link-layer address of a neighboring node to which an IPv6 PDU is being forwarded. Nodes also use ND to determine when the link-layer address of a neighboring node has changed, and whether IPv6 PDUs can be sent to and received from a neighbor.

7.8 Routing and Route Management

Routing is the process of forwarding PDUs between connected network segments (also known as links or subnets). Routing is a primary function of a network layer protocol, whether it is IP version 4 or version 6. IPv6 routers provide the primary means for joining together two or more IPv6 network segments. Network segments are identified by using an IPv6 network prefix and prefix length. Routers pass IPv6 PDUs from one network segment to another. IPv6 routers are attached to two or more IPv6 network segments and enable hosts on those segments to forward IPv6 PDUs. IPv6 PDUs are exchanged and processed on each host by using IPv6 at layer 3 (the Internet Protocol layer).

Datagrams with a source and destination IP address identified in the header are handed to the IP protocol engine/layer. Above the IPv6 layer, transport services on the source host pass data in the form of TCP segments or UDP PDUs down to the IPv6 layer. IPv6 layer services on each sending host examine the destination address of each PDU, compare this address to a locally maintained routing table, and then determine what additional forwarding is required. The IPv6 layer creates IPv6 PDUs with source and destination address information that is used to route the data through the network. The IPv6 layer then passes PDUs down to the link layer, where the PDUs are converted into frames for transmission over network-specific media on a physical network. This process occurs in reverse order on the destination host [MSD200401].

IPv6 hosts utilize routing tables to maintain information about other IPv6 networks and IPv6 hosts. The routing tables provide important information about how to communicate with remote networks and hosts. Every device that implements IPv6 determines how to forward PDUs on the basis of the contents of the IPv6 routing table. The following information is contained in the IPv6 routing table:

- An address prefix
- The interface over which PDUs that match the address prefix are sent
- A forwarding or next-hop address
- A preference value used to select between multiple routes with the same prefix
- The lifetime of the route
- The specification of whether the route is published (advertised in a Routing Advertisement)
- The specification of how the route is aged
- The route type

The IPv6 routing table is built automatically, based on the current IPv6 configuration of the router. When forwarding IPv6 PDUs, the router searches the routing table for an entry that is the most specific match to the destination IPv6 address. A route for the link-local prefix (FE80::/64) is not displayed.

Typically, a default route is used by an end device because it is not practical for an end device to maintain a routing table for each communication device on an IPv6 network. The default route (a route with a prefix of ::/0) is typically used to forward an IPv6 PDU to a default router on the local link. Because the router that corresponds to the default router contains information about the network prefixes of the other IPv6 subnets within the larger IPv6 internetwork, it forwards the PDU to other routers until the PDU is eventually delivered to the destination.

The following steps occur during the routing process [MSD200401]:

1. Before a communication device sends an IPv6 PDU, it inserts its source IPv6 address and the destination IPv6 address (for the recipient) into the IPv6 header.
2. The device then examines the destination IPv6 address, compares it to a locally maintained IPv6 routing table, and takes appropriate action. The device does one of the following:
   – It passes the PDU to a protocol layer above IPv6 on the local host.
   – It forwards the PDU through one of its attached network interfaces.
   – It discards the PDU.
3. IPv6 searches the routing table for the route that is the closest match to the destination IPv6 address. The most specific to the least specific route is determined in the following order:
   – A route that matches the destination IPv6 address (a host route with a 128-bit prefix length).
   – A route that matches the destination with the longest prefix length.
   – The default route (the network prefix ::/0).
4. If a matching route is not found, the destination is determined to be an on-link destination.

7.9 Configuration Methods

IPv6 Stateless Address Autoconfiguration (RFC 2462) specifies procedures by which a node may autoconfigure addresses, based on router advertisements and the use of a valid lifetime to support renumbering of addresses on the Internet. In addition, the protocol interaction by which a node begins stateless or stateful autoconfiguration is specified. DHCP is one vehicle to perform stateful autoconfiguration; compatibility with stateless address autoconfiguration is a design requirement of DHCP [DRO200301].

As we have seen in previous section, the IPv6 protocol can use two address configuration methods: (1) automatic configuration and (2) manual configuration. Autoconfigured addresses exist in one or more of the states depicted in Figure 7.13: tentative, preferred, deprecated, valid (= preferred + deprecated), and, invalid. IPv6 nodes (hosts and routers) automatically create unique link-local addresses for all LAN interfaces that appear to be Ethernet interfaces. IPv6 hosts use received Router Advertisement messages to automatically configure the following parameters [MSD200401].

- A default router
- The default setting for the Hop Limit field in the IPv6 header.
- The timers used in Neighbor Discovery processes.
- The MTU of the local link.

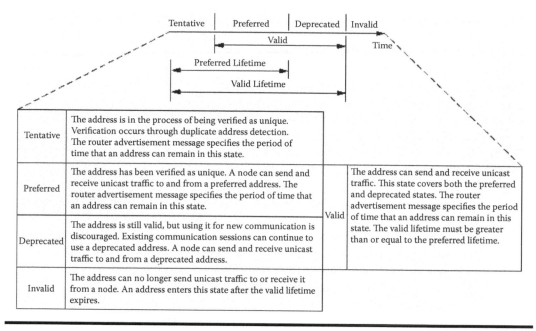

Tentative	The address is in the process of being verified as unique. Verification occurs through duplicate address detection. The router advertisement message specifies the period of time that an address can remain in this state.	
Preferred	The address has been verified as unique. A node can send and receive unicast traffic to and from a preferred address. The router advertisement message specifies the period of time that an address can remain in this state.	Valid: The address can send and receive unicast traffic. This state covers both the preferred and deprecated states. The router advertisement message specifies the period of time that an address can remain in this state. The valid lifetime must be greater than or equal to the preferred lifetime.
Deprecated	The address is still valid, but using it for new communication is discouraged. Existing communication sessions can continue to use a deprecated address. A node can send and receive unicast traffic to and from a deprecated address.	
Invalid	The address can no longer send unicast traffic to or receive it from a node. An address enters this state after the valid lifetime expires.	

Figure 7.13   Address states.

■ The list of network prefixes that are defined for the link. Each network prefix contains both the IPv6 network prefix and its valid and preferred lifetimes. If indicated, a network prefix is combined with the interface identifier to create a stateless IPv6 address configuration for the receiving interface. A network prefix also defines the range of addresses for nodes on the local link.
■ 6to4 addresses on a 6to4 tunneling interface for all public IPv4 addresses that are assigned to the computer (some implementations).
■ Intrasite Automatic Tunnel Addressing Protocol (ISATAP) addresses on an automatic interface for all IPv4 addresses that are assigned to the computer (some implementations).
■ The stack to query for IPv6 ISATAP routers in an IPv4 environment (some implementations).
■ Routes to off-link prefixes, if the off-link address prefix is advertised by a router (some implementations).

DHCP is *not* utilized in IPv6 to configure a link-local scope IP address. The link-local scope of an IPv6 addresses is always configured automatically. Addresses with other scopes are configured by router advertisements. Specifically, unique link-local addresses are automatically configured for each interface on each IPv6 node (host or router). To communicate with IPv6 nodes that are not on attached links, the host must have additional global unicast addresses. Additional addresses for hosts are obtained from router advertisements, while additional addresses for routers must be assigned manually. To communicate with IPv6 nodes on other network segments, IPv6 uses a default router. A default router is automatically assigned based on the receipt of a router advertisement. Alternately, one can add a default route to the IPv6 routing table. Note that one does not need to configure a default router for a network that consists of a single network segment.

The following sequence identifies the address autoconfiguration process for an IPv6 node, such as an IPv6-based VoIP phone or a VSAT node.

1. A tentative link-local address is derived, based on the link-local prefix of FE80::/64 and the 64-bit interface identifier.
2. Duplicate address detection is performed to verify the uniqueness of the tentative link-local address.
3. If duplicate address detection fails, one must manually configure the node, *or*

   If duplicate address detection succeeds, the tentative address is assumed to be valid and unique. The link-local address is initialized for the interface. The corresponding solicited-node multicast link-layer address is registered with the network adapter.

For an IPv6 host, address autoconfiguration continues as follows:

1. The host sends a Router Solicitation message.
2. If a Router Advertisement message is received, the configuration information that is included in the message is set on the host.
3. For each stateless autoconfiguration address prefix that is included, the following processes occur:
   The address prefix and the appropriate 64-bit interface identifier are used to derive a tentative address.
4. Duplicate address detection is used to verify the uniqueness of the tentative address. If the tentative address is in use, the address is not initialized for the interface. If the tentative address is not in use, the address is initialized. This includes setting the valid and preferred lifetimes based on information included in the Router Advertisement message.

Other configuration processes are shown in Table 7.8 [MSD200401].

Table 7.8 Configurations of interest

Configuration	Description
Single subnet with link-local addresses	This configuration supports the installation of the IPv6 protocol on at least two nodes on the same network segment without intermediate routers.
IPv6 traffic between nodes on different subnets of an IPv6 internetwork	This configuration includes two separate network segments (also known as links or subnets), and an IPv6-capable router that connects the network segments and forwards IPv6 Protocol Data Units (PDUs) between the hosts.
IPv6 traffic between nodes on different subnets of an IPv4 internetwork	This configuration supports IPv6 traffic that is carried as the payload of an IPv4 PDU (treating the IPv4 infrastructure as an IPv6 link-layer) without the deployment of IPv6 routers.
IPv6 traffic between nodes in different sites across the internet	This configuration supports the 6to4 tunneling technique. The IPv6 traffic is encapsulated with an IPv4 header before it is sent over an IPv4 internetwork such as the Internet.

7.10 Dynamic Host Configuration Protocol for IPv6

The Dynamic Host Configuration Protocol for IPv6 (DHCPv6 or more simply DHCP) enables DHCP servers to pass configuration parameters such as IPv6 network addresses to IPv6 nodes. DHCP provides both robust stateful autoconfiguration and autoregistration of DNS Host Names.

DHCP offers the capability of automatic allocation of reusable network addresses and additional configuration flexibility. This protocol is a stateful counterpart to "IPv6 Stateless Address Autoconfiguration" (RFC 2462) and can be used separately or concurrently with the latter to obtain configuration parameters. DHCP is a client/server protocol that provides managed configuration of devices. DHCP can provide a device with addresses assigned by a DHCP server and other configuration information. The operational models and relevant configuration information for DHCPv4 and DHCPv6 are significantly different [DRO200301].

Clients and servers exchange DHCP messages using UDP. The client uses a link-local address or an address determined through other mechanisms for transmitting and receiving DHCP messages. DHCP servers receive messages from clients using a reserved, link-scoped multicast address. A DHCP client transmits most messages to this reserved multicast address so that the client need not be configured with the address or addresses of DHCP servers [DRO200301].

DHCP makes use of the following multicast addresses:

- All_DHCP_Relay_Agents_and_Servers (FF02::1:2) A link-scoped multicast address used by a client to communicate with neighboring (i.e., on-link) relay agents and servers. All servers and relay agents are members of this multicast group.
- All_DHCP_Servers (FF05::1:3) A site-scoped multicast address used by a relay agent to communicate with servers, either because the relay agent wants to send messages to all servers or because it does not know the unicast addresses of the servers. Note that in order for a relay agent to use this address, it must have an address of sufficient scope to be reachable by the servers. All servers within the site are members of this multicast group.

To allow a DHCP client to send a message to a DHCP server that is not attached to the same link, a DHCP relay agent on the client's link will relay messages between the client and server. The operation of the relay agent is transparent to the client and the discussion of message exchanges in the remainder of this section will omit the description of message relaying by relay agents. Once the client has determined the address of a server, it may under some circumstances send messages directly to the server using unicast [DRO200301].

The following list provides basic DHCP terminology:

- Appropriate to the link: An address is "appropriate to the link" when the address is consistent with the DHCP server's knowledge of the network topology, prefix assignment, and address assignment policies.
- Identity association (IA): A collection of addresses assigned to a client. Each IA has an associated IAID. A client may have more than one IA assigned to it; for example, one for each of its interfaces. Each IA holds one type of address; for example, an identity association for temporary addresses (IA_TA) holds temporary addresses (see "identity association for temporary addresses"). Throughout this section, "IA" is used to refer to an identity association without identifying the type of addresses in the IA.
- DUID: A DHCP Unique IDentifier for a DHCP participant; each DHCP client and server has exactly one DUID.

- Binding: A binding (or, client binding) is a group of server data records containing the information the server has about the addresses in an IA or configuration information explicitly assigned to the client. Configuration information that has been returned to a client through a policy—for example, the information returned to all clients on the same link—does not require a binding. A binding containing information about an IA is indexed by the tuple <DUID, IA-type, IAID> (where IA-type is the type of address in the IA; for example, temporary). A binding containing configuration information for a client is indexed by <DUID>.
- Configuration parameter: An element of the configuration information set on the server and delivered to the client using DHCP. Such parameters may be used to carry information to be used by a node to configure its network subsystem and enable communication, for example on a link or internetwork.
- DHCP client (or client): A node that initiates requests on a link to obtain configuration parameters from one or more DHCP servers.
- DHCP domain: A set of links managed by DHCP and operated by a single administrative entity.
- DHCP realm: A name used to identify the DHCP administrative domain from which a DHCP authentication key was selected.
- DHCP relay agent (or relay agent): A node that acts as an intermediary to deliver DHCP messages between clients and servers, and is on the same link as the client.
- DHCP server (or server): A node that responds to requests from clients, and may or may not be on the same link as the clients.
- Identity association identifier (IAID): An identifier for an IA, chosen by the client. Each IA has an IAID, which is chosen to be unique among all IAIDs for IAs belonging to that client.
- Identity association for nontemporary addresses (IA_NA): An IA that carries assigned addresses that are not temporary addresses (see "identity association for temporary addresses").
- Identity association for temporary addresses (IA_TA): An IA that carries temporary addresses.
- Message: A unit of data carried as the payload of a UDP datagram, exchanged among DHCP servers, relay agents, and clients.
- Reconfigure key: A key supplied to a client by a server used to provide security for Reconfigure messages.
- Relaying: A DHCP relay agent relays DHCP messages between DHCP participants.
- Transaction ID: An opaque value used to match responses with replies initiated either by a client or server.

Clients listen for DHCP messages on UDP port 546. Servers and relay agents listen for DHCP messages on UDP port 547. DHCP defines and makes use of the following message types [DRO200301]:

1. SOLICIT: A client sends a Solicit message to locate servers.
2. ADVERTISE: A server sends an Advertise message to indicate that it is available for DHCP service, in response to a Solicit message received from a client.
3. REQUEST: A client sends a Request message to request configuration parameters, including IP addresses, from a specific server.

4. CONFIRM: A client sends a Confirm message to any available server to determine whether the addresses it was assigned are still appropriate to the link to which the client is connected.
5. RENEW: A client sends a Renew message to the server that originally provided the client's addresses and configuration parameters to extend the lifetimes on the addresses assigned to the client and to update other configuration parameters.
6. REBIND: A client sends a Rebind message to any available server to extend the lifetimes on the addresses assigned to the client and to update other configuration parameters; this message is sent after a client receives no response to a Renew message.
7. REPLY: A server sends a Reply message containing assigned addresses and configuration parameters in response to a Solicit, Request, Renew, Rebind message received from a client. A server sends a Reply message containing configuration parameters in response to an Information-request message. A server sends a Reply message in response to a Confirm message confirming or denying that the addresses assigned to the client are appropriate to the link to which the client is connected. A server sends a Reply message to acknowledge receipt of a Release or Decline message.
8. RELEASE: A client sends a Release message to the server that assigned addresses to the client to indicate that the client will no longer use one or more of the assigned addresses.
9. DECLINE: A client sends a Decline message to a server to indicate that the client has determined that one or more addresses assigned by the server are already in use on the link to which the client is connected.
10. RECONFIGURE: A server sends a Reconfigure message to a client to inform the client that the server has new or updated configuration parameters and that the client is to initiate a Renew/Reply or Information-Request/Reply transaction with the server to receive the updated information.
11. INFORMATION-REQUEST: A client sends an Information-Request message to a server to request configuration parameters without the assignment of any IP addresses to the client.
12. RELAY-FORW: A relay agent sends a Relay-Forward message to relay messages to servers, either directly or through another relay agent. The received message, either a client message or a Relay-Forward message from another relay agent, is encapsulated in an option in the Relay-Forward message.
13. RELAY-REPL: A server sends a Relay-Reply message to a relay agent containing a message that the relay agent delivers to a client. The Relay-Reply message may be relayed by other relay agents for delivery to the destination relay agent. The server encapsulates the client message as an option in the Relay-Reply message, which the relay agent extracts and relays to the client.

7.11 IPv6 and Related Protocols (Details)

We introduced a number of basic IPv6 concepts in previous sections. The sections that follow focus on a more formal description of Internet Protocol version 6 (IPv6). The discussion is based on IETF RFC 2460 [DEE199801]. There is an extensive body of technical research literature on this topic.

IPv6 is a new version of the Internet Protocol, designed as the successor to IP version 4 (IPv4) described in RFC 791. The changes from IPv4 to IPv6 fall primarily into the following categories:

■ Expanded Addressing Capabilities: IPv6 increases the IP address size from 32 bits to 128 bits to support more levels of addressing hierarchy, a much greater number of addressable nodes, and simpler autoconfiguration addressing scheme. The scalability of multicast routing is improved by adding a "scope" field to multicast addresses and a new type of address called an "anycast address" is defined, used to send a packet to any one of a group of nodes.

- Header Format Simplification. Some IPv4 header fields have been dropped or made optional to reduce the common-case processing cost of packet handling and to limit the bandwidth cost of the IPv6 header.
- Improved Support for Extensions and Options. Changes in the way IP header options are encoded allow for more efficient forwarding, less stringent limits on the length of options, and greater flexibility for introducing new options in the future.
- Flow Labeling Capability. A new capability is added to enable the labeling of packets belonging to particular traffic "flows" for which the sender requests special handling, such as nondefault quality of service or "real-time" service.
- Authentication and Privacy Capabilities. Extensions to support authentication, data integrity, and (optional) data confidentiality are specified for IPv6.

RFC 2460 specifies the basic IPv6 header and the initially defined IPv6 extension headers and options. It also discusses packet size issues, the semantics of flow labels and traffic classes, and the effects of IPv6 on upper-layer protocols. The format and semantics of IPv6 addresses are specified separately in RFC 2373 (now obsoleted by RFC 3513). The IPv6 version of ICMP, which all IPv6 implementations are required to include, is specified in ICMPv6 (RFC 2483). Developers should refer directly to all relevant IETF RFCs for normative guidelines.

The following nomenclature is used in the standard:

- Node—a device that implements IPv6.
- Router—a node that forwards IPv6 packets not explicitly addressed to itself. (See Note below).
- Host—any node that is not a router. (See Note below.)
- Upper layer—a protocol layer immediately above IPv6. Examples are transport protocols such as TCP and UDP, control protocols such as ICMP, routing protocols such as OSPF, and internet or lower-layer protocols being "tunneled" over (i.e., encapsulated in) IPv6.
- Link—a communication facility or medium over which nodes can communicate at the link layer, that is, the layer immediately below IPv6. Examples are Ethernets (simple or bridged), PPP links, X.25, Frame Relay, or ATM networks, and Internet (or higher) layer "tunnels," such as tunnels over IPv4 or IPv6 itself.
- Neighbors—nodes attached to the same link.
- Interface—a node's attachment to a link.
- Address—an IPv6-layer identifier for an interface or a set of interfaces.
- Packet—an IPv6 header plus payload.
- Link MTU—the maximum transmission unit, that is, maximum packet size in octets that can be conveyed over a link.
- Path MTU—the minimum link MTU of all the links in a path between a source node and a destination node.

Note: it is possible, though unusual, for a device with multiple interfaces to be configured to forward non-self-destined packets arriving from some set (fewer than all) of its interfaces, and to discard non-self-destined packets arriving from its other interfaces. Such a device must obey the protocol requirements for routers when receiving packets from, and interacting with neighbors, the former (forwarding) interfaces. It must obey the protocol requirements for hosts when receiving packets from, and interacting with neighbors over, the latter (nonforwarding) interfaces.

7.12 IPv6 Header Format

Figure 7.14 depicts the IPv6 Header format.

The fields in the header have the following meanings:

Version: 4-bit Internet Protocol version number = 6.

Traffic Class: 8-bit traffic class field.

Flow Label: 20-bit flow label.

Payload Length: 16-bit unsigned integer. Length of the IPv6 payload, that is, the rest of the packet following this IPv6 header, in octets. (Note that any extension headers present are considered part of the payload, that is, included in the length count.)

Next Header: 8-bit selector. Identifies the type of header immediately following the IPv6 header. Uses the same values as the IPv4 Protocol field.

Hop Limit: 8-bit unsigned integer. Decremented by 1 by each node that forwards the packet. The packet is discarded if Hop Limit is decremented to zero.

Source Address: 128-bit address of the originator of the packet. This is covered later in more detail.

Destination Address: 128-bit address of the intended recipient of the packet (possibly not the ultimate recipient, if a Routing header is present).

```
+-+-+-+-+-+-+-+-+-+-+-+-+-+-+-+-+-+-+-+-+-+-+-+-+-+-+-+-+-+-+-+-+
|Version| Traffic Class |              Flow Label               |
+-+-+-+-+-+-+-+-+-+-+-+-+-+-+-+-+-+-+-+-+-+-+-+-+-+-+-+-+-+-+-+-+
|         Payload Length        |  Next Header  |   Hop Limit   |
+-+-+-+-+-+-+-+-+-+-+-+-+-+-+-+-+-+-+-+-+-+-+-+-+-+-+-+-+-+-+-+-+
|                                                               |
+                                                               +
|                                                               |
+                        Source Address                         +
|                                                               |
+                                                               +
|                                                               |
+-+-+-+-+-+-+-+-+-+-+-+-+-+-+-+-+-+-+-+-+-+-+-+-+-+-+-+-+-+-+-+-+
|                                                               |
+                                                               +
|                                                               |
+                      Destination Address                      +
|                                                               |
+                                                               +
|                                                               |
+-+-+-+-+-+-+-+-+-+-+-+-+-+-+-+-+-+-+-+-+-+-+-+-+-+-+-+-+-+-+-+-+
```

Figure 7.14 IPv6 header format.

7.13 IPv6 Extension Headers

In IPv6, optional Internet-layer information is encoded in separate headers that may be placed between the IPv6 header and the upper-layer header in a packet. There are a small number of such extension headers, each identified by a distinct Next Header value. As illustrated in the examples of Figure 7.15, an IPv6 packet may carry zero, one, or more extension headers, each identified by the Next Header field of the preceding header:

With one exception, extension headers are not examined or processed by any node along a packet's delivery path, until the packet reaches the node (or each of the set of nodes, in the case of multicast) identified in the Destination Address field of the IPv6 header. There, normal demultiplexing on the Next Header field of the IPv6 header invokes the module to process the first extension header, or the upper-layer header if no extension header is present. The contents and semantics of each extension header determine whether or not to proceed to the next header. Therefore, extension headers must be processed strictly in the order they appear in the packet. A receiver must not, for example, scan through a packet looking for a particular kind of extension header and process that header prior to processing all preceding ones.

The exception referred to in the preceding paragraph is the Hop-by-Hop Options header, which carries information that must be examined and processed by every node along a packet's delivery path, including the source and destination nodes. The Hop-by-Hop Options header, when present, must immediately follow the IPv6 header. Its presence is indicated by the value zero in the Next Header field of the IPv6 header.

If, as a result of processing a header, a node is required to proceed to the next header but the Next Header value in the current header is unrecognized by the node, it should discard the packet and send an ICMP Parameter Problem message to the source of the packet, with an ICMP Code value of 1 ("unrecognized Next Header type encountered") and the ICMP Pointer field containing the offset of the unrecognized value within the original packet. The same action should be taken if a node encounters a Next Header value of zero in any header other than an IPv6 header.

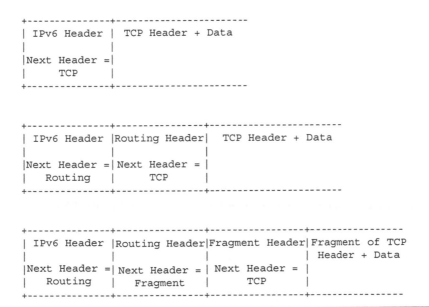

Figure 7.15   Examples of extension headers.

Each extension header is an integer multiple of 8 octets long, to retain 8-octet alignment for subsequent headers. Multi-octet fields within each extension header are aligned on their natural boundaries, that is, fields of width n octets are placed at an integer multiple of n octets from the start of the header, for $n = 1, 2, 4$, or 8.

A full implementation of IPv6 includes implementation of the following extension headers:

Hop-by-Hop Options
Routing (Type 0)
Fragment
Destination Options
Authentication
Encapsulating Security Payload

The first four are specified in this RFC; the last two are specified in RFC 2402 and RFC 2406, respectively.

7.13.1 Extension Header Order

When more than one extension header is used in the same packet, it is recommended that those headers appear in the following order:

IPv6 header
Hop-by-Hop Options header
Destination Options header (Note 1)
Routing header
Fragment header
Authentication header (Note 2)
Encapsulating Security Payload header (Note 2)
Destination Options header (Note 3)
Upper-layer header

Note 1: for options to be processed by the first destination that appears in the IPv6 Destination Address field plus subsequent destinations listed in the Routing header.

Note 2: additional recommendations regarding the relative order of the Authentication and Encapsulating Security Payload headers are given in RFC 2406.

Note 3: for options to be processed only by the final destination of the packet.

Each extension header should occur at most once, except for the Destination Options header which should occur at most twice (once before a Routing header and once before the upper-layer header).

If the upper-layer header is another IPv6 header (in the case of IPv6 being tunneled over or encapsulated in IPv6), it may be followed by its own extension headers, which are separately subject to the same ordering recommendations.

If and when other extension headers are defined, their ordering constraints relative to the above listed headers must be specified.

IPv6 nodes must accept and attempt to process extension headers in any order and occurring any number of times in the same packet, except for the Hop-by-Hop Options header which is restricted to appear immediately after an IPv6 header only.

```
+-+-+-+-+-+-+-+-+-+-+-+-+-+-+-+-+- - - - - - - -
| Option Type |Opt Data Len| Option Data
+-+-+-+-+-+-+-+-+-+-+-+-+-+-+-+-+- - - - - - - -
```

Figure 7.16 Extension headers options.

7.13.2 *Options*

Two of the currently-defined extension headers—the Hop-by-Hop Options header and the Destination Options header—carry a variable number of type-length-value (TLV) encoded "options," of the format shown in Figure 7.16.

Option Type: 8-bit identifier of the type of option.
Opt Data Len: 8-bit unsigned integer. Length of the Option Data field of this option, in octets.
Option Data: Variable-length field. Option-Type-specific data.

The sequence of options within a header must be processed strictly in the order they appear in the header; a receiver must not, for example, scan through the header looking for a particular kind of option and process that option prior to processing all preceding ones.

The Option Type identifiers are internally encoded such that their highest-order two bits specify the action that must be taken if the processing IPv6 node does not recognize the Option Type:

- 00—skip over this option and continue processing the header.
- 01—discard the packet.
- 10—discard the packet and, regardless of whether or not the packet's Destination Address was a multicast address, send an ICMP Parameter Problem, Code 2, message to the packet's Source Address, pointing to the unrecognized Option Type.
- 11—discard the packet and, only if the packet's Destination Address was not a multicast address, send an ICMP Parameter Problem, Code 2, message to the packet's Source Address, pointing to the unrecognized Option Type.

The third-highest-order bit of the Option Type specifies whether or not the Option Data of that option can change enroute to the packet's final destination. When an Authentication header is present in the packet, for any option whose data may change en route, its entire Option Data field must be treated as zero-valued octets when computing or verifying the packet's authenticating value.

- 0—Option Data does not change en route
- 1—Option Data may change en route

The three high-order bits described above are to be treated as part of the Option Type, not independent of the Option Type. That is, a particular option is identified by a full 8-bit Option Type, not just the low-order 5 bits of an Option Type.

The same Option Type numbering space is used for both the Hop-by-Hop Options header and the Destination Options header. However, the specification of a particular option may restrict its use to only one of those two headers.

Individual options may have specific alignment requirements, to ensure that multi-octet values within Option Data fields fall on natural boundaries. The alignment requirement of an option is specified using the notation $xn + y$, meaning the Option Type must appear at an integer multiple of x octets from the start of the header, plus y octets. For example:

$2n$ means any 2-octet offset from the start of the header;

$8n + 2$ means any 8-octet offset from the start of the header, plus 2 octets.

There are two padding options which are used when necessary to align subsequent options and to pad out the containing header to a multiple of 8 octets in length. These padding options must be recognized by all IPv6 implementations:

Pad1 option (alignment requirement: none)

```
+-+-+-+-+-+-+-+-+
|       0       |
+-+-+-+-+-+-+-+-+
```

Note: the format of the Pad1 option is a special case—it does not have length and value fields.

The Pad1 option is used to insert one octet of padding into the Options area of a header. If more than one octet of padding is required, the PadN option, described next, should be used, rather than multiple Pad1 options.

PadN option (alignment requirement: none)

The PadN option is used to insert two or more octets of padding into the Options area of a header. For N octets of padding, the Opt Data Len field contains the value N-2, and the Option Data consists of N-2 zero-valued octets.

7.13.3 Hop-by-Hop Options Header

The Hop-by-Hop Options header is used to carry optional information that must be examined by every node along a packet's delivery path. The Hop-by-Hop Options header is identified by a Next Header value of 0 in the IPv6 header, and has the format of Figure 7.17.

```
+-+-+-+-+-+-+-+-+-+-+-+-+-+-+-+-+-+-+-+-+-+-+-+-+-+-+-+-+-+-+-+-+
| Next Header   | Hdr Ext Len   |                               |
+-+-+-+-+-+-+-+-+-+-+-+-+-+-+-+-+                               +
|                                                               |
.                                                               .
.                            Options                            .
.                                                               .
|                                                               |
+-+-+-+-+-+-+-+-+-+-+-+-+-+-+-+-+-+-+-+-+-+-+-+-+-+-+-+-+-+-+-+-+
```

Figure 7.17 Hop-by-Hop Options header.

The fields are as follows:

Next Header: 8-bit selector. Identifies the type of header immediately following the Hop-by-Hop Options header. Uses the same values as the IPv4 Protocol field [RFC-1700 et seq.].

Hdr Ext Len: 8-bit unsigned integer. Length of the Hop-by-Hop Options header in 8-octet units, not including the first 8 octets.

Options: Variable-length field, of length such that the complete Hop-by-Hop Options header is an integer multiple of 8 octets long. Contains one or more TLV-encoded options.

7.13.4 *Routing Header*

The Routing header is used by an IPv6 source to list one or more intermediate nodes to be "visited" on the way to a packet's destination. This function is very similar to IPv4's Loose Source and Record Route option. The Routing header is identified by a Next Header value of 43 in the immediately preceding header, and has the format of Figure 7.18.

The fields are as follows:

Next Header: 8-bit selector. Identifies the type of header immediately following the Routing header. Uses the same values as the IPv4 Protocol field.

Hdr Ext Len: 8-bit unsigned integer. Length of the Routing header in 8-octet units, not including the first 8 octets.

Routing Type: 8-bit identifier of a particular Routing header variant.

Segments Left: 8-bit unsigned integer. Number of route segments remaining, that is, number of explicitly listed intermediate nodes still to be visited before reaching the final destination.

Type-Specific Data: Variable-length field, of format determined by the Routing Type, and of length such that the complete Routing header length is an integral multiple of 8 octets.

If, while processing a received packet, a node encounters a Routing header with an unrecognized Routing Type value, the required behavior of the node depends on the value of the Segments Left field, as follows:

If Segments Left is zero, the node must ignore the Routing header and proceed to process the next header in the packet, whose type is identified by the Next Header field in the Routing header.

If Segments Left is nonzero, the node must discard the packet and send an ICMP Parameter Problem, Code 0, message to the packet's Source Address, pointing to the unrecognized Routing Type.

If, after processing a Routing header of a received packet, an intermediate node determines that the packet is to be forwarded onto a link whose link MTU is less than the size of the packet,

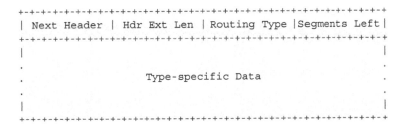

Figure 7.18 Routing header.

the node must discard the packet and send an ICMP Packet Too Big message to the packet's Source Address.

The Type 0 Routing header has the format shown in Figure 7.19.

The fields are as follows:

Next Header: 8-bit selector. Identifies the type of header immediately following the Routing header. Uses the same values as the IPv4 Protocol field (RFC-1700).

Hdr Ext Len: 8-bit unsigned integer. Length of the Routing header in 8-octet units, not including the first 8 octets. For the Type 0 Routing header, Hdr Ext Len is equal to two times the number of addresses in the header.

Routing Type: 0.

Segments Left: 8-bit unsigned integer. Number of route segments remaining, that is, number of explicitly listed intermediate nodes still to be visited before reaching the final destination.

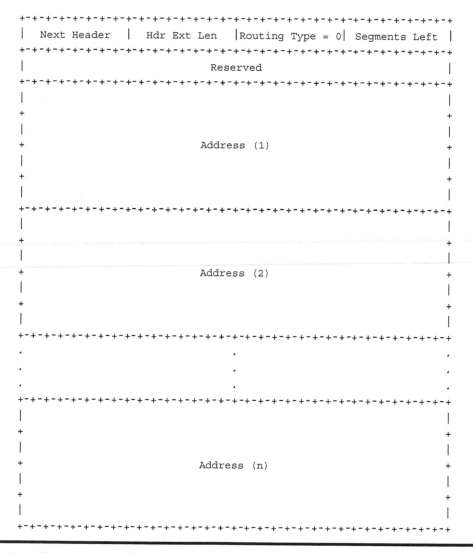

Figure 7.19 Type 0 Routing header.

Reserved: 32-bit reserved field. This is initialized to zero for transmission and ignored on reception.

Address[1, 2, ..., *n*]: Vector of 128-bit addresses, numbered 1 to *n*.

Multicast addresses must not appear in a Routing header of Type 0 or in the IPv6 Destination Address field of a packet carrying a Routing header of Type 0.

A Routing header is not examined or processed until it reaches the node identified in the Destination Address field of the IPv6 header. In that node, dispatching on the Next Header field of the immediately preceding header causes the Routing header module to be invoked, which, in the case of Routing Type 0, performs the following algorithm:

```
if Segments Left = 0 {
    proceed to process the next header in the packet, whose type is
    identified by the Next Header field in the Routing header
}
else if Hdr Ext Len is odd {
        send an ICMP Parameter Problem, Code 0, message to the Source
        Address, pointing to the Hdr Ext Len field, and discard the
        packet
}
else {
compute n, the number of addresses in the Routing header, by
dividing Hdr Ext Len by 2

if Segments Left is greater than n {
    send an ICMP Parameter Problem, Code 0, message to the Source
    Address, pointing to the Segments Left field, and discard the
    packet
}
else {
    decrement Segments Left by 1;
    compute i, the index of the next address to be visited in
    the address vector, by subtracting Segments Left from n
    if Address [i] or the IPv6 Destination Address is multicast {
        discard the packet
    }
    else {
        swap the IPv6 Destination Address and Address[i]
        if the IPv6 Hop Limit is less than or equal to 1 {
            send an ICMP Time Exceeded -- Hop Limit Exceeded in
            Transit message to the Source Address and discard the
            packet
        }
        else {
            decrement the Hop Limit by 1
            resubmit the packet to the IPv6 module for transmission
            to the new destination
        }
    }
  }
 }
}
```

As an example of the effects of the above algorithm, consider the case of a source node S sending a packet to destination node D, using a Routing header to cause the packet to be routed via intermediate nodes I1, I2, and I3. The values of the relevant IPv6 header and Routing header fields on each segment of the delivery path would be as follows:

As the packet travels from S to I1:

```
Source Address = S                Hdr Ext Len = 6
Destination Address = I1          Segments Left = 3
                                  Address[1] = I2
                                  Address[2] = I3
                                  Address[3] = D
```
As the packet travels from I1 to I2:
```
Source Address = S                Hdr Ext Len = 6
Destination Address = I2          Segments Left = 2
                                  Address[1] = I1
                                  Address[2] = I3
                                  Address[3] = D
```
As the packet travels from I2 to I3:
```
Source Address = S                Hdr Ext Len = 6
Destination Address = I3          Segments Left = 1
                                  Address[1] = I1
                                  Address[2] = I2
                                  Address[3] = D
```
As the packet travels from I3 to D:
```
Source Address = S                Hdr Ext Len = 6
Destination Address = D           Segments Left = 0
                                  Address[1] = I1
                                  Address[2] = I2
                                  Address[3] = I3
```

7.13.5 Fragment Header

The Fragment header is used by an IPv6 source to send a packet larger than would fit in the path MTU to its destination. (Note: Unlike IPv4, fragmentation in IPv6 is performed only by source nodes, not by routers along a packet's delivery path.) The Fragment header is identified by a Next Header value of 44 in the immediately preceding header, and has the format shown in Figure 7.20.

The fields are as follows:

Next Header: 8-bit selector. Identifies the initial header type of the Fragmentable Part of the original packet (defined below). Uses the same values as the IPv4 Protocol field (RFC-1700).

```
+-+-+-+-+-+-+-+-+-+-+-+-+-+-+-+-+-+-+-+-+-+-+-+-+-+-+-+-+-+-+
| Next Header |   Reserved   |    Fragment Offset   |Res|M|
+-+-+-+-+-+-+-+-+-+-+-+-+-+-+-+-+-+-+-+-+-+-+-+-+-+-+-+-+-+-+
|                       Identification                     |
+-+-+-+-+-+-+-+-+-+-+-+-+-+-+-+-+-+-+-+-+-+-+-+-+-+-+-+-+-+-+
```

Figure 7.20   Fragment header.

Reserved: 8-bit reserved field. This field is initialized to zero for transmission and ignored on reception.

Fragment Offset: 13-bit unsigned integer. The offset, in 8-octet units, of the data following this header, relative to the start of the Fragmentable Part of the original packet.

Res: 2-bit reserved field. This field is initialized to zero for transmission and ignored on reception.

M flag: 1 = more fragments; 0 = last fragment.

Identification: 32 bits. See description below.

To send a packet that is too large to fit in the MTU of the path to its destination, a source node may divide the packet into fragments and send each fragment as a separate packet, to be reassembled at the receiver.

For every packet that is to be fragmented, the source node generates an Identification value. The Identification must be different than that of any other fragmented packet sent recently* with the same Source Address and Destination Address. If a Routing header is present, the Destination Address of concern is that of the final destination.

The initial, large, unfragmented packet is referred to as the "original packet," and it is considered to consist of two parts, as seen in Figure 7.21.

The Unfragmentable Part consists of the IPv6 header plus any extension headers that must be processed by nodes en route to the destination, that is, all headers up to and including the Routing header if present, else the Hop-by-Hop Options header if present, else no extension headers.

The Fragmentable Part consists of the rest of the packet, that is, any extension headers that need be processed only by the final destination node(s), plus the upper-layer header and data.

The Fragmentable Part of the original packet is divided into fragments, each, except possibly the last ("rightmost") one, being an integer multiple of 8 octets long. The fragments are transmitted in separate "fragment packets" as illustrated in Figure 7.22.

Each fragment packet is composed of:

(1) The Unfragmentable Part of the original packet, with the Payload Length of the original IPv6 header changed to contain the length of this fragment packet only (excluding the length of the IPv6 header itself), and the Next Header field of the last header of the Unfragmentable Part changed to 44.

---

* "Recently" means within the maximum likely lifetime of a packet, including transit time from source to destination and time spent awaiting reassembly with other fragments of the same packet. However, it is not required that a source node know the maximum packet lifetime. Rather, it is assumed that the requirement can be met by maintaining the Identification value as a simple, 32-bit, "wrap-around" counter, incremented each time a packet must be fragmented. It is an implementation choice whether to maintain a single counter for the node or multiple counters, for example, one for each of the node's possible source addresses, or one for each active (source address, destination address) combination.

```
Original Packet:
+-----------------+-------------------//----------------------+
| Unfragmentable  |        Fragmentable                       |
|     Part        |            Part                           |
+-----------------+-------------------//----------------------+
```

Figure 7.21 Original packet.

(2) A Fragment header containing:

The Next Header value that identifies the first header of the Fragmentable Part of the original packet.

A Fragment Offset containing the offset of the fragment, in 8-octet units, relative to the start of the Fragmentable Part of the original packet. The Fragment Offset of the first ("leftmost") fragment is 0.

An M flag value of 0 if the fragment is the last ("rightmost") one, else an M flag value of 1.

The Identification value generated for the original packet.

(3) The fragment itself.

The lengths of the fragments must be chosen such that the resulting fragment packets fit within the MTU of the path to the packets' destination(s).

At the destination, fragment packets are reassembled into their original, unfragmented form, as illustrated in Figure 7.23.

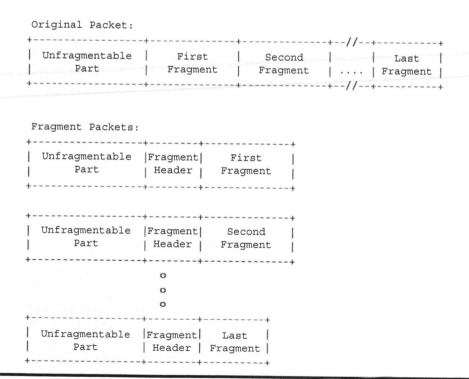

Figure 7.22 Fragmentable parts.

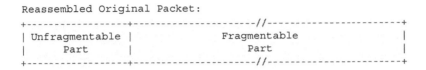

```
Reassembled Original Packet:

+-----------------+--------------------//----------------------+
| Unfragmentable  |      Fragmentable                           |
|     Part        |         Part                                |
+-----------------+--------------------//----------------------+
```

Figure 7.23   Reassembled original packet.

The following rules govern reassembly:

- An original packet is reassembled only from fragment packets that have the same Source Address, Destination Address, and Fragment Identification.
- The Unfragmentable Part of the reassembled packet consists of all headers up to, but not including, the Fragment header of the first fragment packet (that is, the packet whose Fragment Offset is zero), with the following two changes:

  The Next Header field of the last header of the Unfragmentable Part is obtained from the Next Header field of the first fragment's Fragment header.

  The Payload Length of the reassembled packet is computed from the length of the Unfragmentable Part and the length and offset of the last fragment. For example, a formula for computing the Payload Length of the reassembled original packet is:

$$PL.orig = PL.first - FL.first - 8 + (8 * FO.last) + FL.last$$

  where

  PL.orig = Payload Length field of reassembled packet.
  PL.first = Payload Length field of first fragment packet.
  FL.first = length of fragment following Fragment header of first fragment packet.
  FO.last = Fragment Offset field of Fragment header of last fragment packet.
  FL.last = length of fragment following Fragment header of last fragment packet.

- The Fragmentable Part of the reassembled packet is constructed from the fragments following the Fragment headers in each of the Fragment packets. The length of each fragment is computed by subtracting from the packet's Payload Length the length of the headers between the IPv6 header and fragment itself; its relative position in Fragmentable Part is computed from its Fragment Offset value.
- The Fragment header is not present in the final, reassembled packet.

The following error conditions may arise when reassembling fragmented packets:

- If insufficient fragments are received to complete reassembly of a packet within 60 s of the reception of the first-arriving fragment of that packet, reassembly of that packet must be abandoned and all the fragments that have been received for that packet must be discarded. If the first fragment (i.e., the one with a Fragment Offset of zero) has been received, an ICMP Time Exceeded—Fragment Reassembly Time Exceeded message should be sent to the source of that fragment.
- If the length of a fragment, as derived from the fragment packet's Payload Length field, is not a multiple of 8 octets and the M flag of that fragment is 1, then that fragment must be discarded and an ICMP Parameter Problem, Code 0, message should be sent to the source of the fragment, pointing to the Payload Length field of the fragment packet.

■ If the length and offset of a fragment are such that the Payload Length of the packet reassembled from that fragment would exceed 65,535 octets, then that fragment must be discarded and an ICMP Parameter Problem, Code 0, message should be sent to the source of the fragment, pointing to the Fragment Offset field of the fragment packet.

The following conditions are not expected to occur, but are not considered errors if they do:

■ The number and content of the headers preceding the Fragment header of different fragments of the same original packet may differ. Whatever headers are present, preceding the Fragment header in each fragment packet, are processed when the packets arrive, prior to queuing the fragments for reassembly. Only those headers in the Offset zero fragment packet are retained in the reassembled packet.

■ The Next Header values in the Fragment headers of different fragments of the same original packet may differ. Only the value from the Offset zero fragment packet is used for reassembly.

7.13.6 Destination Options Header

The Destination Options header is used to carry optional information that need be examined only by a packet's destination node(s). The Destination Options header is identified by a Next Header value of 60 in the immediately preceding header, and has the format shown in Figure 7.24.

Next Header: 8-bit selector. Identifies the type of header immediately following the Destination Options header. Uses the same values as the IPv4 Protocol field.

Hdr Ext Len: 8-bit unsigned integer. Length of the Destination Options header in 8-octet units, not including the first 8 octets.

Options: Variable-length field, of length such that the complete Destination Options header is an integer multiple of 8 octets long. Contains one or more TLV-encoded options.

Note that there are two possible ways to encode optional destination information in an IPv6 packet: either as an option in the Destination Options header or as a separate extension header. The Fragment header and the Authentication header are examples of the latter approach. Which approach can be used depends on what action is desired of a destination node that does not understand the optional information:

■ If the desired action is for the destination node to discard the packet and, only if the packet's Destination Address is not a multicast address, send an ICMP Unrecognized Type

```
+-+-+-+-+-+-+-+-+-+-+-+-+-+-+-+-+-+-+-+-+-+-+-+-+-+-+-+-+-+-+-+-+
| Next Header | Hdr Ext Len |                                  |
+-+-+-+-+-+-+-+-+-+-+-+-+-+-+-+                                  +
|                                                               |
.                                                               .
.                           Options                             .
.                                                               .
|                                                               |
+-+-+-+-+-+-+-+-+-+-+-+-+-+-+-+-+-+-+-+-+-+-+-+-+-+-+-+-+-+-+-+-+
```

Figure 7.24   Destination options header.

message to the packet's Source Address, then the information may be encoded either as a separate header or as an option in the Destination Options header whose Option Type has the value 11 in its highest-order two bits. The choice may depend on such factors as which takes fewer octets, or which yields better alignment or more efficient parsing.

■ If any other action is desired, the information must be encoded as an option in the Destination Options header whose Option Type has the value 00, 01, or 10 in its highest-order two bits, specifying the desired action.

7.13.7 No Next Header

The value 59 in the Next Header field of an IPv6 header or any extension header indicates that there is nothing following that header. If the Payload Length field of the IPv6 header indicates the presence of octets past the end of a header whose Next Header field contains 59, those octets must be ignored, and passed on unchanged if the packet is forwarded.

7.14 Packet Size Issues

IPv6 requires that every link in the internet have an MTU of 1280 octets or greater. On any link that cannot convey a 1280-octet packet in one piece, link-specific fragmentation, and reassembly must be provided at a layer below IPv6.

Links that have a configurable MTU (e.g., PPP links defined in RFC 1661) must be configured to have an MTU of at least 1280 octets. It is recommended that they be configured with an MTU of 1500 octets or greater, to accommodate possible encapsulations (i.e., tunneling) without incurring IPv6-layer fragmentation.

From each link to which a node is directly attached, the node must be able to accept packets as large as that link's MTU.

It has been recommended that IPv6 nodes implement Path MTU Discovery (RFC-1981), to discover and take advantage of path MTUs greater than 1280 octets. However, a minimal IPv6 implementation (e.g., in a boot ROM) may simply restrict itself to sending packets no larger than 1280 octets, and omit implementation of Path MTU Discovery.

To send a packet larger than a path's MTU, a node may use the IPv6 Fragment header to fragment the packet at the source and have it reassembled at the destinations. However, the use of such fragmentation is discouraged in any application that is able to adjust its packets to fit the measured path MTU (i.e., down to 1280 octets).

A node must be able to accept a fragmented packet that, after reassembly, is as large as 1500 octets. A node is permitted to accept fragmented packets that reassemble to more than 1500 octets. An upper-layer protocol or application that depends on IPv6 fragmentation to send packets larger than the MTU of a path should not send packets larger than 1500 octets unless it has assurance that the destination is capable of reassembling packets of that larger size.

In response to an IPv6 packet that is sent to an IPv4 destination (i.e., a packet that undergoes translation from IPv6 to IPv4), the originating IPv6 node may receive an ICMP Packet Too Big message reporting a Next-Hop MTU less than 1280. In that case, the IPv6 node is not required to reduce the size of subsequent packets to less than 1280, but must include a fragment header in those packets so that the IPv6-to-IPv4 translating router can obtain a suitable Identification value to use in resulting IPv4 fragments. Note that this means the payload may have to be reduced to 1232 octets (1280 minus 40 for the IPv6 header and 8 for the fragment header), and smaller still if additional extension headers are used.

7.15 Flow Labels

The 20-bit Flow Label field in the IPv6 header may be used by a source to label sequences of packets for which it requests special handling by the IPv6 routers, such as nondefault quality of service or "real-time" service. This aspect of IPv6 is still experimental to a large degree and subject to change as the requirements for flow support in the Internet become clearer (RFC 3697, March 2004, and RFC 3595, September 2003, provide some current thinking on the topic). Hosts or routers that do not support the functions of the Flow Label field are required to set the field to zero when originating a packet, pass the field on unchanged when forwarding a packet, and ignore the field when receiving a packet.

7.16 Traffic Classes

The 8-bit Traffic Class field in the IPv6 header is available for use by originating nodes and/or forwarding routers to identify and distinguish between different classes or priorities of IPv6 packets. There are a number of experiments under way in the use of the IPv4 Type of Service and/or Precedence bits to provide various forms of "differentiated service" for IP packets, other than through the use of explicit flow setup. The Traffic Class field in the IPv6 header is intended to allow similar functionality to be supported in IPv6.

The expectation is that experimentation will eventually lead to agreement on what sorts of traffic classifications are most useful for IP packets. Detailed definitions of the syntax and semantics of all or some of the IPv6 Traffic Class bits, whether experimental or intended for eventual standardization, are to be provided in separate documents.

The following general requirements apply to the Traffic Class field:

- The service interface to the IPv6 service within a node must provide a means for an upper-layer protocol to supply the value of the Traffic Class bits in packets originated by that upper-layer protocol. The default value must be zero for all 8 bits.
- Nodes that support a specific (experimental or eventual standard) use of some or all of the Traffic Class bits are permitted to change the value of those bits in packets that they originate, forward, or receive, as required for that specific use. Nodes should ignore and leave unchanged any bits of the Traffic Class field for which they do not support a specific use.
- An upper-layer protocol must not assume that the value of the Traffic Class bits in a received packet are the same as the value sent by the packet's source.

7.17 Upper-Layer Protocol Issues

7.17.1 Upper-Layer Checksums

Any transport or other upper-layer protocol that includes the addresses from the IP header in its checksum computation must be modified for use over IPv6, to include the 128-bit IPv6 addresses instead of 32-bit IPv4 addresses. In particular, Figure 7.25 shows the TCP and UDP "pseudo-header" for IPv6:

- If the IPv6 packet contains a Routing header, the Destination Address used in the pseudo-header is that of the final destination. At the originating node, that address will be in the last element of the Routing header; at the recipients, that address will be in the Destination Address field of the IPv6 header.

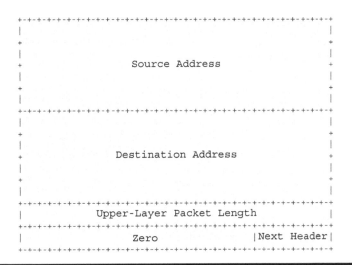

Figure 7.25   TCP and UDP "pseudo-header" for IPv6.

■ The Next Header value in the pseudo-header identifies the upper-layer protocol (e.g., 6 for TCP, or 17 for UDP). It will differ from the Next Header value in the IPv6 header if there are extension headers between the IPv6 header and the upper- layer header.

■ The Upper-Layer Packet Length in the pseudo-header is the length of the upper-layer header and data (e.g., TCP header plus TCP data). Some upper-layer protocols carry their own length information (e.g., the Length field in the UDP header); for such protocols, that is the length used in the pseudo-header. Other protocols (such as TCP) do not carry their own length information, in which case the length used in the pseudo-header is the Payload Length from the IPv6 header, minus the length of any extension headers present between the IPv6 header and the upper-layer header.

■ Unlike IPv4, when UDP packets are originated by an IPv6 node, the UDP checksum is not optional. That is, whenever originating a UDP packet, an IPv6 node must compute a UDP checksum over the packet and the pseudo-header, and, if that computation yields a result of zero, it must be changed to hex FFFF for placement in the UDP header. IPv6 receivers must discard UDP packets containing a zero checksum, and should log the error.

The IPv6 version of ICMP includes the above pseudo-header in its checksum computation; this is a change from the IPv4 version of ICMP, which does not include a pseudo-header in its checksum. The reason for the change is to protect ICMP from misdelivery or corruption of those fields of the IPv6 header on which it depends, which, unlike IPv4, are not covered by an IP-layer checksum. The Next Header field in the pseudo-header for ICMP contains the value 58, which identifies the IPv6 version of ICMP.

7.17.2 Maximum Packet Lifetime

Unlike IPv4, IPv6 nodes are not required to enforce maximum packet lifetime. That is the reason the IPv4 "Time to Live" field was renamed "Hop Limit" in IPv6. In practice, very few, if any, IPv4 implementations conform to the requirement that they limit packet lifetime, so this is not a change in practice. Any upper-layer protocol that relies on the internet layer (whether IPv4 or IPv6) to limit packet lifetime ought to be upgraded to provide its own mechanisms for detecting and discarding obsolete packets.

7.17.3 Maximum Upper-Layer Payload Size

When computing the maximum payload size available for upper-layer data, an upper-layer protocol must take into account the larger size of the IPv6 header relative to the IPv4 header. For example, in IPv4, TCP's Maximum Segment Size (MSS) option is computed as the maximum packet size (a default value or a value learned through Path MTU Discovery) minus 40 octets (20 octets for the minimum-length IPv4 header and 20 octets for the minimum-length TCP header). When using TCP over IPv6, the MSS must be computed as the maximum packet size minus 60 octets, because the minimum-length IPv6 header (i.e., an IPv6 header with no extension headers) is 20 octets longer than a minimum-length IPv4 header.

7.17.4 Responding to Packets Carrying Routing Headers

When an upper-layer protocol sends one or more packets in response to a received packet that included a Routing header, the response packets must not include a Routing header that was automatically derived by "reversing" the received Routing header unless the integrity and authenticity of the received Source Address and Routing header have been verified (e.g., via the use of an Authentication header in the received packet). In other words, only the following kinds of packets are permitted in response to a received packet bearing a Routing header.

- Response packets that do not carry Routing headers;
- Response packets that carry Routing headers that were NOT derived by reversing the Routing header of the received packet (e.g., a Routing header supplied by local configuration); and
- Response packets that carry Routing headers that were derived by reversing the Routing header of the received packet if and only if the integrity and authenticity of the Source Address and Routing header from the received packet have been verified by the responder.

7.18 Semantics and Usage of the Flow Label Field

A flow is a sequence of packets sent from a particular source to a particular (unicast or multicast) destination for which the source desires special handling by the intervening routers. The nature of that special handling might be conveyed to the routers by a control protocol, such as a resource reservation protocol, or by information within the flow's packets themselves, for example, in a Hop-by-Hop Option.

There may be multiple active flows from a source to a destination, as well as traffic that is not associated with any flow. A flow is uniquely identified by the combination of a source address and a nonzero Flow Label. Packets that do not belong to a flow carry a Flow Label of zero.

A Flow Label is assigned to a flow by the flow's source node. New Flow Labels must be chosen (pseudo-) randomly and uniformly from the range 1 to FFFFF hex. The purpose of the random allocation is to make any set of bits within the Flow Label field suitable for use as a hash key by routers, for looking up the state associated with the flow.

All packets belonging to the same flow must be sent with the same source address, destination address and flow label. If any of those packets includes a Hop-by-Hop Options header, then they all must be originated with the same Hop-by-Hop Options header contents (excluding the Next Header field of the Hop-by-Hop Options header). If any of those packets includes a Routing header, then they all must be originated with the same contents in all extension headers up to and including the Routing header (excluding the Next Header field in the Routing header). The routers or destinations are permitted, but not required, to verify that these conditions are satisfied. If a violation is detected, it should be reported to the source by an ICMP Parameter Problem message, Code 0, pointing to the high-order octet of the Flow Label field (i.e., offset 1 within the IPv6 packet).

The maximum lifetime of any flow-handling state established along a flow's path must be specified as part of the description of the state-establishment mechanism, for example, the resource reservation protocol or the flow-setup hop-by-hop option. A source must not reuse a Flow Label for a new flow within the maximum lifetime of any flow-handling state that might have been established for the prior use of that Flow Label.

When a node stops and restarts (e.g., as a result of a "crash"), it must be careful not to use a Flow Label that it might have used for an earlier flow whose lifetime may not have expired yet. This may be accomplished by recording flow label usage on stable storage so that it can be remembered across crashes, or by refraining from using any flow labels until the maximum lifetime of any possible previously established flows has expired. If the minimum time for rebooting the node is known, that time can be deducted from the necessary waiting period before starting to allocate flow labels.

There is no requirement that all, or even most, packets belong to flows, that is, carry nonzero Flow Labels. This observation is placed here to remind protocol designers and implementers not to assume otherwise. For example, it would be unwise to design a router whose performance would be adequate only if most packets belonged to flows, or to design a header compression scheme that only worked on packets that belonged to flows.

7.19 Formatting Guidelines for Options

This section addresses how to lay out the fields when designing new options to be used in the Hop-by-Hop Options header or the Destination Options header. These guidelines are based on the following assumptions:

- One desirable feature is that any multi-octet fields within the Option Data area of an option be aligned on their natural boundaries, that is, fields of width n octets should be placed at an integer multiple of n octets from the start of the Hop-by-Hop or Destination Options header, for $n = 1, 2, 4,$ or 8.
- Another desirable feature is that the Hop-by-Hop or Destination Options header take up as little space as possible, subject to the requirement that the header be an integer multiple of 8 octets long.
- It may be assumed that, when either of the option-bearing headers is present, it carries a very small number of options, usually only one.

These assumptions suggest the following approach to laying out the fields of an option: order the fields from smallest to largest, with no interior padding, then derive the alignment requirement for the entire option based on the alignment requirement of the largest field (up to a maximum alignment of 8 octets). This approach is illustrated in the following examples:

Example 1:

If an option X required two data fields, one of length 8 octets and one of length 4 octets, it would be laid out as follows:

```
                                  +-+-+-+-+-+-+-+-+-+-+-+-+-+-+-+-+
                                  | Option Type=X | Opt Data Len=12 |
 +-+-+-+-+-+-+-+-+-+-+-+-+-+-+-+-+-+-+-+-+-+-+-+-+-+-+-+-+-+-+-+-+
 |                        4-octet field                           |
 +-+-+-+-+-+-+-+-+-+-+-+-+-+-+-+-+-+-+-+-+-+-+-+-+-+-+-+-+-+-+-+-+
 |                                                                |
 +                        8-octet field                          +
 |                                                                |
 +-+-+-+-+-+-+-+-+-+-+-+-+-+-+-+-+-+-+-+-+-+-+-+-+-+-+-+-+-+-+-+-+
```

Its alignment requirement is 8n+2, to ensure that the 8-octet field starts at a multiple-of-8 offset from the start of the enclosing header. A complete Hop-by-Hop or Destination Options header containing this one option would look as follows:

```
 +-+-+-+-+-+-+-+-+-+-+-+-+-+-+-+-+-+-+-+-+-+-+-+-+-+-+-+-+-+-+-+-+
 | Next Header | Hdr Ext Len=1 | Option Type=X |Opt Data Len=12 |
 +-+-+-+-+-+-+-+-+-+-+-+-+-+-+-+-+-+-+-+-+-+-+-+-+-+-+-+-+-+-+-+-+
 |                        4-octet field                          |
 +-+-+-+-+-+-+-+-+-+-+-+-+-+-+-+-+-+-+-+-+-+-+-+-+-+-+-+-+-+-+-+-+
 |                                                                |
 +                        8-octet field                          +
 |                                                                |
 +-+-+-+-+-+-+-+-+-+-+-+-+-+-+-+-+-+-+-+-+-+-+-+-+-+-+-+-+-+-+-+-+
```

Example 2:

If an option Y required three data fields, one of length 4 octets, one of length 2 octets, and one of length 1 octet, it would be laid out as follows:

```
                                         +-+-+-+-+-+-+-+-+
                                         | Option Type=Y |
 +-+-+-+-+-+-+-+-+-+-+-+-+-+-+-+-+-+-+-+-+-+-+-+-+-+-+-+-+-+-+-+-+
 | Opt Data Len=7 | 1-octet field |        2-octet field         |
 +-+-+-+-+-+-+-+-+-+-+-+-+-+-+-+-+-+-+-+-+-+-+-+-+-+-+-+-+-+-+-+-+
 |                        4-octet field                          |
 +-+-+-+-+-+-+-+-+-+-+-+-+-+-+-+-+-+-+-+-+-+-+-+-+-+-+-+-+-+-+-+-+
```

Its alignment requirement is 4n+3, to ensure that the 4-octet field starts at a multiple-of-4 offset from the start of the enclosing header. A complete Hop-by-Hop or Destination Options header containing this one option would look as follows:

```
 +-+-+-+-+-+-+-+-+-+-+-+-+-+-+-+-+-+-+-+-+-+-+-+-+-+-+-+-+-+-+-+-+
 | Next Header | Hdr Ext Len=1 | Pad1 Option=0 | Option Type=Y |
 +-+-+-+-+-+-+-+-+-+-+-+-+-+-+-+-+-+-+-+-+-+-+-+-+-+-+-+-+-+-+-+-+
 | Opt Data Len=7 |1-octet field |       2-octet field           |
 +-+-+-+-+-+-+-+-+-+-+-+-+-+-+-+-+-+-+-+-+-+-+-+-+-+-+-+-+-+-+-+-+
```

```
+-+-+-+-+-+-+-+-+-+-+-+-+-+-+-+-+-+-+-+-+-+-+-+-+-+-+-+-+-+-+-+-+
|                          4-octet field                        |
+-+-+-+-+-+-+-+-+-+-+-+-+-+-+-+-+-+-+-+-+-+-+-+-+-+-+-+-+-+-+-+-+
| PadN Option=1 | Opt Data Len=2 |       0        |      0       |
+-+-+-+-+-+-+-+-+-+-+-+-+-+-+-+-+-+-+-+-+-+-+-+-+-+-+-+-+-+-+-+-+
```

Example 3:

A Hop-by-Hop or Destination Options header containing both options X and Y from Examples 1 and 2 would have one of the two following formats, depending on which option appeared first:

```
+-+-+-+-+-+-+-+-+-+-+-+-+-+-+-+-+-+-+-+-+-+-+-+-+-+-+-+-+-+-+-+-+
| Next Header | Hdr Ext Len=3 | Option Type=X | Opt Data Len=12 |
+-+-+-+-+-+-+-+-+-+-+-+-+-+-+-+-+-+-+-+-+-+-+-+-+-+-+-+-+-+-+-+-+
|                          4-octet field                        |
+-+-+-+-+-+-+-+-+-+-+-+-+-+-+-+-+-+-+-+-+-+-+-+-+-+-+-+-+-+-+-+-+
|                                                               |
+                          8-octet field                        +
|                                                               |
+-+-+-+-+-+-+-+-+-+-+-+-+-+-+-+-+-+-+-+-+-+-+-+-+-+-+-+-+-+-+-+-+
| PadN Option=1 | Opt Data Len=1 |       0       | Option Type=Y |
+-+-+-+-+-+-+-+-+-+-+-+-+-+-+-+-+-+-+-+-+-+-+-+-+-+-+-+-+-+-+-+-+
| Opt Data Len=7 | 1-octet field |       2-octet field          |
+-+-+-+-+-+-+-+-+-+-+-+-+-+-+-+-+-+-+-+-+-+-+-+-+-+-+-+-+-+-+-+-+
|                          4-octet field                        |
+-+-+-+-+-+-+-+-+-+-+-+-+-+-+-+-+-+-+-+-+-+-+-+-+-+-+-+-+-+-+-+-+
| PadN Option=1 | Opt Data Len=2 |       0       |       0       |
+-+-+-+-+-+-+-+-+-+-+-+-+-+-+-+-+-+-+-+-+-+-+-+-+-+-+-+-+-+-+-+-+
```

```
+-+-+-+-+-+-+-+-+-+-+-+-+-+-+-+-+-+-+-+-+-+-+-+-+-+-+-+-+-+-+-+-+
| Next Header   | Hdr Ext Len=3 | Pad1 Option=0 | Option Type=Y |
+-+-+-+-+-+-+-+-+-+-+-+-+-+-+-+-+-+-+-+-+-+-+-+-+-+-+-+-+-+-+-+-+
| Opt Data Len=7 | 1-octet field |       2-octet field          |
+-+-+-+-+-+-+-+-+-+-+-+-+-+-+-+-+-+-+-+-+-+-+-+-+-+-+-+-+-+-+-+-+
|                          4-octet field                        |
+-+-+-+-+-+-+-+-+-+-+-+-+-+-+-+-+-+-+-+-+-+-+-+-+-+-+-+-+-+-+-+-+
| PadN Option=1 | Opt Data Len=4 |       0       |       0       |
+-+-+-+-+-+-+-+-+-+-+-+-+-+-+-+-+-+-+-+-+-+-+-+-+-+-+-+-+-+-+-+-+
|       0       |       0       | Option Type=X | Opt Data Len=12|
+-+-+-+-+-+-+-+-+-+-+-+-+-+-+-+-+-+-+-+-+-+-+-+-+-+-+-+-+-+-+-+-+
|                          4-octet field                        |
+-+-+-+-+-+-+-+-+-+-+-+-+-+-+-+-+-+-+-+-+-+-+-+-+-+-+-+-+-+-+-+-+
|                                                               |
+                          8-octet field                        +
|                                                               |
+-+-+-+-+-+-+-+-+-+-+-+-+-+-+-+-+-+-+-+-+-+-+-+-+-+-+-+-+-+-+-+-+
```

7.20 Introduction to Addressing

This section defines the addressing architecture of the IPv6 protocol (RFC 3513). This expands some of the details already provided on this topic in the previous sections. The RFC includes the basic formats for the various types of IPv6 addresses (unicast, anycast, and multicast). This section covers the IPv6 addressing model, text representations of IPv6 addresses, and the definition of IPv6 unicast addresses, anycast addresses, and multicast addresses, based on the RFC [HIN200301]. The discussion is for pedagogical purposes and developers should refer to the latest IETF documentation.

7.21 IPv6 Addressing

IPv6 addresses are 128-bit identifiers for interfaces and sets of interfaces (where "interface" is as defined earlier). There are three types of addresses:

- Unicast: An identifier for a single interface. A packet sent to a unicast address is delivered to the interface identified by that address.
- Anycast: An identifier for a set of interfaces (typically belonging to different nodes). A packet sent to an anycast address is delivered to one of the interfaces identified by that address (the "nearest" one, according to the routing protocols' measure of distance).
- Multicast: An identifier for a set of interfaces (typically belonging to different nodes). A packet sent to a multicast address is delivered to all interfaces identified by that address.

There are no broadcast addresses in IPv6, their function being superseded by multicast addresses.

In RFC 3513, fields in addresses are given a specific name, for example, "subnet." When this name is used with the term "ID" for identifier after the name (e.g., "subnet ID"), it refers to the contents of the named field. When it is used with the term "prefix" (e.g., "subnet prefix"), it refers to all of the address from the left up to and including this field. In IPv6, all zeros and all ones are legal values for any field, unless specifically excluded. Specifically, prefixes may contain, or end with, zero-valued fields.

7.21.1 Addressing Model

IPv6 addresses of all types are assigned to interfaces, not nodes. An IPv6 unicast address refers to a single interface. Since each interface belongs to a single node, any of that node's interfaces' unicast addresses may be used as an identifier for the node.

All interfaces are required to have at least one link-local unicast address. A single interface may also have multiple IPv6 addresses of any type (unicast, anycast, and multicast) or scope. Unicast addresses with scope greater than link-scope are not needed for interfaces that are not used as the origin or destination of any IPv6 packets to or from non-neighbors. This is sometimes convenient for point-to-point interfaces. There is one exception to this addressing model:

> A unicast address or a set of unicast addresses may be assigned to multiple physical interfaces if the implementation treats the multiple physical interfaces as one interface when presenting it to the internet layer. This is useful for load-sharing over multiple physical interfaces.

Currently IPv6 continues the IPv4 model that a subnet prefix is associated with one link. Multiple subnet prefixes may be assigned to the same link.

7.21.2 Text Representation of Addresses

There are three conventional forms for representing IPv6 addresses as text strings:

1. The preferred form is x:x:x:x:x:x:x:x, where the "x's" are the hexadecimal values of the eight 16-bit pieces of the address.
   Examples:

   ```
   FEDC:BA98:7654:3210:FEDC:BA98:7654:3210
   1080:0:0:0:8:800:200C:417A
   ```

   Note that it is not necessary to write the leading zeros in an individual field, but there must be at least one numeral in every field (except for the case described in point 2.).
2. Due to some methods of allocating certain styles of IPv6 addresses, it will be common for addresses to contain long strings of zero bits. To make writing addresses containing zero bits easier, a special syntax is available to compress the zeros. The use of "::" indicates one or more groups of 16 bits of zeros. The "::" can only appear once in an address. The "::" can also be used to compress leading or trailing zeros in an address.
   For example, the following addresses:

   ```
   1080:0:0:0:8:800:200C:417A       a unicast address
   FF01:0:0:0:0:0:0:101             a multicast address
   0:0:0:0:0:0:0:1                  the loopback address
   0:0:0:0:0:0:0:0                  the unspecified addresses
   ```

   may be represented as:

   ```
   1080::8:800:200C:417A            a unicast address
   FF01::101                        a multicast address
   ::1                              the loopback address
   ::                               the unspecified addresses
   ```

3. An alternative form that is sometimes more convenient when dealing with a mixed environment of IPv4 and IPv6 nodes is x:x:x:x:x:x:d.d.d.d, where the "x's" are the hexadecimal values of the six high-order 16-bit pieces of the address, and the "d's" are the decimal values of the four low-order 8-bit pieces of the address (standard IPv4 representation).
   Examples:

   ```
   0:0:0:0:0:0:13.1.68.3
   0:0:0:0:0:FFFF:129.144.52.38
   ```

   or in compressed form:

   ```
   ::13.1.68.3
   ::FFFF:129.144.52.38
   ```

7.21.3 *Text Representation of Address Prefixes*

The text representation of IPv6 address prefixes is similar to the way IPv4 addresses prefixes are written in CIDR notation. An IPv6 address prefix is represented by the notation:

```
ipv6-address/prefix-length
```

where

    ipv6-address is an IPv6 address in any of the notations listed earlier.

    prefix-length is a decimal value specifying how many of the leftmost contiguous bits of the address comprise the prefix.

For example, the following are legal representations of the 60-bit prefix 12AB00000000CD3 (hexadecimal):

```
12AB:0000:0000:CD30:0000:0000:0000:0000/60
12AB::CD30:0:0:0:0/60
12AB:0:0:CD30::/60
```

The following are *not* legal representations of the above prefix:

12AB:0:0:CD3/60	may drop leading zeros, but not trailing zeros, within any 16-bit chunk of the address
12AB::CD30/60	address to left of "/" expands to 12AB:0000:0000:0000:0000:000:0000:CD30
12AB::CD3/60	address to left of "/" expands to 12AB:0000:0000:0000:0000:000:0000:0CD3

When writing both a node address and a prefix of that node address (e.g., the node's subnet prefix), the two can combined as follows:

The node address	12AB:0:0:CD30:123:4567:89AB:CDEF
and its subnet number	12AB:0:0:CD30::/60

can be abbreviated as 12AB:0:0:CD30:123:4567:89AB:CDEF/60.

7.21.4 *Address Type Identification*

The type of an IPv6 address is identified by the high-order bits of the address, as follows:

Address type	Binary prefix	IPv6 notation
Unspecified	00...0 (128 bits)	::/128
Loopback	00...1 (128 bits)	::1/128
Multicast	11111111	FF00::/8
Link-local unicast	1111111010	FE80::/10
Site-local unicast[a]	1111111011	FEC0::/10
Global unicast	(everything else)	

[a] *Deprecated*

Anycast addresses are taken from the unicast address spaces (of any scope) and are not syntactically distinguishable from unicast addresses.

7.21.5 Unicast Addresses

IPv6 unicast addresses are aggregable with prefixes of arbitrary bit-length similar to IPv4 addresses under Classless Interdomain Routing.

There are several types of unicast addresses in IPv6, in particular global unicast, site-local unicast, and link-local unicast. There are also some special-purpose subtypes of global unicast, such as IPv6 addresses with embedded IPv4 addresses or encoded Network Service Access Point (NSAP) addresses.

IPv6 nodes may have considerable or little knowledge of the internal structure of the IPv6 address, depending on the role the node plays (for instance, host versus router). At a minimum, a node may consider that unicast addresses (including its own) have no internal structure:

```
|                              128 bits                              |
+-------------------------------------------------------------------+
|                            node address                           |
+-------------------------------------------------------------------+
```

A slightly sophisticated host (but still rather simple) may additionally be aware of subnet prefixes for the links it is attached to, where different addresses may have different values for *n:*

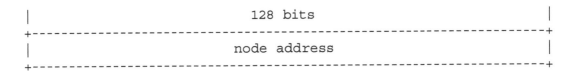

```
|              n bits                |         128-n bits        |
+------------------------------------+---------------------------+
|            subnet prefix           |        interface ID       |
+------------------------------------+---------------------------+
```

Though a very simple router may have no knowledge of the internal structure of IPv6 unicast addresses, routers will more generally have knowledge of one or more of the hierarchical boundaries for the operation of routing protocols. The known boundaries will differ from router to router, depending on what positions the router holds in the routing hierarchy.

7.21.5.1 Interface Identifiers

Interface identifiers in IPv6 unicast addresses are used to identify interfaces on a link. They are required to be unique within a subnet prefix. It is recommended that the same interface identifier not be assigned to different nodes on a link. They may also be unique over a broader scope. In some cases an interface's identifier will be derived directly from that interface's link-layer address. The same interface identifier may be used on multiple interfaces on a single node, as long as they are attached to different subnets.

Note that the uniqueness of interface identifiers is independent of the uniqueness of IPv6 addresses.

For all unicast addresses, except those that start with binary value 000, Interface IDs are required to be 64 bits long and to be constructed in Modified EUI-64 format.

Modified EUI-64 format based Interface identifiers may have global scope when derived from a global token (e.g., IEEE 802 48-bit MAC or IEEE EUI-64 identifiers) or may have local scope where a global token is not available (e.g., serial links, tunnel endpoints, etc.) or where global tokens are undesirable (e.g., temporary tokens for privacy.

Modified EUI-64 format interface identifiers are formed by inverting the "u" bit (universal/local bit in IEEE EUI-64 terminology) when forming the interface identifier from IEEE EUI-64 identifiers. In the resulting Modified EUI-64 format the "u" bit is set to one (1) to indicate global scope, and it is set to zero (0) to indicate local scope. The first three octets in binary of an IEEE EUI-64 identifier are as follows:

```
 0               0  0              1  1              2
|0               7  8              5  6              3|
+------+------+------+------+------+------+
| cccc | ccug | cccc | cccc | cccc | cccc |
+------+------+------+------+------+------+
```

written in Internet standard bit order, where "u" is the universal/local bit, "g" is the individual/group bit, and "c" are the bits of the company_id.

The motivation for inverting the "u" bit when forming an interface identifier is to make it easy for system administrators to manually configure nonglobal identifiers when hardware tokens are not available. This is expected to be case for serial links, tunnel endpoints, etc. The alternative would have been for these to be of the form 0200:0:0:1, 0200:0:0:2, etc., instead of the much simpler 1, 2, etc.

The use of the universal/local bit in the Modified EUI-64 format identifier is to allow development of future technology that can take advantage of interface identifiers with global scope.

7.21.5.2 The Unspecified Address

The address 0:0:0:0:0:0:0:0 is called the *unspecified address*. It must never be assigned to any node. It indicates the absence of an address. One example of its use is in the Source Address field of any IPv6 packets sent by an initializing host before it has learned its own address.

The unspecified address must not be used as the destination address of IPv6 packets or in IPv6 Routing Headers. An IPv6 packet with a source address of unspecified must never be forwarded by an IPv6 router.

7.21.5.3 The Loopback Address

The unicast address 0:0:0:0:0:0:0:1 is called the *loopback address*. It may be used by a node to send an IPv6 packet to itself. It may never be assigned to any physical interface. It is treated as having link-local scope, and may be thought of as the link-local unicast address of a virtual interface (typically called "the loopback interface") to an imaginary link that goes nowhere.

The loopback address must not be used as the source address in IPv6 packets that are sent outside of a single node. An IPv6 packet with a destination address of loopback must never be sent outside of a single node and must never be forwarded by an IPv6 router. A packet received on an interface with destination address of loopback must be dropped.

7.21.5.4 Global Unicast Addresses

The general format for IPv6 global unicast addresses is as follows:

```
|           n bits           |    m bits    |      128-n-m bits        |
+----------------------------+--------------+--------------------------+
| global routing prefix      | subnet ID    |      interface ID        |
+----------------------------+--------------+--------------------------+
```

where the global routing prefix is a (typically hierarchically-structured) value assigned to a site (a cluster of subnets/links), the subnet ID is an identifier of a link within the site, and the interface ID is as defined earlier.

All global unicast addresses other than those that start with binary 000 have a 64-bit interface ID field (i.e., $n + m = 64$). Global unicast addresses that start with binary 000 have no such constraint on the size or structure of the interface ID field.

Examples of global unicast addresses that start with binary 000 are the IPv6 address with embedded IPv4 addresses and the IPv6 address containing encoded NSAP addresses.

7.21.5.5 IPv6 Addresses with Embedded IPv4 Addresses

The IPv6 transition mechanisms include a technique for hosts and routers to dynamically tunnel IPv6 packets over an IPv4 routing infrastructure. IPv6 nodes that use this technique are assigned special IPv6 unicast addresses that carry a global IPv4 address in the low-order 32 bits. This type of address is termed an "IPv4-compatible IPv6 address" and has the format:

```
|           80 bits          |      16      |         32 bits          |
+----------------------------+--------------+--------------------------+
| 0000...................0000 |     0000     |      IPv4 address         |
+----------------------------+--------------+--------------------------+
```

Note: The IPv4 address used in the "IPv4-compatible IPv6 address" must be a globally unique IPv4 unicast address.

A second type of IPv6 address which holds an embedded IPv4 address is also defined. This address type is used to represent the addresses of IPv4 nodes as IPv6 addresses. This type of address is termed an "IPv4-mapped IPv6 address" and has the format:

```
|           80 bits          |      16      |         32 bits          |
+----------------------------+--------------+--------------------------+
| 0000...................0000 |     FFFF     |      IPv4 address         |
+----------------------------+--------------+--------------------------+
```

7.21.5.6 Local-Use IPv6 Unicast Addresses

There are two types of local-use unicast addresses defined. These are link-local and site-local. The link-local is for use on a single link and the site-local is for use in a single site (recall that site-local addresses have been deprecated). Link-Local addresses have the following format:

```
|          10 bits          |   54 bits    |            64 bits            |
+---------------------------+--------------+-------------------------------+
|        1111111010         |      0       |         interface ID          |
+---------------------------+--------------+-------------------------------+
```

Link-local addresses are designed to be used for addressing on a single link for purposes such as automatic address configuration, neighbor discovery, or when no routers are present.

Routers must not forward any packets with link-local source or destination addresses to other links.

7.21.6 Anycast Addresses

An IPv6 anycast address is an address that is assigned to more than one interface (typically belonging to different nodes), with the property that a packet sent to an anycast address is routed to the "nearest" interface having that address, according to the routing protocols' measure of distance.

Anycast addresses are allocated from the unicast address space, using any of the defined unicast address formats. Thus, anycast addresses are syntactically indistinguishable from unicast addresses. When a unicast address is assigned to more than one interface, thus turning it into an anycast address, the nodes to which the address is assigned must be explicitly configured to know that it is an anycast address.

For any assigned anycast address, there is a longest prefix P of that address that identifies the topological region in which all interfaces belonging to that anycast address reside. Within the region identified by P, the anycast address must be maintained as a separate entry in the routing system (commonly referred to as a "host route"); outside the region identified by P, the anycast address may be aggregated into the routing entry for prefix P.

Note that in the worst case, the prefix P of an anycast set may be the null prefix, that is, the members of the set may have no topological locality. In that case, the anycast address must be maintained as a separate routing entry throughout the entire Internet, which presents a severe scaling limit on how many such "global" anycast sets may be supported. Therefore, it is expected that support for global anycast sets may be unavailable or very restricted.

One expected use of anycast addresses is to identify the set of routers belonging to an organization providing internet service. Such addresses could be used as intermediate addresses in an IPv6 Routing header, to cause a packet to be delivered via a particular service provider or sequence of service providers.

Some other possible uses are to identify the set of routers attached to a particular subnet, or the set of routers providing entry into a particular routing domain.

There is little experience with widespread, arbitrary use of internet anycast addresses, and some known complications and hazards when using them in their full generality. Until more experience has been gained and solutions are specified, the following restrictions are imposed on IPv6 anycast addresses:

■ An anycast address must not be used as the source address of an IPv6 packet.
■ An anycast address must not be assigned to an IPv6 host; that is, it may be assigned to an IPv6 router only.

7.21.6.1 Required Anycast Address

The Subnet-Router anycast address is predefined. Its format is as follows:

```
|            n bits              |          128-n bits            |
+-------------------------------+--------------------------------+
|          subnet prefix         |         00000000000000         |
+-------------------------------+--------------------------------+
```

The "subnet prefix" in an anycast address is the prefix which identifies a specific link. This anycast address is syntactically the same as a unicast address for an interface on the link with the interface identifier set to zero.

Packets sent to the Subnet-Router anycast address will be delivered to one router on the subnet. All routers are required to support the Subnet-Router anycast addresses for the subnets to which they have interfaces.

The subnet-router anycast address is intended to be used for applications where a node needs to communicate with any one of the set of routers.

7.21.7 Multicast Addresses

An IPv6 multicast address is an identifier for a group of interfaces (typically on different nodes). An interface may belong to any number of multicast groups. Multicast addresses have the following format:

```
|    8     |   4    |   4    |            112 bits                |
+----------+--------+--------+-----------------------------------+
| 11111111 |  flgs  |  scop  |             group ID              |
+----------+--------+--------+-----------------------------------+
```

binary 11111111 at the start of the address identifies the address as being a multicast address.

flgs is a set of 4 flags: |0|0|0|T|

> The high-order 3 flags are reserved, and must be initialized to 0.
> T = 0 indicates a permanently assigned ("well-known")
> multicast address, assigned by the Internet Assigned Number Authority (IANA).
> T = 1 indicates a nonpermanently assigned ("transient") multicast address.

scop is a 4-bit multicast scope value used to limit the scope of the multicast group. The values are:

> 0 reserved
> 1 interface-local scope
> 2 link-local scope
> 3 reserved
> 4 admin-local scope
> 5 site-local scope (deprecated)
> 6 (unassigned)
> 7 (unassigned)
> 8 organization-local scope
> 9 (unassigned)
> A (unassigned)
> B (unassigned)
> C (unassigned)

D (unassigned)
E global scope
F reserved

The **interface-local scope** spans only a single interface on a node, and is useful only for loopback transmission of multicast.

Link-local (and **site-local**) multicast scopes span the same topological regions as the corresponding unicast scopes.

Admin-local scope is the smallest scope that must be administratively configured, that is, not automatically derived from physical connectivity or other, non- multicast-related configuration.

Organization-local scope is intended to span multiple sites belonging to a single organization.

Scopes labeled **(unassigned)** are available for administrators to define additional multicast regions.
Group ID identifies the multicast group, either permanent or transient, within the given scope.

The "meaning" of a permanently assigned multicast address is independent of the scope value. For example, if the "NTP servers group" is assigned a permanent multicast address with a group ID of 101 (hex), then:

FF01:0:0:0:0:0:0:101 means all NTP servers on the same interface (i.e., the same node) as the sender.
FF02:0:0:0:0:0:0:101 means all NTP servers on the same link as the sender.
FF05:0:0:0:0:0:0:101 means all NTP servers in the same site as the sender.
FF0E:0:0:0:0:0:0:101 means all NTP servers in the internet.

Non-permanently assigned multicast addresses are meaningful only within a given scope.

Multicast addresses must not be used as source addresses in IPv6 packets or appear in any Routing header. Routers must not forward any multicast packets beyond of the scope indicated by the scop field in the destination multicast address.

Nodes must not originate a packet to a multicast address whose scop field contains the reserved value 0; if such a packet is received, it must be silently dropped. Nodes should not originate a packet to a multicast address whose scop field contains the reserved value F; if such a packet is sent or received, it must be treated the same as packets destined to a global (scop E) multicast address.

7.21.7.1 Predefined Multicast Addresses

The following well-known multicast addresses are predefined. The group ID's defined in this section are defined for explicit scope values.

Use of these group IDs for any other scope values, with the T flag equal to 0, is not allowed.

```
Reserved Multicast Addresses:      FF00:0:0:0:0:0:0:0
                                   FF01:0:0:0:0:0:0:0
                                   FF02:0:0:0:0:0:0:0
                                   FF03:0:0:0:0:0:0:0
                                   FF04:0:0:0:0:0:0:0
                                   FF05:0:0:0:0:0:0:0
                                   FF06:0:0:0:0:0:0:0
                                   FF07:0:0:0:0:0:0:0
                                   FF08:0:0:0:0:0:0:0
                                   FF09:0:0:0:0:0:0:0
                                   FF0A:0:0:0:0:0:0:0
```

```
FF0B:0:0:0:0:0:0:0
FF0C:0:0:0:0:0:0:0
FF0D:0:0:0:0:0:0:0
FF0E:0:0:0:0:0:0:0
FF0F:0:0:0:0:0:0:0
```

The above multicast addresses are reserved and shall never be assigned to any multicast group.

```
All Nodes Addresses:        FF01:0:0:0:0:0:0:1
                            FF02:0:0:0:0:0:0:1
```

The above multicast addresses identify the group of all IPv6 nodes, within scope 1 (interface-local) or 2 (link-local).

```
All Routers Addresses:      FF01:0:0:0:0:0:0:2
                            FF02:0:0:0:0:0:0:2
                            FF05:0:0:0:0:0:0:2
```

The above multicast addresses identify the group of all IPv6 routers, within scope 1 (interface-local), 2 (link-local), or 5 (site-local).

```
Solicited-Node Address:     FF02:0:0:0:0:1:FFXX:XXXX
```

Solicited-node multicast address are computed as a function of a node's unicast and anycast addresses. A solicited-node multicast address is formed by taking the low-order 24 bits of an address (unicast or anycast) and appending those bits to the prefix FF02:0:0:0:0:1:FF00::/104 resulting in a multicast address in the range

```
FF02:0:0:0:0:1:FF00:0000
```
to
```
FF02:0:0:0:0:1:FFFF:FFFF
```

For example, the solicited node multicast address corresponding to the IPv6 address 4037:: 01:800:200E: 8C6C is FF02::1:FF0E:8C6C. IPv6 addresses that differ only in the high-order bits, for example, due to multiple high-order prefixes associated with different aggregations, will map to the same solicited-node address thereby, reducing the number of multicast addresses a node must join.

A node is required to compute and join (on the appropriate interface) the associated Solicited-Node multicast addresses for every unicast and anycast address it is assigned.

7.21.8 A Node's Required Addresses

A host is required to recognize the following addresses as identifying itself:

- Its required Link-Local Address for each interface.
- Any additional Unicast and Anycast Addresses that have been configured for the node's interfaces (manually or automatically).
- The loopback address.
- The All-Nodes Multicast Addresses.
- The Solicited-Node Multicast Address for each of its unicast and anycast addresses.
- Multicast Addresses of all other groups to which the node belongs.

A router is required to recognize all addresses that a host is required to recognize, plus the following addresses as identifying itself:

- The Subnet-Router Anycast Addresses for all interfaces for which it is configured to act as a router.
- All other Anycast Addresses with which the router has been configured.
- The All-Routers Multicast Addresses.

7.22 IANA Considerations: The Initial Assignment of IPv6 Address Space

Allocation	Prefix (binary)	Fraction of address space
Unassigned (see Note 1 below)	0000 0000	1/256
Unassigned	0000 0001	1/256
Reserved for NSAP Allocation	0000 001	1/128 (RFC 1888)
Unassigned	0000 01	1/64
Unassigned	0000 1	1/32
Unassigned	0001	1/16
Global Unicast	001	1/8 (per RFC 2374)
Unassigned	010	1/8
Unassigned	011	1/8
Unassigned	100	1/8
Unassigned	101	1/8
Unassigned	110	1/8
Unassigned	1110	1/16
Unassigned	1111 0	1/32
Unassigned	1111 10	1/64
Unassigned	1111 110	1/128
Unassigned	1111 1110 0	1/512
Link—Local Unicast addresses	1111 1110 10	1/1024
Site—Local Unicast addresses	1111 1110 11	1/1024
Multicast addresses	1111 1111	1/256

Notes:

1. The "unspecified address," the "loopback address," and the IPv6 addresses with embedded IPv4 addresses are assigned out of the 0000 0000 binary prefix space.
2. For now, IANA should limit its allocation of IPv6 unicast address space to the range of addresses that start with binary value 001. The rest of the global unicast address space (approximately 85 percent of the IPv6 address space) is reserved for future definition and use, and is not to be assigned by IANA at this time.

7.23 Creating Modified EUI-64 Format Interface Identifiers

Depending on the characteristics of a specific link or node there are a number of approaches for creating Modified EUI-64 format interface identifiers. This appendix describes some of these approaches.

EUI is defined in IEEE, "Guidelines for 64-bit Global Identifier (EUI-64) Registration Authority," March 1997.

7.23.1 Links or Nodes with IEEE EUI-64 Identifiers

The only change needed to transform an IEEE EUI-64 identifier to an interface identifier is to invert the "u" (universal/local) bit. For example, a globally unique IEEE EUI-64 identifier of the form:
    identifier are as follows:

```
|0                1|1              3|3              4|4              6|
|0                5|6              1|2              7|8              3|
+-----------------+----------------+----------------+----------------+
| cccccc0gcccccccc | cccccccmmmmmmmm | mmmmmmmmmmmmmmmm | mmmmmmmmmmmmmmmm |
+-----------------+----------------+----------------+----------------+
```

where "c" are the bits of the assigned company_id, "0" is the value of the universal/local bit to indicate global scope, "g" is individual/group bit, and "m" are the bits of the manufacturer-selected extension identifier. The IPv6 interface identifier would be of the form:

```
|0                1|1              3|3              4|4              6|
|0                5|6              1|2              7|8              3|
+-----------------+----------------+----------------+----------------+
| cccccc1gcccccccc | cccccccmmmmmmmm | mmmmmmmmmmmmmmmm | mmmmmmmmmmmmmmmm |
+-----------------+----------------+----------------+----------------+
```

The only change is inverting the value of the universal/local bit.

7.23.2 Links or Nodes with IEEE 802 48-Bit MACs

EUI64 defines a method to create a IEEE EUI-64 identifier from an IEEE 48-bit MAC identifier. This is to insert two octets, with hexadecimal values of 0xFF and 0xFF, in the middle of the 48- bit MAC (between the company_id and vendor-supplied ID). For example, the 48-bit IEEE MAC with global scope:

```
|0                1|1              3|3              4|
|0                5|6              1|2              7|
+-----------------+-------------------+-------------------+
| cccccc0gcccccccc | cccccccmmmmmmmm | mmmmmmmmmmmmmmmm |
+-----------------+-------------------+-------------------+
```

where "c" are the bits of the assigned company_id, "0" is the value of the universal/local bit to indicate global scope, "g" is individual/group bit, and "m" are the bits of the manufacturer-selected extension identifier. The interface identifier would be of the form:

```
|0                1|1              3|3              4|4              6|
|0                5|6              1|2              7|8              3|
+-----------------+----------------+----------------+----------------+
| cccccc1gcccccccc | cccccccc11111111 | 11111110mmmmmmmm | mmmmmmmmmmmmmmmm |
+-----------------+----------------+----------------+----------------+
```

When IEEE 802 48bit MAC addresses are available (on an interface or a node), an implementation may use them to create interface identifiers due to their availability and uniqueness properties.

7.23.3 Links with Other Kinds of Identifiers

There are a number of types of links that have link layer interface identifiers other than IEEE EIU-64 or IEEE 802 48-bit MACs. Examples include LocalTalk and Arcnet. The method to create an Modified EUI-64 format identifier is to take the link identifier (e.g., the LocalTalk 8-bit node identifier) and zero fill it to the left. For example, a LocalTalk 8-bit node identifier of hexadecimal value 0x4F results in the following interface identifier:

```
|0                 1|1                 3|3                 4|4                 6|
|0                 5|6                 1|2                 7|8                 3|
+------------------+------------------+------------------+------------------+
| 0000000000000000 | 0000000000000000 | 0000000000000000 | 0000000001001111 |
+------------------+------------------+------------------+------------------+
```

Note that this results in the universal/local bit set to "0" to indicate local scope.

7.23.4 Links without Identifiers

There are a number of links that do not have any type of built-in identifier. The most common of these are serial links and configured tunnels. Interface identifiers must be chosen that are unique within a subnet-prefix.

When no built-in identifier is available on a link, the preferred approach is to use a global interface identifier from another interface or one which is assigned to the node itself. When using this approach, no other interface connecting the same node to the same subnet-prefix may use the same identifier.

If there is no global interface identifier available for use on the link the implementation needs to create a local-scope interface identifier. The only requirement is that it be unique within a subnet prefix. There are many possible approaches to select a subnet-prefix-unique interface identifier. These include:

- Manual configuration
- Node serial number
- Other node-specific token

The subnet-prefix-unique interface identifier should be generated in a manner that does not change after a reboot of a node or if interfaces are added or deleted from the node. The selection of the appropriate algorithm is link and implementation dependent.

7.24 64-Bit Global Identifier (EUI-64) Registration Authority

The IEEE-defined 64-bit extended unique identifier (EUI-64) is a concatenation of the 24-bit company_id value by the IEEE Registration Authority and a 40-bit extension identifier assigned by the organization with that company_id assignment. The IEEE administers the assignment of 24-bit *company_id* values. The assignments of these values are public, so that a user of an EUI-64 value can identify the manufacturer that provided any value. The IEEE/RAC has no control over the assignments of 40-bit extension identifiers and assumes no liability for assignments of duplicate EUI-64 identifiers assigned by manufacturers.

7.24.1 Application Restrictions

Given the minimal probability of consuming all the EUI-64 identifiers, the IEEE/RAC places minimal restrictions on their use within standards. However, if used within the context of an IEEE standard, the documentation shall be reviewed by the IEEE/RAC for correctness and clarity. The IEEE/RAC shall not otherwise restrict the use of EUI-64 identifiers within standards. If the EUI-64 is referenced within non-IEEE standards, there shall not be any reference to IEEE unless approved by the IEEE/RAC.

7.24.2 Distribution Restrictions

Given the minimal probability of consuming all the EUI-64 identifiers, the IEEE/RAC places minimal restrictions on their redistribution through third parties, as follows:

1. Allocation. The EUI-64 values shall be sold within electronically-readable parts; no more than one EUI-64 value shall be contained within each component that is manufactured.
2. Packaging. A component containing the EUI-64 value shall have a distinguishing characteristic (such as color or shape) to distinguish it from other commonly used identifier components.
3. Documentation. Readily available documentation.
4. Legal indemnification. Any organization producing EUI-64 values shall indemnify the IEEE for damages arising from duplicate number assignments.

The term EUI-64 is trademarked by the IEEE. Companies are allowed to use this term for commercial purposes, but only if their use of this term has been reviewed by the IEEE/RAC and the proposed products using the EUI-64 conform to these restrictions.

7.24.3 Application Documentation

As a condition for receiving a company_id assignment, a manufacturer of EUI-64 values accepts the following responsibilities:

1. This documentation shall be readily available (at no cost) to any purchaser of EUI-64 values.
2. The manufacturer's part specification should include an unambiguous description of how the EUI-64 value is accessed (pin and/or address descriptions).

7.24.4 Manufacturer-Assigned Identifiers

The manufacturer identifier assignment allows the assignee to generate approximately 1 trillion (10^{12}) unique EUI-64 values, by varying the last 40 bits. The IEEE intends not to assign another OUI/company_id value to a manufacturer of EUI-64 values until the manufacturer has consumed, in product, the preponderance (more than 90 percent) of this block of potential unique words. It is incumbent upon the manufacturer to ensure that large portions of the unique word block are not left unused in manufacturing.

7.25 More on Transition Approaches and Mechanisms

Although most technical aspects of IPv6 have been defined for some time, deployment of IPv6 is occurring gradually. Initially, IPv6 is being deployed within isolated islands with interconnectivity among the islands being achieved by the existing IPv4 infrastructure, and a number of transition mechanisms have been defined to interconnect such islands.

There is an additional need for support for IPv6 hosts and routers that need to interoperate with legacy IPv4 hosts and an overview of such mechanisms for this purpose is provided in RFC 2893. That RFC defines the following types of nodes with respect to the transition to IPv6.

IPv4-only node: A host or router that implements only IPv4. An IPv4-only node does not understand IPv6. The installed base of IPv4 hosts and routers are examples of IPv4-only nodes.

IPv6/IPv4 node: A host or router that implements both the IPv4 and IPv6 protocols.

IPv6-only node: A host or router that implements IPv6 and does not implement IPv4.

IPv6 node: Any host or router that implements IPv6. IPv6/IPv4 and IPv6-only nodes are both IPv6 nodes.

IPv4 node: Any host or router that implements IPv4. IPv6/IPv4 and IPv4-only nodes are both IPv4 nodes.

The RFC also defines the IPv4-compatible IPv6 address, for example, ::156.55.23.5. IPv4-compatible IPv6 addresses are used to implement a simple automatic tunneling mechanism.

In addition to connectivity issues at the IP layer, the transition to IPv6 is also not entirely transparent to the networking layers above IP. As discussed previously, IPv6 addresses are significantly longer in size than IPv4 addresses and thus will require a change in application programming interfaces (APIs) or service primitive parameters that include IP addresses. Applications must also be extended to select the appropriate protocol, IPv4 or IPv6, when a DNS lookup returns both types of addresses. In general, legacy applications written for IPv4 either need to be rewritten or amended to support IPv6. For example, the application layer file transfer protocol (FTP) embeds IP addresses in its protocol fields, and could thus require changes to both the client and server FTP applications.

The IETF has defined a number of specific mechanisms to assist in transitioning to IPv6. These mechanisms are generally classified as belonging to the following categories.

Dual-stack—The principal building block for transitioning is the *dual-stack* approach. Dual-stack nodes, as the name suggests, maintain two protocol stacks that operate in parallel and thus allow the end system or router to operate via either protocol. In end systems they enable both IPv4 and IPv6 capable applications to operate on the same node. Dualstack capabilities in routers allow handling of both IPv4 and IPv6 packet types.

Translation—Translation refers to the direct conversion of protocols (e.g., between IPv4 and IPv6) and may include transformation of both the protocol header and the protocol payload. Translation can occur at several layers in the protocol stack, including IP, transport, and application layers. Note that protocol translation can result in feature loss where there is no clear mapping between the features provided by translated protocols. For instance translation of an IPv6 header into an IPv4 header will lead to the loss of the IPv6 flow label and its accompanying functionality.

Tunneling (or encapsulation)—Tunneling is used to interconnect compatible networking nodes or domains across incompatible networks. It can be viewed technically as the transfer of a payload protocol data unit by an encapsulating carrier protocol. For IPv6 transition, the IPv6 protocol data unit is generally carried as the payload of an IPv4 packet. Encapsulation of the payload protocol data unit is performed at the tunnel entrance and de-encapsulation is performed at the tunnel exit point.

Note that a transition mechanism may employ techniques from more than one of these categories. For example, when an end system or router creates an IPv6 in an IPv4 tunnel this could be classified as both dual stack (having both an IPv4 and IPv6 address) and tunneling.

References

[6NE200501] 6NET, "D2.2.4: Final IPv4 to IPv6 Transition Cookbook for Organizational/ISP (NREN) and Backbone Networks," Version: 1.0 (4th February 2005), Project Number: IST-2001-32603, CEC Deliverable Number: 32603/UOS/DS/2.2.4/A1.

[ATT200801] AT&T, "Promotional Literature on IPv6, 2008," http://www.corp.att.com/gov/ipv6/

[DAV200201] J. Davies, *Understanding IPv6*, Microsoft Press, Redmond, WA, 2002.

[DEE199801] S. Deering, R. Hinden, "Internet Protocol, Version 6 (IPv6) Specification," RFC 2460, December 1998. (C) The Internet Society (1998). All Rights Reserved. This document and translations of it may be copied and furnished to others, and derivative works that comment on or otherwise explain it or assist in its implementation may be prepared, copied, published and distributed, in whole or in part, without restriction of any kind, provided that the above copyright notice and this paragraph are included on all such copies and derivative works.

[DEM200301] R. Desmeules, *Cisco Self-Study: Implementing IPv6 Networks (IPV6)*, Pearson Education, May 2003.

[DIR200801] Directorate-Generals Information Society "IPv6: Enabling the Information Society," European Commission Information Society, Europe Information Society Portal, February 18, 2008.

[DRO200301] R. Droms, Ed., J. Bound, B. Volz, T. Lemon, C. Perkins, M. Carney, Dynamic Host Configuration Protocol for IPv6 (DHCPv6), RFC 3315, July 2003. (C) The Internet Society (2003). All Rights Reserved. This document and translations of it may be copied and furnished to others, and derivative works that comment on or otherwise explain it or assist in its implementation may be prepared, copied, published and distributed, in whole or in part, without restriction of any kind, provided that the above copyright notice and this paragraph are included on all such copies and derivative works.

[ERT200401] E. Ertekin, C. Christou, "IPv6 Header Compression," North American IPv6 Summit, Santa Monica, CA, June 2004.

[GIL200001] R. Gilligan, E. Nordmark, Transition Mechanisms for IPv6 Hosts and Routers, RFC 2893, August 2000. (C) The Internet Society (2000). All Rights Reserved. This document and translations of it may be copied and furnished to others, and derivative works that comment on or otherwise explain it or assist in its implementation may be prepared, copied, published and distributed, in whole or in part, without restriction of any kind, provided that the above copyright notice and this paragraph are included on all such copies and derivative works.

[GON199801] M. Goncalves, K. Niles, *IPv6 Networks*, McGraw-Hill Osborne, New York, 1998.

[GOS200301] S. Goswami, *Internet Protocols: Advances, Technologies, and Applications*, Kluwer Academic Publishers, New York, May 2003.

[GRA200001] B. Graham, *TCP/IP Addressing: Designing and Optimizing your IP Addressing Scheme* (2nd edition), Morgan Kaufmann, San Francisco, 2000.

[HAG200201] S. Hagen, *IPv6 Essentials*, O'Reilly, Sebastopol, CA, 2002.

[HIN200301] R. Hinden, S. Deering, "Internet Protocol Version 6 (IPv6) Addressing Architecture" RFC 3513, April 2003. (C) The Internet Society (2003). All Rights Reserved. This document and translations of it may be copied and furnished to others, and derivative works that comment on or otherwise explain it or assist in its implementation may be prepared, copied, published and distributed, in whole or in part, without restriction of any kind, provided that the above copyright notice and this paragraph are included on all such copies and derivative works.

[HUI199701] C. Huitema, *IPv6 the New Internet Protocol* (2nd edition), Prentice Hall, Englewood Cliffs, NJ, 1997.

[HUI200401] C. Huitema, B. Carpenter, RFC3879: Deprecating Site Local Addresses, September 2004.

[IPV200401] IPv6Forum, "IPv6 Vendors Test Voice, Wireless and Firewalls on Moonv6," November 15, 2004, http://www. ipv6forum.com/

[IPV200501] IPv6 Portal, http://www.ipv6tf.org/meet/faqs.php

[ITO200401] J. Itojun Hagino, *IPv6 Network Programming*, Butterworth-Heinemann, Oxford, UK, 2004.

[LAD200601] L. Ladid, "European IPv6 Roadmap 2006 Recommendations," European IPv6 Task Force, IPv6 Task Force Steering Committee, IST-2-004572-CA IPv6 TF-SC Euro-v6-Roadmap, 8/10/2006.

[LEE200501] H. K. Lee, *Understanding IPv6*, Springer-Verlag, New York, 2005.

[LOS200301] P. Loshin, *IPv6: Theory, Protocol, and Practice* (2nd edition), Elsevier Science & Technology Books, New York, 2003.

[MIL199701] M. A. Miller, *Implementing IPv6: Migrating to the Next Generation Internet Protocol*, Wiley, New York, 1997.

[MIL200001] M. Miller, P. E. Miller, *Implementing IPv6: Supporting the Next Generation Internet Protocols* (2nd edition), Hungry Minds, New York, 2000.

[MIN200601] D. Minoli, *VoIP over IPv6*, Elsevier, New York, 2006.

[MSD200401] Microsoft Corporation, MSDN Library, Internet Protocol, 2004, http://msdn.microsoft.com.

[MUR200501] N. R. Murphy, D. Malone, *IPv6 Network Administration*, O'Reilly & Associates, Sebastopol, CA, 2005.

[SHI200501] M.-K. Shin, Ed., Y.-G. Hong, J. Hagino, P. Savola, E. M. Castro, "Application Aspects of IPv6 Transition," RFC 4038, March 2005.

[SOL200401] H. S. Soliman, *Mobile IPv6*, Pearson Education, Englewood Cliffs, NJ, 2004.

[SRI200101] P. Srisuresh, K. Egevang "Traditional IP Network Address Translator (Traditional NAT)," RFC 3022, January 2001.

[TEA200401] D. Teare, C. Paquet, "CCNP Self-Study: Advanced IP Addressing," Cisco Press, San Jose, CA, June 11, 2004.

[WEG199901] J. D. Wegner, *IP Addressing and Subnetting, Including IPv6*, 1999, Elsevier Science & Technology Books, New York.

Appendix 7A: Header Compression

Implementation of IPv6 raises concerns related to new packet headers because, as we have seen in this chapter, the packet header size for an IP datagram doubles from 20 bytes (IPv4) to at least 40 bytes (IPv6). Furthermore, incorporation of network-layer encryption mechanism (i.e., IPSec) nearly doubles the IP operational overhead. This predicament is undesirable for many wireless and satellite networks because they are typically bandwidth-constrained or bandwidth is (relatively) expensive. This issue also impacts delay-sensitive applications such as VoIP and IP-based video-conferencing; this is the reason Header Compression (HC) was deemed critical for VoIP even in an IPv4 environment. HC is, therefore, particularly important in IPv6-over-satellite environments. This appendix, based on reference [ERT200401] and used with permission, provides a short overview of this important topic.

HC methods that reduce the expanded overhead of IPv6 are able to increase user throughput and the number of users a network can support. Consider a case where two routers are connected using a line operating at 1 Mbps and where the packet payload is 20 bytes (constant). With the addition of the 40 byte IPv6 Header, 2083 packets can be sent per second transmitted across the link. Over a one second period about 666 kb transmitted is IPv6 overhead and about 333 kb transmitted is actual user data (i.e., 66 percent of data transmitted is overhead). If the header is compressed to 2 bytes, about 5680 packets sent per second; over a one second period, about 90 kb transmitted is IPv6 overhead and about 910 kb transmitted is actual user data (i.e., 9 percent of data transmitted is overhead). This shows that, under the best case scenario, HC can theoretically decrease header overhead by 95 percent. Obviously, the shorter the packets, the more onerous is the impact of the overhead. VoIP and video-over-IP use relatively short packers. Furthermore, studies show that 40 percent of packets that traverse the Internet are 40 bytes in length. It follows that in IPv6 the 40 byte header on a datagram carrying a 40-byte payload results in a situation where 50 percent of the bandwidth is consumed for overhead. The same studies show that the average packet length is around 355 bytes (Header and Payload).

Assuming that no header compression is applied, IP header overhead is calculated as follows:

Percentage Overhead = IP Header Bytes/Total Bytes Transmitted = 40 Bytes/375 Bytes = 10.67% overhead

Assuming that a header compression algorithm is applied to the ESP/IP header, overhead calculations can be made as follows (assuming header size is reduced to 2 bytes per packet):

Percentage Overhead = Compressed Header Bytes/Total Bytes Transmitted = 2 Bytes/337 Bytes = 0.59% overhead

Recognizing the need to reduce the size of IP headers, the IETF has led the development of HC algorithms. In these algorithms, compression is applied to Layer 3 (IP) and several Layer 4 protocol headers. HC solutions can reduce the additional overhead introduced by network-layer encryption mechanisms (e.g., IPSec). Compression algorithms instantiated on encryption/decryption devices have the ability to (a) compress inner headers before encryption; and (b) compress outer ESP/IP headers after encryption. See Figure 7A. HC algorithms exploit protocol inter-packet header field redundancies to improve overall efficiency.

Compression is applied over a link between a source node (i.e., compressor) and a destination node (i.e., decompressor). Both compressor and decompressor store header fields of each packet stream and associate each stream with a context identifier (CID). Upon receipt of a packet with an associated context, the compressor removes the IPv6 header fields from the packet header and appends a CID. Upon receipt of a packet with a CID, the decompressor inserts IPv6 header fields back into the packet header and transmits the packet. The reduction in the number of full headers transmitted can result in an overall decrease in overhead. See Figure 7B.

Two compression protocols have emerged from the IETF: (1) Internet Protocol Header Compression (IPHC), and (2) ROHC Working Group: Robust Header Compression (ROHC).

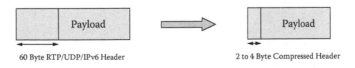

Figure 7A   Goal of compression algorithms in an IPv6 environment.

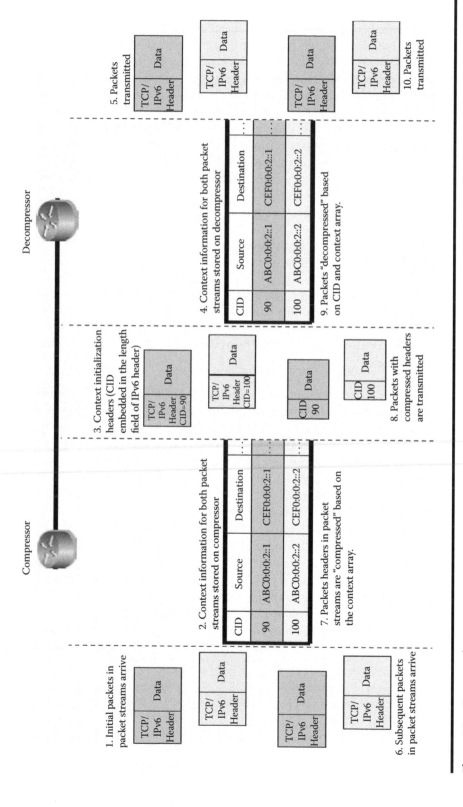

Figure 7B Compression process.

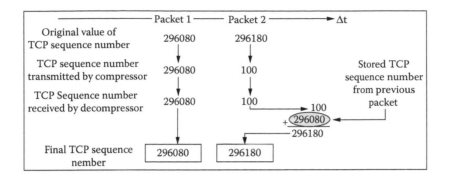

Figure 7C IPCH mechanisms.

- IPHC is an HC scheme designed for low bit error rate links. Compression profiles are defined in RFC 2507 and RFC 2508. IPHC supports compression of TCP/IP, UDP/IP, RTP/UDP/IP, and ESP/IP headers; "Enhanced" Compression of RTP/UDP/IP (ECRTP) headers is defined in RFC 3545.
- ROHC is an HC scheme designed for wireless links. It provides greater compression compared to IPHC at the cost of greater implementation complexity. ROHC is more suitable for high bit error rate (BER) and long round-trip time (RTT) links. Compression profiles are defined in RFC 3095 and RFC 3096. ROHC supports the compression of ESP/IP, UDP/IP, RTP/UDP/IP headers.

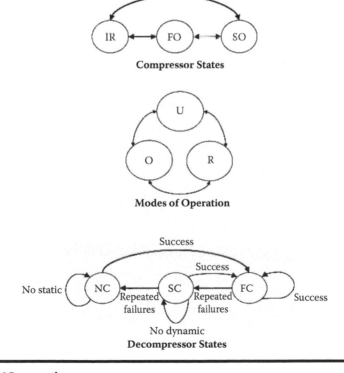

Figure 7D ROHC operation.

IPHC mechanisms facilitate compression of dynamic header fields (e.g., TCP sequence numbers.) Delta-based differential encoding mechanisms provide compression of incremental header fields, as depicted in Figure 7C. However, if sequential packets are lost, compressor¬-decompressor contexts will desynchronize; resynchronization depends on how quickly the compressor notices desynchonization; the compressor reestablishes context by transmitting a full header. IPHC algorithm performs poorly over high BER, long-RTT links.

ROHC provides improved header compression over IPHC in high BER and high RTT wireless links. These benefits, however, come at a cost, because ROHC implementation is significantly more complex than IPHC. The operation of ROHC is shown in Figure 7D (for details of state machines,see RFC 3095.) ROHC incorporates enhanced mechanisms compared to IPHC; for example, implementation flexibility (capability to compress headers with or without a feedback mechanism), and improved compression ratios (specifically, the ability to compress, e.g., TCP SYN and FIN messages and timestamps). In summary, it provides greater compression compared to IPHC at the cost of greater implementation complexity.

Note: Cisco Systems OS provides IPHC implementation; furthermore, both IPHC and ROHC are specified in Release 4 and Release 5 of the 3rd Generation Partnership Project (3GPP).

Chapter 8

Carrying IPv4, IPv6, and TCP over Satellite Links

This chapter looks at the issue of supporting upper-layer protocols over satellite links, particularly Transmission Control Protocol (TCP). At this juncture, TCP is the protocol used by most data applications, including Internet access; VoIP and video conferencing/streaming use User Datagram Protocol (UDP), while traditional video broadcast operates typically directly over the bearer channel. The performance of TCP over high-latency networks—for example, over a network using GEO satellites—has a direct impact on the perceived quality of such two-way communications.

As noted, some satellite applications, such as traditional (digital) video broadcasting, make use of satellite links as a PHY (physical) layer service, without explicit higher-layer protocols. Here, a transponder (or a portion of it) is used to support a digitally modulated signal that comprises a single or several multiplexed digital video channels. Some Data Link Layer framing such as MultiProtocol Encapsulation (MPE) and/or Moving Picture Expert Group (MPEG-2) Transport Stream (TS) may be employed in digital video transmission, but these mechanisms are typically used in a simplex mode, not in a duplex handshake mode. Heavy use is made of Forward Error Correction (FEC) techniques. Similarly, IPTV broadcasting over a satellite backbone and data casting may use network layer protocols (IPv4, IPv6), but also tend to use simplex links over the satellite backbone distribution network. However, there are many applications that require two-way duplex communications which make use of higher layer protocols requiring a two-way handshake, such as TCP ([RFC793], [RFC2581]), or protocols that make use of TCP (e.g., Hypertext Transfer protocol (HTTP), FTP, among others.) The more "chatty" an application is, the more visible the impact of TCP will be. Internet access via satellites, particularly GEO satellites, is a basic two-way application of interest.

We start the discussion with a brief overview of satellite-based communications at the IP layer and then examine approaches to support TCP effectively.

8.1 IP Networking

Typically, users employ a satellite channel as a PHY (physical) layer service, perhaps connecting a dispersed set of routers, for either an intranet or Internet access application. A certain amount of transponder bandwidth is allocated to the user—say, the entire 36 MHz at C-band or Ku-band—and some modulation scheme is used to achieve digital carriage—say, DVB-S2 to support 90 Mbps raw and 75 Mbps net (accounting for FEC). This bandwidth can be used to support one or more backbone trunks in the (core) network. Figure 8.1 depicts a basic "intranet" application where terrestrial links are substituted with satellite-based links; in this case, all links are substituted and all locations have satellite antennas.

Figure 8.2 depicts the same "intranet" application as Figure 8.1, but where terrestrial links are substituted with satellite-based links; in this case, all links are substituted, but the routers use

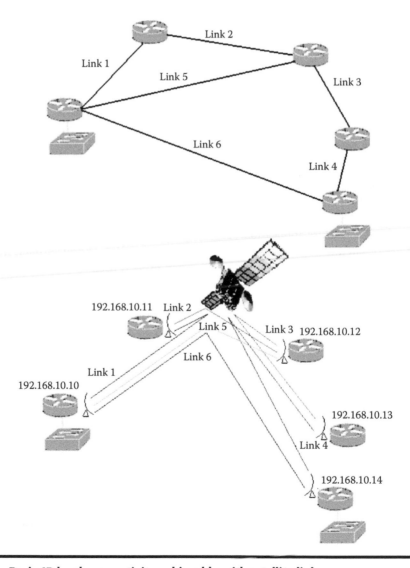

Figure 8.1   Basic IP-level connectivity achievable with satellite links.

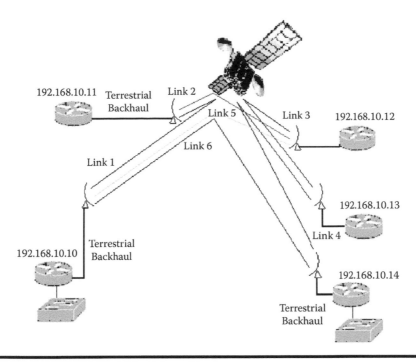

Figure 8.2   Use of backhaul to support IP-level connectivity achievable with satellite links.

some terrestrial links to reach the teleport location of a service provider, thereby obviating the need for satellite antennas at the customer location. Figure 8.3 depicts an example of a basic "intranet" application also with Internet access where terrestrial links are substituted with satellite-based links; in this case, all links are substituted and all locations have satellite antennas. Figure 8.4 shows a case where some links are replaced with satellite links, but others use terrestrial links. This could be the case when some routers are in one continent and the other routers are in another continent, or in the case where some routers are on the East Coast of the United States and the other routers are on the West Coast. Figure 8.5 depicts a future environment where routing decisions can be made (more efficiently) at the satellite level; this technology may become prevalent in the next decade. As an example in this arena, on March 29, 2007, a Cisco Systems router, flying in low Earth orbit onboard the UK-DMC satellite built by Surrey Satellite Technology Ltd (SSTL), was successfully configured by NASA Glenn Research Center to use IPSec and IPv6 technologies in space.

Finally, Figure 8.6 shows an example of the use of IPv6 for satellite-based access to remote wireless sensors networks. Given the large number of nodes involved in these types of applications, IPv6 addressability is highly desirable. There are many applications that (will increasingly) use networked sensors (e.g., see [SOH200701]). These sensors are invariably IP-enabled. Some high-end applications (e.g., military theater applications, homeland security/border control, even medical Body Area Networks) use a very large number of sensors (in aggregate), such that the use of IPv6 is highly desirable. The IETF recently started a Work Group (WG) called IPv6 over low power Wireless Personal Area Networks (6lowpan). As another somewhat related example, during 2004–2006 NATO sponsored the Interoperable Networks for Secure Communication (INSC) project which aimed at supporting theater MANETs (mobile ad networks) with native IPv6 services over a DVB-S-based satellite link. During Phase 2 (2006), vehicles are connected directly via satellite to the remote core network using DVB-S/DVB-RCS (Return Channel via Satellite) technology.

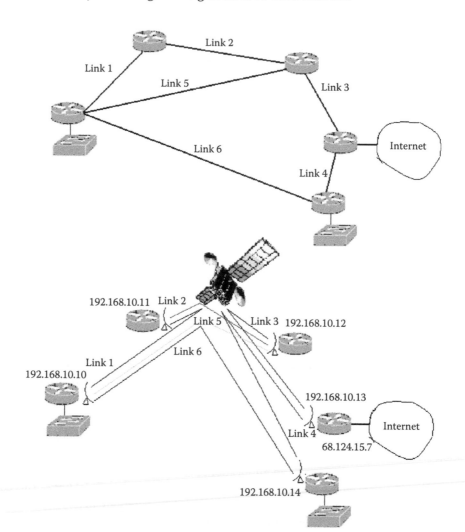

Figure 8.3 Internet access connectivity achievable with satellite links.

8.2 Very Small Aperture Terminal (VSAT) Systems

A very small aperture terminal (VSAT) is a self-contained, usually two-way earth station support-ing business voice/data/Internet communications. A VSAT system comprises the following:

■ A small-sized antenna (typically 1.2 to 2.4 m in diameter; see Figure 8.7 for an example)
■ Send and receive amplifiers
■ A satellite modem/router for medium-speed business communications almost invariably in an IP-networking environment
■ A supportive dedicated or shared satellite link
■ A central large-antenna earth station managing reception/transmission for the remotes, along with hub electronics and TCP acceleration

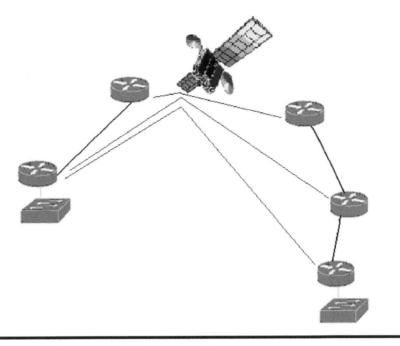

Figure 8.4   Mixed satellite/terrestrial IP connectivity.

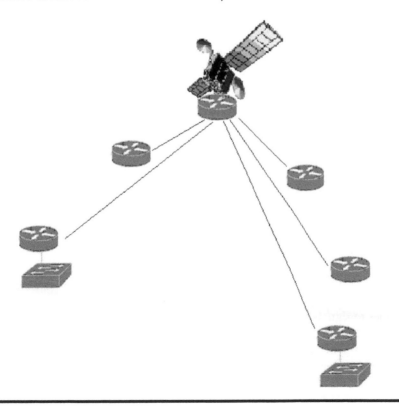

Figure 8.5   Future goal: Onboard-based routing functionality.

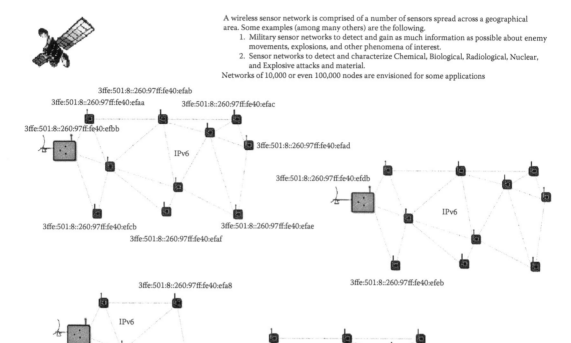

A wireless sensor network is comprised of a number of sensors spread across a geographical area. Some examples (among many others) are the following.

1. Military sensor networks to detect and gain as much information as possible about enemy movements, explosions, and other phenomena of interest.
2. Sensor networks to detect and characterize Chemical, Biological, Radiological, Nuclear, and Explosive attacks and material.

Networks of 10,000 or even 100,000 nodes are envisioned for some applications

Figure 8.6   Use of IPv6 for satellite-based access to remote wireless sensor networks.

The antenna for Ku operation can range in diameter from 75 cm to 1.8 m, and for C-band operation from 1.8 m to 2.4 m. One-way systems can use antennas as small as 45 cm. At the practical level, VSATs operating at C-band (which suffers less from rain attenuation, but requires larger antennas) are used in Asia, Africa, and Latin America, while VSATs operating at Ku-band (which can use smaller antennas, but suffers from rain fade) are used in Europe and North America. For any given system, the size of the antenna and the BUC need to be determined using the link budget analysis discussed in Chapter 6—also taking into account the desired throughput and link availability.)

Communication is usually from the remote terminals to the hub (and vice versa), effectively in a star configuration (logical meshing is supported using a dual hop: remote_a- hub- remote_b.) (The *outbound* direction applies to signals transmitted from the hub to the VSAT; the *inbound* direction applies to signals transmitted from the VSAT to the hub.) As the Internet user community becomes increasingly peer-to-peer, the demand for bidirectional capacity will increase. VSAT satellite services are generally asymmetrical with more bandwidth outbound than inbound; however, evolving applications may soon require larger inbound bandwidth. At press time there were reportedly more than 500,000 terminals installed in more than 120 countries around the world.

Figure 8.8 depicts a typical example of a network arrangement. Generally, the remote devices are supported on a shared service-provider network, where these remote devices comprise an IPv4 subnet. IP traffic (IP packets) is transmitted over the network. Nonregistered IP addresses are

Figure 8.7   Example of VSAT antenna.

typically used for the subnet, and Network Address Translation (NAT) is employed at the hub to reach either the Internet or a firm's corporate intranet. In the future, IPv6 addresses and IP packets may be used (either in a native mode or in a tunneled mode.)

Typical VSAT applications include the following:

Carrier applications
> VoIP
> Broadband rural telephony
> Virtual network operator (VNO)
> GSM backhaul
> Internet access

Corporate applications
> VoIP
> Broadband Internet access
> Disaster recovery/business continuity
> Distance learning

Government applications
> Disaster preparedness
> First responders
> Comms on the Move (COTM)
> Foreign service (embassy connectivity)
> Internet access
> Field training

Industry applications
> Oil and gas
> Maritime
> Retail

Corporate Infrastructure

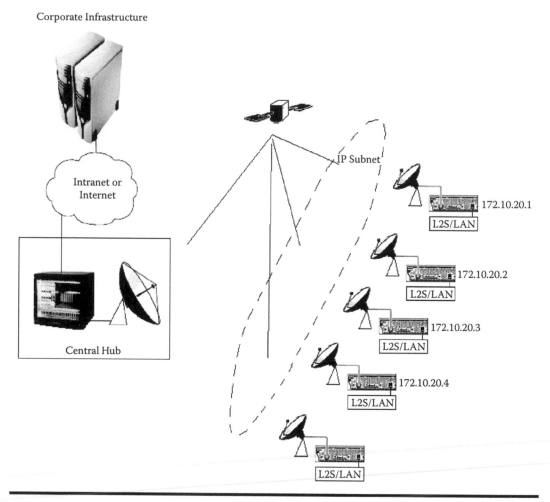

Figure 8.8   Typical VSAT environment.

Media entertainment
Healthcare
Financial education
Internet access

The hub-to-remote channel (return channel) is usually operated in a TDMA modem, while the remote-to-hub channel is operated in a random access/contention mode (see Chapter 4; a number of commercial systems also use TDMA in the return channel). The performance of the system depends on how much bandwidth is allocated for the inbound and outbound channels, the number of users, the oversubscription allowed, and the application.

A VSAT has an outdoor unit (ODU) and an indoor unit (IDU). The outdoor unit (ODU) includes the RF equipment required to transmit and receive from the antenna. The ODU consists of these elements:

■ Low noise block (LNB), which is a down converter and receiver
■ Block-up converter (BUC); this is the up converter and transmitter

- Ortho-mode transducer (OMT); transmits and receives waveguide joint
- Microwave filters, which protect the LNB from the transmit signals.

The indoor unit (IDU) consists of the following elements:

- Integrated receiver decoder (IRD)
- Modem
- Switch/router

Usually two coax cables carry the signals between the IDU and the ODU (power for the LNB and BUC as well as control signals are also carried along these cables). See Figure 8.9.

Mid-range satellite equipment can transmit data to the network hub at speeds up to 1.544 Mbps. Such VSAT terminal modems may have specs such as these:

- Burst rates: 156, 312, 625, 1250, 2500 ksym/s
- Bit rates (rate 2/3 FEC): 208, 416, 833, 1667, 3333 Kbps
- FEC: DVB-compliant R/S (204, 188) and convolutional (R = 1/2, 2/3, 3/4, 5/6, 7/8)
- Modulation: QPSK
- Network: (1) 10/100 BaseT Ethernet (RJ-45)
- Antenna diameters: 0.96, 1.2, 1.8, and 2.4 M
- ODU power: 1- and 2-W Ku-band, 4-, and 5-W C-band

Higher-end system may have typical parameters such as these (e.g., see Figure 8.10):

- IP router, TCP optimization over satellite, 3DES Encryption, and QoS/prioritization
- Data throughput: 18 Mbps downstream, and 4.2 Mbps upstream (scalable for both inbound and outbound data rates)
- TDM (downstream), multifrequency TDMA (upstream)
- Channel rates outroute: 11.5 Mbps or 5.75 Msps
- Inroute: 5.75 Mbps or 2.875 Msps
- Modulation: QPSK
- IP data rates: outroute—64 kbps–9.1Mbps; inroute—64 Kbps–4.2 Mbps
- FEC: outroute—turbo product coding (TPC) rate 0.793; inroute: TPC rate 0.793 or TPC rate 0.66
- Eb/No: 4.6 Eb/No for 10^{-9} quasi error free @ 0.793 FEC; 5.4 Eb/No for 10^{-9} quasi error free @ 0.66 FEC
- Protocols supported: TCP, UDP, ICMP, IGMP, RIP Ver2, Static Routes, NAT, DHCP, DNS Caching, cRTP; Security: 3DES Link Encryption; Traffic Engineering: QoS (CBWFQ), CIR, Rate Limiting, Dynamic Allocation

The central hub (also known as the "master earth station") is typically comprised of the following components [STA200802]:

- Large dish antenna (4.5 to 11 m in diameter)
- Satellite network management system (NMS) and provisioning stations, from which a network operator can monitor and control all components of the enterprise satellite communications network

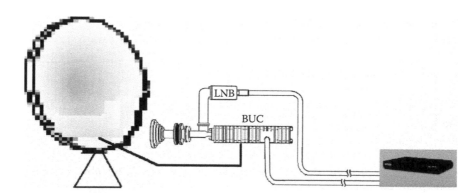

Note: Drawing not to Scale

Figure 8.9   VSAT ODU/IDU.

■ Baseband equipment that handles satellite access, routing between the hub and remote earth stations, dial backup, quality of service (QoS), TCP acceleration, and HTTP acceleration

■ Optional components: Web caches, MPEG transport coder/decoder, application server farms, and audio/video broadcast programming devices

VSAT services can be provided by a facilities-based operator (an entity that owns satellites) or by a Virtual Network Operator (VNO). A VNO leases hub services from an operator that owns satellites (or collocates its own equipment at the teleport site of such operator), and also leases

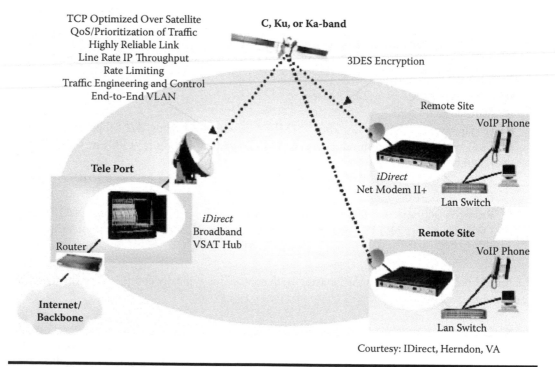

Courtesy: IDirect, Herndon, VA

Figure 8.10   A robust VSAT network.

transponder capacity from the same satellite provider. The VNO is then able to resell the services to a variety of end users. Traffic management is important in this context, especially if heavy over-subscription is employed by the VNO.

8.3 Issues Related to TCP support

TCP-based applications have technical challenges that surface in both terrestrial and satellite high-speed networks; fortunately a number of workarounds exist, as we discuss below. Because of the handshake/windowing mechanism, TCP requires special treatment to operate effectively over a satellite link. In general, there is a lot of concern related to the performance of TCP in a network that consists of a satellite link or uses both satellite and terrestrial components. One classical method to improve the performance of data transfers over satellites is to use a performance-enhancing proxy often dubbed "spoofing." Spoofing involves the transparent splitting of a TCP connection between the source and destination by some entity within the network path [ISH200101]. This concept is expanded in the sections that follow. Figure 8.11 depicts the basic TCP PDU, and Figure 8.12 shows the TCP protocol state machine showing three-way handshake for opening a connection.

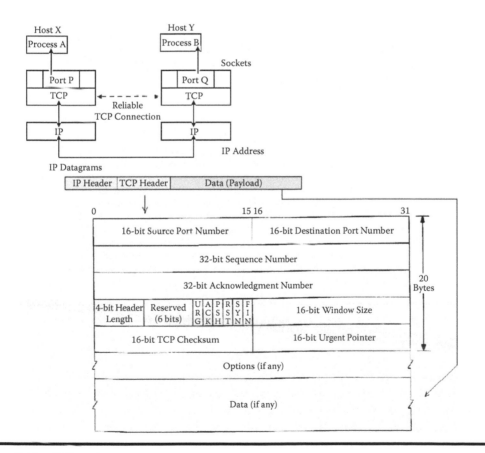

Figure 8.11   TCP environment.

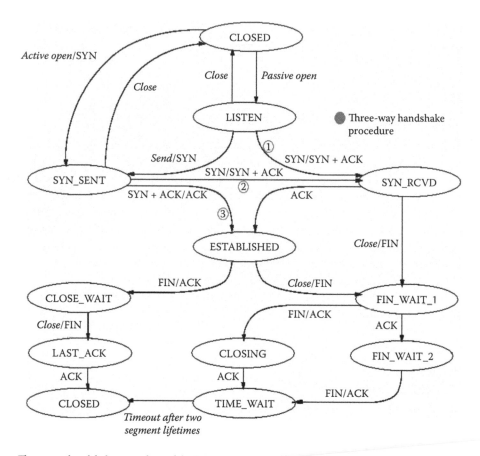

Three-way handshake procedure of the TCP process:

The ESTABLISHED state is where data transfer can occur between the two ends in both directions. The two transitions leading to the ESTABLISHED state correspond to the opening of a connection. Consider a client/server environment. If a connection is in the LISTEN state and a SYN segment arrives, the connection makes a transition to the SYN_RCVD state and takes the action of replying with an ACK+SYN segement. The client does an "active open" which causes its end of the connection to send a SYN segment to the server and to move to the SYN_SENT state. The arrival of the SYN+ACK segment causes the client to move to the ESTABLISHED state and to send an acknowledgement back to the server. When this ACK arrives the server moves to the ESTABLISHED state. This is the three-way handshake procedure of the TCP process.

Figure 8.12   TCP state machine showing three-way handshake for opening a connection.

At face value, satellite links present a challenge to two-way interactive traffic because of the *latency* between two earth stations due to electromagnetic signal propagation on its way to the satellite and back. As we saw in passing in Chapter 1, for GEO satellite systems the latency is at least 238 ms; some earth station–satellite pairs have higher latency because the slant range is longer than the minimal equatorial straight-up distance. Protocol artifacts (e.g., framing) and multiplexing-related queuing for a transmission slot can add extra delays, making the end-to-end

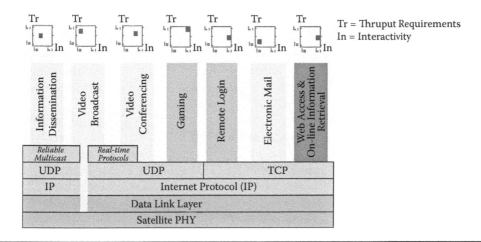

Figure 8.13   Application profiles.

latency as high as 400 ms. This is about an order of magnitude *higher* than the latency on a cross-continental point-to-point fiber-optic link.

Various applications differ in bandwidth requirement and interactivity. The latency usually does not degrade bulk data transfer and broadcast-type applications, but it impacts interactive applications that require continuous handshaking between two endpoints, such as is the case with native TCP. Common Internet applications include Web browsing, remote login (Telnet), video teleconferencing, e-mail, and streaming. The requirements of these applications on network bandwidth and interactivity, their tolerance to communication noise, and their implementation techniques are very different. (See Figure 8.13 for a graphical perspective of these requirements.) Some applications require guaranteed delivery, so they use TCP and, consequently, are sensitive to latency; other applications use UDP or other real-time multicast protocols that can tolerate delays. The sections that follow cover these issues.

8.4 TCP Performance-Affecting Mechanisms

8.4.1 Background

A TCP connection and the flow control of such connections make use of the "window size" mechanism. The windowing mechanism establishes how much data is allowed to be transmitted before an acknowledgement is received. TCP specifies a maximum segment size (MSS). The transmitting entity maintains a *congestion window*, limiting the total number of unacknowledged Protocol Data Units (PDUs; packets) that may be in transit end to end. In high-latency networks, there typically would be many unacknowledged Transport Layer PDUs. The issue is particularly acute in high-latency, high-bandwidth paths.

There are two issues of interest: *performance* and *reliability*. TCP performance depends not upon the transfer rate itself, but rather upon the product of the transfer rate and the round-trip delay. This "bandwidth × delay product" measures the amount of data that would "fill the pipe"; it is the buffer space required at sender and receiver to obtain maximum throughput on the TCP connection over the path, that is, the amount of unacknowledged data that TCP must handle to keep the pipeline full. TCP performance problems arise when the bandwidth × delay product is large. A path operating in this region is referred to as a "long, fat pipe," and a network containing this path as an "LFN." High-capacity packet satellite channels are LFNs. For example, a DS1-speed satellite channel has a bandwidth × delay product of 10^6 bits or more; this corresponds to 100 outstanding TCP segments of 1200 bytes each. There are three fundamental *performance* challenges for TCP over LFN paths [JAC199201]:

1. *Window size limit.* The TCP header uses a 16-bit field to report the receive window size to the sender. Therefore, the largest window that can be used in classical TCP is $2^{16} = 64$ kilobytes. To circumvent this problem one could define an extension to TCP to allow a larger window size, say $2^{32} = 128$ kilobytes. A TCP extension known as TCP-LW ("large-window") was proposed in the late 1990s (as described in RFC 1323 and RFC 2018) to increase the maximum window size to 2^{32}; this allows better utilization of links with large-bandwidth delay products.

2. *Recovery from losses.* Packet losses in an LFN can have a significant effect on through-put. Original TCP implementations caused the data pipeline to drain with every packet loss, and require a slow-start action to recover. The Fast Retransmit and Fast Recovery algorithms address this issue. Their combined effect is to recover from one packet loss per window, without draining the pipeline. However, more than one packet loss per window typically results in a retransmission timeout and the resulting pipeline drain and slow start. Expanding the window size to match the capacity of an LFN results in a corresponding increase of the probability of more than one packet per window being dropped. This could have a serious effect on the throughput of TCP over an LFN. The standard TCP acknowledgment scheme is coarse: if a segment is lost, TCP senders will retransmit all data sent starting from the lost segment without regard to the successful transmission of later segments. To generalize the Fast Retransmit/Fast Recovery mecha-nism to handle multiple packets dropped per window, selective acknowledgments would be required. Unlike the normal cumulative acknowledgments of TCP, selective acknowl-edgments give the sender a complete picture of which segments are queued at the receiver and which have not yet arrived. The selective acknowledgment feature is important in the LFN regime.

3. *Round-trip measurement.* TCP implements reliable data delivery by retransmitting segments that are not acknowledged within some retransmission timeout (RTO) interval. Accurate dynamic determination of an appropriate RTO is essential to TCP performance. RTO is determined by estimating the mean and variance of the measured round-trip time (RTT), that is, the time interval between sending a segment and receiving an acknowledgment for it.

High transfer rate can also impact TCP *reliability* by violating the assumptions behind the TCP mechanism for duplicate detection and sequencing. An especially serious kind of error may result from an accidental reuse of TCP sequence numbers in data segments (see [JAC199201] for exam-ples of how this may occur).

8.4.2 Classical TCP Mechanisms

Studies show that latency is a critical factor in a satellite network only for the response times of small transfers; namely, if an interactive application repeatedly opens a TCP connection only to transfer a small amount of data, the application and the underlying protocol stack will operate inefficiently. This section highlights classical TCP mechanisms, with an eye to establishing what changes can be made to support TCP sessions over a satellite link. Many commercial TCP implementations now support TCP-SACK (Selective ACKnowledge) (RFC 1072 and RFC 2018) for selective acknowledgment* and other high-performance TCP options such as Large Windows (LW) (RFC 1323) (however, it may be necessary to explicitly enable them on some systems). This section is based on reference [ZHA199701].

Bandwidth adaptation. TCP adapts to the available bandwidth of the network by increasing its window size as congestion decreases, and reducing the window size as it increases. The speed of the adaptation is proportional to the latency, or the round trip time of the acknowledgment. In a satellite network with longer latency, bandwidth adaptation takes longer and, as a result, TCP congestion control is not as effective as would be the case in a terrestrial network. Also, it will take much longer for TCP's linear increase to recover the window size after a packet loss if a TCP "large-window" extension is used.

Selective acknowledgment. The standard TCP acknowledgment scheme is coarse, namely, if a segment is lost, the transmitting TCP entity will retransmit all data sent starting from the lost segment without regard to the successful transmission of later segments. TCP considers this lost segment as an indication of congestion and reduce its window size by half. A relatively new TCP-SACK option allows the receiver to explicitly inform the sender of the loss. It follows that a sender can retransmit the lost segments immediately rather than waiting for a timeout, reacting to supposed congestion, and multiplicatively decreasing its window. If lost segments are not caused by congestion, or if the congestion is transient, throughput in TCP-SACK should be much better. This is helpful in satellite networks because anything that triggers timeouts and window size reduction will force a lengthy recovery in TCP.

TCP slow start. When a TCP connection first starts up or is idle for a long time, it needs to quickly determine the available bandwidth on the network. It does so by starting with an initial window size of one segment (usually 512 bytes), then increasing the window size as packets are delivered successfully and acknowledgments arrive, until reaching the network saturation state (indicated by a packet drop). On the one hand, slow start avoids congesting the network before it has a good assessment of the available bandwidth; on the other hand, TCP bandwidth utilization is suboptimal during the procedure. Therefore, the shorter the TCP slow start process is, the better performance that can be achieved. The total time of a TCP slow start period is approximately $RTT \times \log_2(B/MSS)$, where RTT is the round trip time (twice the latency), B the available bandwidth, and MSS the TCP segment size. Although the growth is exponential, for high-bandwidth and high-latency networks (e.g., satellite links), this can still take a significant relative amount of time.

Congestion avoidance. TCP congestion control responds to congestion slowly because of latency; if such congestion can be avoided before it happens, it is advantageous for high-speed and high-latency networks. In recent years new techniques have been introduced in TCP to avoid congestion before it happens. One such approach is Random Early Detection (RED). RED mechanisms require each router/gateway to monitor its own queue length; when imminent congestion is detected the TCP sender is notified. By dropping a packet earlier than would otherwise be the case, RED sends an implicit notification of congestion. The sender is effectively notified by the timeout of this packet. The principle behind the RED approach is that a few earlier-than-usual

---

* There is a limit on the number of SACK blocks that can be carried by the acknowledgment packets. Under some error conditions, this limitation can force the TCP sender to retransmit packets that have already been received successfully by the receiver [SRI200201].

drops may help avoid more packet drops later on. The TCP sender can then reduce its window before serious congestion occurs. Although this and similar approaches have not been widely adopted as yet, it holds promise for satellite networks.

TCP for transactions. Many TCP applications involve only simple communications between the client and the server. The interaction is called a transaction: a client sends a request to a server and the server replies. HTTP for WWW browsing applications is a typical example of TCP with transactional behavior. Under standard TCP, even a small transaction involving a single request segment and a reply must undergo TCP's three-way handshake in preparation for bidirectional data transfer, as we saw in Figure 8.12. If the request is larger than a segment, TCP must also undergo the slow start procedure. It is fairly inefficient to establish such a TCP connection, send and receive a small amount of data, and then tear it down. Transaction TCP (T/TCP) is an extension to TCP designed to make such behavior more efficient [BRA199401]. T/TCP does this by bypassing the three-way handshake and slow start, using the cached state information from previous connections. T/TCP uses a monotonically increasing variable CC (Connection Counts) to bypass the three-way handshake and reduce TIME_WAIT period; hence, T/TCP greatly decreases the overhead standard TCP introduces when dealing with transaction-oriented connections. Although T/TCP is designed mainly for short client-server interaction applications, it can be used to reduce the impact of latency on the beginning of a TCP connection. If slow start can be avoided, significant performance improvement can be achieved in a satellite-based network. (Some implementations of T/TCP have emerged over the years for Sun Solaris and Linux, but its overall penetration is limited.)

TCP for lossy networks. TCP was designed to optimize its performance to deal with packet losses in the network due to congestion. TCP is unable to determine if a packet loss is due to congestion or corruption of the packet due to errors in the network. As a result, TCP generally performs poorly in lossy environments because it interprets packet corruption as congestion in the network. Thus, instead of increasing or, at least, maintaining its sending rate to overcome these errors due to corruption, TCP will decrease its sending rate to reduce what it perceives as congestion in the network. This reduction in sending rate results in low throughputs for bulk transfers. HeAder ChecKsum option (HACK) is a proposed modification to the TCP protocol that allows it to perform better in lossy environments [BAL200101].

8.4.3 Performance Improvements for TCP over Satellite Links

Three classes of techniques have evolved over the years to mitigate the impact of (satellite) latency when using TCP protocols:

- Implementation of new versions of TCP (TCP extensions) that perform better over the satellite networks;
- Deployment of gateways at the satellite network boundaries to perform special functions to speed up TCP sessions (this action also known as TCP acceleration and as middleware);
- Enhancements to applications to use TCP more efficiently.

8.4.3.1 TCP Extensions

As we have seen, issues such as small window size, prolonged slow-start period, and ineffective bandwidth adaptation affect network performance over satellite links. As discussed above, possible TCP extensions to deal with performance degradation include (among others):

- TCP-LW
- TCP-SACK
- T/TCP
- RED gateways
- HACK

Of these, the first two have seen the most deployment to date.

8.4.3.2 Gateways/Performance Enhancement Proxies (PEP)

Major performance improvements can be achieved by working at the network infrastructure level without the need to modify the TCP protocol and/or the endsystems. A Performance Enhancement Proxy operates on a router (or other appropriate device) along the path of a TCP connection, for example, at the edge of the satellite hops. When a data packet arrives at the PEP, the PEP forwards the packet to the destination host, transmits the corresponding ACK (premature ACK) to the source host on behalf of the destination host, and stores a copy of the packet in a local buffer (PEP buffer) in case the packet needs to be retransmitted [WAN200701]. Two variants of the PEP concept exist as follows [ZHA199701]:

Split-TCP. The basic concept of the split-TCP approach (sometimes also referred to as indirect TCP), is to segment an end-to-end TCP connection into multiple segments. Each segment is itself a complete TCP connection, and data streams are forwarded from one segment to another (using buffering if necessary). See Figure 8.14. In a satellite environment, the middle segment spans the high-latency satellite link, and the other two segments connect the routers that join the terrestrial Internet and the satellite link to the original endpoints. The splitting action isolates the effects of

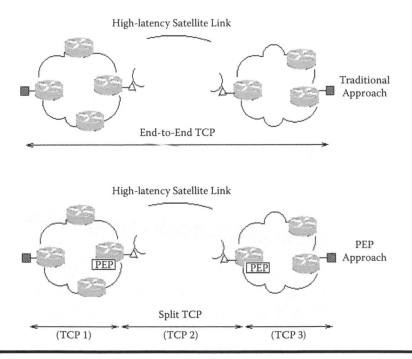

Figure 8.14   Split TCP.

high latency. If the first and the last TCP segments span a low-latency network, TCP slow start can speed up more quickly and the normal window size (without TCP-LW) will operate as expected. The middle segment, however, could implement special features such as TCP-LW to deal with long latency. Using split-TCP allows TCP performance to be improved without changes to application software. Note however, that this approach "stretches" the concept of end-to-end semantics because the sender may think a segment has arrived at the destination while it is actually still in transit.

TCP spoofing. TCP spoofing (also known as TCP acceleration) compensates for the space-link transit time. TCP spoofing is accomplished by deploying special equipment at the carrier's main

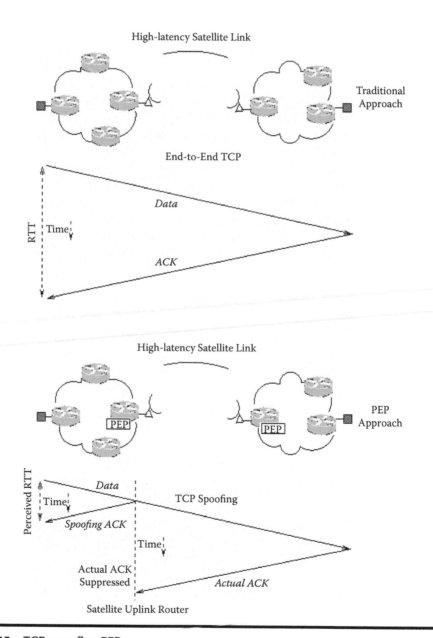

Figure 8.15 TCP-spoofing PEP.

satellite hub site. This equipment appears to TCP as if it were the remote location, while acting as a relay or forwarder for data packets going to and from the remote satellite location. In TCP spoofing, the more common PEP implementation, an intermediate gateway (usually at the satellite uplink) prematurely acknowledges a TCP segment without waiting for the actual acknowledgment from the receiver (see Figure 8.15). This gives the transmitting entity the appearance of a low-latency network, so the TCP slow start phase can progress more rapidly. The intermediate PEP gateway buffers TCP segments in transit. When the actual acknowledgment from the receiver arrives at the gateway, it is suppressed to prevent duplicate acknowledgments from reaching the sender. If the receiver's acknowledgment never arrives and the gateway times out, it retransmits the lost segment from its local buffer.

When the spoofing equipment receives Internet traffic destined for a remote satellite location, it acknowledges receipt of the packet immediately on behalf of the remote site so that more data packets will follow immediately. In this manner, the latency is "hidden" because the acknowledgments are returned rapidly. As a result, TCP moves out of slow-start mode quickly and builds to the highest possible speed. The acceleration equipment also tracks real acknowledgments coming back from the remote site and suppresses them. If the acknowledgment is not received from the remote site, the system automatically resends the packet from its buffer [STA200801].

As shown in Figure 8.16 (developed in reference [ISH200101]), there is generally a 100 percent to 250 percent improvement in throughput when using PEP (the percent difference is calculated by taking the throughput difference of spoofing and end-to-end TCP over the throughput of end-to-end TCP). Studies have shown that [ISH200101]:

- Spoofing is beneficial for large file transfers.
- For small transfer sizes, spoofing increases the throughput as observed by the data sender.
- Spoofing was much less beneficial for throughput observed at the receiver, which is the vantage. point perceived by the end-user. Because a majority of data sent across networks is small and sent to the user, spoofing will likely not provide a large advantage from the user's perspective.
- Spoofing's benefit to Web servers and other content providers is significant.
- Spoofing allows for data to accumulate at the spoofer, creating a second bottleneck and increasing the number of dropped data packets, which also degrades the receivers perceived performance.
- The performance benefits of spoofing increase as the amount of network congestion grows.

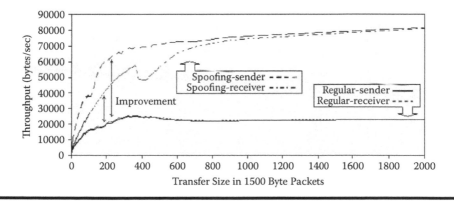

Figure 8.16 Performance improvements with TCP-spoofing PEP.

Again, PEP-based TCP spoofing "stretches" the concept of end-to-end semantics because the sender may think a segment has arrived at the destination while it is actually still in transit. This is acceptable in many applications, such as WWW browsing through proxy, but it may cause problems if an application is built upon end-to-end semantics. The concept of network-oriented processing in general is gaining more acceptance lately (e.g., [MIN200601]. [MIN200801].)

8.4.4 Application-Level Approaches

Finally, some changes could be made to applications to optimize their use over a satellite link. For example, applications use TCP more effectively by avoiding small and short transfers, because, as noted earlier, an interactive application that repeatedly opens a TCP connection only to transfer a small amount of data will operate inefficiently. Such potential changes include the following [ZHA199701].

Persistent TCP connections. Client/server-type applications operate with the client program sending a request to the server and the server returning appropriate data. If each request/reply is implemented in a separate TCP session, the overall performance of the application will be impacted. One can speed up the performance the application by redesigning the application to

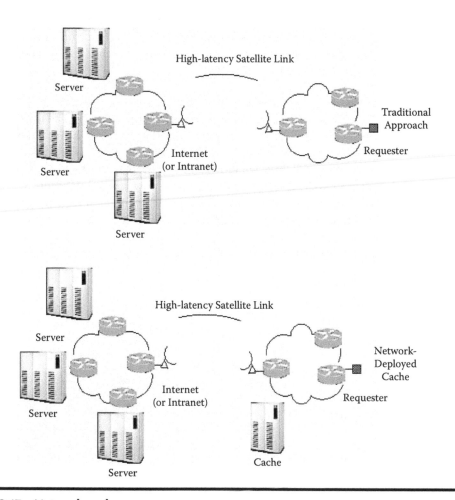

Figure 8.17   Network caches.

use a single TCP session for all the small requests/replies. For example, HTTP protocol performs poorly because a Web client gets each Web object (page of text, icons, images, etc.) in a separate TCP connection (a typical Web page consists of many such small objects). HTTP 1.1 alleviates the problem by utilizing a persistent connection to combine many small transfers into a single request and to pipeline the transfers so that the transmission delays overlap with each other.

Caching. Caching can reduce network utilization and latency as perceived by the user. Caches can be used, for example, in the context of Internet access (WWW/HTTP); caches would be deployed by service providers and/or incorporated in the terrestrial satellite platform that supports the satellite link (say, in a VSAT environment). See Figure 8.17. Commonly requested Web pages and/or documents are sent (and refreshed) to local caches. By (proper) design, most of the requests to a Web page are able to be satisfied by a nearby cache (perhaps avoiding the satellite link), with only occasional retrievals from remote servers.

References

[BAL200101] R. K. Balan, B. P. Lee, K. R. R. Kumar, L. Jacob, W. K. G. Seah, A. L. Ananda, "TCP HACK: TCP Header Checksum Option to Improve Performance over Lossy Links," INFOCOM 2001. *Proceedings of the Twentieth Annual Joint Conference of the IEEE Computer and Communications Societies,* IEEE, Vol. 1, 2001, pp. 309–318, Digital Object Identifier 10.1109/INFCOM.2001.916713.

[BRA199401] R. Braden, "T/TCP–TCP Extensions for Transactions Functional Specification," Request for Comments: 1644, July 1994, IETF.

[ISH200101] J. Ishac, M. Allman, "On the performance of TCP spoofing in satellite networks," *Proceedings of IEEE Military Communications Conference on Communications for Network-Centric Operations: Creating the Information Force (MILCOM 2001),* IEEE, Vol. 1, Issue, 2001 pp. 700–704, Digital Object Identifier 10.1109/MILCOM.2001.985925.

[JAC199201] V. Jacobson, R. Braden, D. Borman, Request for Comments: 1323, TCP Extensions for High Performance, May 1992.

[MIN200601] D. Minoli, "Do We Need Computing in the Network?" *Network World,* 10/02/2006.

[MIN200801] D. Minoli, *Enterprise Architecture A to Z: Frameworks, Business Process Modeling, SOA, and Infrastructure Technology,* Taylor & Francis, New York, 2008.

[SOH200701] K. Sohraby, D. Minoli, T. Znati, *Wireless Sensor Networks,* Wiley, New York, 2007.

[SRI200201] K. N. Srijith, L. Jacob, A. L. Ananda, "Worst-Case Performance Limitation of TCP SACK and a Feasible Solution," *ICCS 2002: The 8th International Conference on Communication Systems,* November 25–28, 2002, Vol. 2, pp. 1152–1156.

[STA200801] Staff, "Satellite Internet How Broadband Satellite Internet Acceleration Works," VSAT Systems, LLC. http://www.vsat-systems.com

[STA200802] Staff, "Connecting Cisco IP VSAT Satellite WAN Network Modules," Cisco Systems, http://ciscosystems.com/univercd/cc/td/doc/product/access/acs_mod/cis2600/hw_inst/nm_inst/nm-doc/satnm.htm

[WAN200701] H. Wang, T. Yokohira, and N. Yamai, "Effect of Premature ACK Transmission Timing on Throughput in TCP with a Performance Enhancing Proxy," *IEICE Transactions on Communications* 2007 E90-B(1): 31–41; doi:10.1093/ietcom/e90-b.1.31.

[ZHA199701] Y. Zhang, D. De Lucia, B. Ryu, S. K. Dao, "Satellite Communications in the Global Internet: Issues, Pitfalls, and Potential," NASA Lewis Research Center, Whitepaper, *INET 1997 Proceedings.*

Chapter 9

Satellite Communication in IPv6 Environments

In Chapter 8 we provided an implicit view to the support of IPv6 over satellite links. This chapter gives a more detailed survey of press-time industry activities aimed at advancing the understanding of how to optimally and effectively support IPv6 over satellite systems. There is keen interest in this topic, especially outside the United States. Some specific implementation approaches and suggestions are included in Section 9.4. Appendix A following this chapter provides some additional information.

While this information has a certain time horizon, it is included here to give a sense to the reader of the activities that have taken place in the past few years (2003–2009) in the area of IPv6 over satellite, and to give a sense of the foundation that this work lays for more concrete activities expected in the next few years (2010–2015.)

9.1 Motivations for IPv6 Support

As of press time vendors of satellite equipment appeared in general to continue to focus on topics such as new modulation techniques that can realize more sellable bandwidth out of transponders' capacity. Value-added features such as IP support in satellite environments (e.g., IP-based VSATs, IPTV over satellite, etc.) have seen implementation in the past few years, but they are still somewhat secondary and always seem to be a step behind the terrestrial world, according to some observers [CAM200801].

As we saw earlier in the book, IPv6 offers the potential of achieving the scalability, reachability, end-to-end interworking, quality of service (QoS), and commercial-grade robustness for data as well as for VoIP/triple-play applications that next-generation networks require. According to some observers, the satellite market could in fact be the first to actually capitalize on IPv6 because a large part of the market consists of regions underserved by fiber; these tend to be the very same countries who are short on IPv4 addresses and whose Internet capabilities are hampered as a result. These regions also tend to be the fastest-growing market segments and the ones with the largest population growth. Those same markets may be the first to embrace IPv6 and its large address space. Satellite providers also serve a large market of mobile customers such as the military or emergency response units that must set up and communicate quickly when responding to disasters. Some of IPv6's features of interest include its inherent support for mobility and its auto-configuration capabilities that make deploying an ad hoc network straightforward. Also, there are significant requirements coming from the U.S. DoD. Additionally, many new mobile phones are being released with built-in IPv6 capabilities. These kinds of mobile customers may soon require IPv6 support [CAM200801].

The planning required by service providers should always be ahead of the actual end-user requirements for a given feature by some business-optimal market window (which should neither be "way too early" or "way too late.") At the end-user level, organizations have been attempting to develop business cases that would show some short-term financial benefit for a conversion to IPv6. Such business cases may be difficult to be developed in the immediate future, reminiscent of the desire for a business case to validate a move to VoIP in the late 1990s. However, at this juncture it is clear that VoIP is less expensive compared to traditionally priced telephony with built-in 20th century sociophilosophical pricing fee structures* and may be less costly when one considers the possibility of unified messaging and computer-telephony integration. The introduction of IPv6 may need to parallel an even more basic transition: the introduction of PC-based automation in the mid-1980s. It was difficult to anticipate the eventual macroeconomic productivity gains that were finally realized in the late 1990s and early 2000s when this technology was first being introduced. It took over a decade for the intrinsic and synergistic value of the deployment of PC/server technology to be manifestly and explicitly evident. The expectation of observers is that at some point in time, the technology will simply have self-initiated pull, as more and more vendors bring IPv6 to the market. The transition may occur without much fanfare, initially on a case-by-case, application-by-application, island-by-island basis, until such time when critical mass is achieved. The macro-level advantages of IPv6 will be evident perhaps a decade out from when the technology made its first deployment appearance. If the "technology pull" we mentioned in the previous paragraph will occur around 2010, then the full value of IPv6 will be more fully evident around 2020. However, DoD requirements may pull this through at an earlier time. One should not forget that, after all, it was the military requirements that drove the development of the technology that led to the deployment of the Internet, as we know it today. Also, in consideration of the relatively long life cycle for the deployment of satellite technology (perhaps 2–3 years on the drawing board, and a deployed life of 15–20 years), it is important to start planning at this juncture if the service is expected to be needed 5 years out.

As we saw in Chapter 7, in addition to being backward compatible with IPv4's differentiated services ("diffserv"), IPv6 includes a QoS-supporting feature that allows for additional capabilities. The IPv6 "Flow Label" allows for the mapping of flows (individual transaction sessions) and other information into the IPv6 header for visibility and processing at layer 3. This eliminates the burden on network systems to process flows by using information at multiple layers; in effect, this obviates the need to deploy an overlay network function such as MultiProtocol Label Switching (MPLS), which includes a similar flow function at later "2.5." IPv6 also reduces the security issue of leaving data integrity (encryption) as an optional external processing function. This is important

---

* For example, that "long-distance service" is a "luxury" item and, therefore, people making long-distance calls should subsidize people making local calls, when, in fact, the economics are exactly the opposite; the implication of our statement is that VoIP may or may not be actually cheaper in a true-cost comparison, but it is cheaper in a price comparison.

for satellite communication, where encryption is routine. Furthermore, because satellite-based IP services delivered in remote areas are often used for VoIP, these advanced QoS capabilities will be useful especially in conjunction with header compression. IPv6 also facilitates the end-to-end communications model, a concept that has largely been lost due to the issues that private IP addressing and Network Address Translation (NAT) present [CAM200801]. The macro issue of IPv4 address availability has temporarily been solved by the industry at large with the heavy use of NAT. However, this use of NAT limits applications such as end-to-end public VoIP, global IP-based cellular service, large sensor networks, and true IP addressability of billion of appliances ranging from automobiles to PDAs, home appliances, and so on.

As was implied in the previous chapters, the migration of IPv4 to IPv6 will not happen instantaneously; the expectation is that there will be a period of transition when both protocols are in use over the same infrastructure. This, in fact, gives service providers some time to prepare for the eventual demand in Europe, in Asia, by the U.S. DoD, and by North American businesses. As noted in earlier chapters, to deal with this transition period, the designers of IPv6 have created technologies and address types so that IPv6 nodes can communicate with each other in a mixed environment, even if they are supported at the core by an IPv4-only infrastructure [MIC200701]. Protocol transitions are never trivial and the transition from IPv4 to IPv6 is particularly complex. Protocol transitions are typically achieved by installing and configuring the new protocol on all nodes within the network [AMO200801]. Although this might be possible in a small- or medium-sized organization, the challenge of making a rapid protocol transition in a large organization or a carrier of any size (including an Internet service provider) is challenging. Therefore, while migration is the aspirational goal, consideration must be given to the short-term coexistence of IPv4 and IPv6 nodes.

9.2 Initiatives for IPv6 Support

Satellite services are perceived as being critical to the global information infrastructure (GII) and next-generation networks (NGNs). To fulfill these roles, support for IPv6 is required on all communication systems, including satellites. As of 2008 most satellite vendors asked about IPv6 support state that they have it on their roadmaps; however, introduction of products appears to be less than imminent. A number of initiatives, especially in Europe, however, have been set in motion in the recent past to facilitate the deployment of IPv6-based technology in satellite environments. While these projects all point to "work in progress," they show the keen interest in addressing the need and opportunities for IPv6 over satellite links.

9.2.1 The European Space Agency (ESA) Project

One such initiative is The European Space Agency (ESA) project "Preparation for IPv6 in Satellite Communications." The key objective of this project was to investigate the influence of an introduction of the new Internet Protocol IPv6 on today's satellite communication. Subtending objectives were as follows [ESA200701]:

- Gain a first overview of the influence of link layer, network layer, transport layer, and network management protocol IPv6 issues on satellite networks. Knowing and understanding these IPv6 issues is the required basis before starting the introduction of IPv6 in satellite networks.
- Specify transition scenarios for different satellite architectures, which allow a smooth integration of IPv6.

■ Perform first real demonstrations of IPv6 over satellite; this will allow illustration of the subject of IPv6 over satellite to a broad audience, to raise the awareness for it, and to foster collaboration, for example, between service provider, manufacturer, and future user. The goal is to utilize IPv6 capable DVB-S equipment.

■ Develop a set of recommendations for the next steps required towards a full integration of IPv6 in satellite networks. Such a roadmap can be used by service provider and manufacturer, but also by ESA itself to assist in the planning of future work.

The following tasks are performed in this project:

■ Identification of satellite specific link layer, network layer, transport layer, and management protocol issues for IPv6

■ Investigation of the impact of IPv6 on existing and future satellite network architectures and services

■ Specification of IPv6 transition scenarios for satellite network architectures

■ Identification of possible demonstration scenarios for IPv6 over satellite

■ Selection of and specification of IPv6 pilot demonstration

■ Integration and execution of IPv6 pilot demonstration

■ Identification of an IPv6 roadmap and recommendation for satellite networks

■ Dissemination of project activities and results

The major achievements of the project were [ESA200701]:

■ The majority of satellite specific link layer, network layer, transport layer, and management protocol issues for IPv6 have been identified.

■ The impact of IPv6 on existing and future satellite network architectures has been investigated, architectural shortcomings have been addressed, and potential solutions have been outlined. These solutions include mechanisms to cope with problems resulting from unidirectional links as well as suitable transition methods for networks deploying not IPv6-ready devices. Furthermore, the impact of IPv6 on services has been identified and new or modified service possibilities have been described.

■ For two teleport networks, the first representing a DVB-S/SCPC architecture and the second using a DVB-RCS (digital video broadcast-return channel via satellite) system, detailed transition plans have been investigated, each covering short term solutions using transition methods, long-term solutions requiring the replacement or update of not IPv6-enabled devices, and efforts and costs for transition.

■ To illustrate native IPv6 deployment over satellite, two pilot demonstrations have been specified, implemented, and performed. Both demonstrations used IPv6-capable DVB-S equipment developed in other ESA projects. The first demonstration was to show the usability of advanced services such as IPSec, Mobile IPv6, and audio and video conferencing in a native IPv6 satellite network. For the second demonstration, the SILK network, connecting academic and educational institutions residing in the Central Asian and Caucasian region to a hub station in Hamburg via DVB-S forward and SCPC return links, has been enhanced to support native IPv6 communication.

■ The project and its results have been disseminated in various fora and working groups related to satellite communication and IPv6, such as the ESA IP networking workshop, IPv6 Forum and Cluster events or the Asian Pacific Advanced Networking Conference.

- Finally, a roadmap for the introduction of IPv6 in satellite networks has been identified, containing recommendations for future standardization, dissemination, and deployment activities.

Appendix A provides some results from this initiative.

9.2.2 The SATSIX Project

Another initiative is the SATSIX Project. The SATSIX Project (satellite-based communications systems within IPv6 networks) is part of the European Sixth Framework Program, which aims at implementing innovative concepts and cost-effective solutions for broadband satellite systems and services using IPv6. This project is promoting the introduction of the IPv6 protocol into satellite-based communication systems. The goal of SATSIX is to assist the industry implement IPv6. It also focuses on satellite broadband multimedia systems by exploiting the common components defined by the DVB-S2 (digital video broadcast satellite) and DVB-RCS satellite broadband standards [LOU200701].

The intrinsic objectives of SATSIX are to lower the cost of broadband satellite access, through the development of new satellite access techniques and the integration of wireless local loops (Wi-Fi and WiMax) and to develop recommendations, testbeds, trial networks showing how satellite broadband access will integrate Next Generation Networks, based on IPv6, and support new multimedia applications. The SATSIX project will thus focus on satellite systems that offer attractive solutions to the access segment of wider networks in several main scenarios.

The SATSIX project also looked at approaches for a multicast architecture. As one of the key elements in an IPv6-supported DVB-RCS satellite network, the multicast architecture design is facing the challenges of interworking between the IPv6 multicast protocols and the satellite signaling while efficiently using the satellite bandwidth. One needs to establish how to enable the multicast group management functions for satellite end users with and without direct IPv6 Multicast Listener Discovery (MLD) router support and how to translate the IP multicast routing protocol messages to the satellite lower layer signaling to establish the satellite channels between two spot beams need to be answered. With the regenerating satellite space segment, dynamic multicast routing is possible, and making an efficient use of the satellite bandwidth has to be considered [YUN200701].

9.2.3 European Commission's Seventh Framework Program

The European Commission has established an Action Plan for accelerating the rollout of IPv6; it calls for Europe's research efforts to be matched by political commitment, with a concerted effort to (1) develop the skills base; (2) accelerate standards and specifications work; and (3) promote awareness throughout the economy. Under the previous Fifth and Sixth Research Framework Program, and now the Seventh Framework Program, the goal is to support an array of research projects (including IPv6 over satellite) to [DIR200801]:

- Ensure that the new protocol can be deployed and implemented over a wide array of devices and communication systems
- Study and investigate some specific aspects related to IPv6, such as:
  - IPv6 knowledge dissemination and training (6DISS) and (6DEPLOY—under negotiation)

- IPv6 mobility mechanisms testbed (ANEMONE)
- The large scale international IPv6 pilot network (6NET)
- Advanced tools and services for IPv6 testing (GO4IT)
- The transition from IPv4 to IPv6 (the LONG project)
- Quality of Service aspects (6QM, GCAP)
- IPv6 over satellite (SatIP6)
- IPv6 over power lines (6POWER)
- IPv6 in wireless (6WINIT, 6HOP) or mobile (OverDRiVE) environments
- IPv6, Autonomic Networks (EFIPSAN)
- IPv6 for Africa (IRMA—under negotiation)
- Europe India IPv6 collaboration (6CHOICE)
- IPv6, Emergency (U2010)

9.2.4 The European IPv6 Task Force

Yet another initiative was sponsored by the European IPv6 Task Force. It recently made the following recommendations to European industry in reference to IPv6 in general and IPv6-satellite-based services in particular [LAD200601] (direct quote follows):

- Promote IPv6 over Broadband: as a benchmark, the Taiwanese and Japanese success story with broadband access using IPv6 is the first visible service where IPv6 can be deployed immediately and in larger scale. Taiwan will deploy IPv6 broadband access for 6 Mio users by 2008 and Japan's Softbank will deliver IPv6 by end of 2006 to its 5 Mio users. These are two examples for European ISPs to look into and win experience from. The Korean strategy is to drive WiBro with IPv6. The EUv6TF has published a Communication for this potential deployment.
- Promote VoIP over IPv6: The other immediate and strategic area where IPv6 could be introduced immediately is in VoIP. An effort in convincing the European Telecom industry and operators is key since in the U.S. corporate operators are deploying VoIP to [… reduce costs]. The European operators need to be convinced to have a new approach to VoIP using IPv6.
- Promote European IPv6-ready technologies and the European companies working in the ICT domains, facilitating the development and growth of SMEs working in new innovative ICT fields and promote the use of SMEs products by the large European groups. One domain we should focus in Europe is Software. Innovation comes mainly from software. A case in point is the unique success story of 6WIND. 6WIND provided for example its software to Samsung, Mercury and Ibit in Korea to let them develop new ranges of IPv6 ready equipments in a few months. Off-the-shelf networking software reduces drastically the time to market and costs.
- Promote open source Linux implementation of IPv6.
- Promote IPv6 over Satellite and HDTV over IPv6: One of the areas where Europe has developed leadership is in satellite communications. With the advent of the all-digital TV by 2010 in Europe, there is a clear potential for Europe to retain its leadership in this strategic market. SES Astra being based in Luxembourg, an EU project has been proposed to SES to work with industry. It would be highly recommended to promote High Definition Video Delivery Service over IPv6 Internet by:

- Establishing operation and extension of IPv6 network infra for HDV contents delivery service.
- Applying network-monitoring tools for analyzing the number of users and IPv6 traffics with VoD service.
- Developing HDV contents service techniques based on VoD and its management schemes.
- Building VoD server and its Web site for HDV contents (e.g., cultural, medical, educational multimedia contents) service and Testing operation and by
- Developing multiuser remote videoconference system based on HD video delivery service and encouraging it.
- Europe could take leadership in researching and investing in:
  - Two-way satellite communications
  - Mobile satellite services
  - DVB-S2 usage (SES Astra has a test carrier already up and running) for IP data delivery,
  - Contribution to the emerging standard for IP over DVB-S2
  - Use of DVB-S (2) and IP for television contribution links
  - Delivering HDTV over IPv6 over DVB-S (or DVB-S2).
  - As a benchmark, NTTPC Communications announced that it started offering an IPv6 Enabled High Quality Video Conference System ("ViPr") in Japan. "ViPr" is manufactured by and imported from Marconi Corporation plc ("Marconi," based in London, UK).

  The all-in-one system ViPr overcomes the challenges of conventional videoconference systems, such as high installation cost and complex operation, and enables a high-end videoconference system at a lower cost. ViPr supports not only conventional IP networks but also IP multicast, IPv6, SIP control, and MPEG2.

  Marconi provides a wide range of network equipments for telcos to enterprises with proven records in the high level of technology. ViPr allows videoconference with clear voice quality and DVD-equivalent video quality using MPEG2 CODEC, with ease-of-use of one-touch operation on the touch panel. Multi-point (up to 15 sites) videoconference is possible without an MCU (multipoint control unit), where participants to the conference can scan and distribute what are on their PC screen or use applications simultaneously. ViPr is already in use in the United States for such applications as remote trials, remote medicine, and distance learning.

■ Promote IPv6 in the home networking. The EUv6TF communication addressed to CENELEC outlines the technical guidelines and practices to achieve successful use of IPv6 in the home connectivity market: http://www.european-ipv6-tf.org/Whitepapers/Forms/AllItems.aspx
■ Fully participate in the R&D activities to be supported in the context of the 6-7th Framework program, with a view to put in place an integrated and structured set of IPv6 activities, covering the full range of IPv6 aspects, from basic research through the development of service enablers and associated software suites, to the large-scale trialing and testing of IPv6 features, for a diversity of applications, in a European-wide environment.
■ Actively contribute towards the acceleration and alignment of ongoing IPv6 work within standards and specifications bodies and urgently develop key guidelines permitting the rapid integration of IPv6 infrastructures and interoperability of IPv6 services and

applications, especially in the ETSI testing events ETSI Plugtests* and within the auspices of the GO4IT IST project in which ETSI is a key member.

■ Where appropriate, develop roadmaps for the design, development and deployment of IPv6 services, equipment and networks, to include technologies such as AAA, DNS, xDSL, etc.

■ Contribute actively to the work of the National IPv6 Task Forces, ensure the collectively increase of IPv6 awareness and permit their members to individually derive their own perspective of the IPv6 business case and their own IPv6 integration strategy.

■ Devote efforts towards the establishment of a European-wide, vendor-independent, training and education program on IPv6, in cooperation with 6DISS (which is funded to do so in developing regions around the world but including southeast Europe)

■ Consider in their manufacturing plans that the majority of mobile devices, and a growing number of household and consumer-electronic devices will require some form of IP connectivity and that the simplest way to offer these devices the fullest range of services is to have a unique globally routable IPv6 address available for all network-enabled components.

■ Seek to develop innovative IPv6-enabled devices, for example, biometric security devices, "IP in a chip" embedded systems components, in-car sensor devices. Seek to design and implement innovative peer-to-peer applications where appropriate, for example, peer-to-peer gaming in the entertainment industry.

■ Take early steps to obtain adequate IPv6 address allocations and where appropriate, and to either accelerate the offer of IPv6 capable services or consider on a priority basis how best to rapidly evolve toward IPv6

■ Address the multi-vendor interoperability issues impeding the wide-scale deployment of PKI and to conduct extensive trials with IP security in IPv6 and the parallel implementation of a PKI.

9.2.5 The "Anywhere, Anytime Internet Access" Project

"Anywhere, anytime Internet access" is a project that looks at multicast and mobility over satellite using IPv6. It is a collaborative project, funded by CNES (French National Space Agency), led by 6WIND (private French developer of embedded IP networking software), Astrium (a designer and manufacturer of satellites) and LSIIT (Strasbourg University IPv6 research team). The objective of the project is to identify applicative scenarios that could best benefit from multicast and mobility features over satellite. From the potential user needs, optimization of the communication needs is studied through system architecture comparisons and identifies standard evolutions to ease these improvements [6WI200801].

9.2.6 Satellite Broadband Multimedia System for IPv6 (SATIP6) Project

The aim of this early project (2002, with completion 2004) was to evaluate and demonstrate key issues of the integration of satellite-based access networks into the Internet to offer multimedia services over wide areas. SATIP6 examined the most immediate technical issues facing satellite

* ETSI Plugtests™ refer to events where engineers get together to test the interoperability of their implementations between each other. Plugtests™ increase the probability of achieving interoperability by debugging the standard and companies' implementations at an early stage. Plugtests™ are part of the standardization process. The implementations tested are prototypes. The events are open to every developer. They are usually short duration events (1–5 days). They take place within the time frame of the standards drafting.

broadband access in the coming years. The project sought to define solutions for the short-to-medium term focusing on better adaptation of DVB-RCS access for IP services and for the longer term in which protocols more optimized for IPv6. Issues such as mobility, QoS, multicast and security were included in the study [FP5200701].

During the project the main requirements of the satellite system were to be identified and the overall system architecture defined. Simulation of these architectural choices including aspects such as scalability, optimal utilization of network resources, minimization of packet loss, and queuing delay will be assessed. In parallel to the simulation activity, a testbed for validating the most significant architectural aspects will be set up to provide, together with the simulation activity, inputs to the architecture specification activity.

An area of interest was the functions to be implemented in the protocol layer between physical medium access and the IP layer (i.e., layer 2). The project aimed at defining solutions for two stages:

1. For the short-to-medium term, focusing on better adaptation of DVB-RCS access for IP services with current satellites (transparent payload)
2. For the long term in which protocols more optimized for IPv6 will be introduced with next-generation satellites as a possible option. Simulation of these architectural choices including aspects such as scalability, optimal utilization of network resources, minimization of packet loss, and queuing delay will be assessed.

9.3 Actual IPv6 Satellite Demos and Services

Some satellite-based IPv6 demos have already taken place. This section identifies a few.

Working together, Cisco Systems, NASA Glenn Research Center and Surrey Satellite Technology Ltd (SSTL) recently configured and tested IPSec and IPv6 on a satellite. On Thursday, March 29, 2007, a Cisco Systems router, flying in low earth orbit onboard the UK-DMC satellite built by SSTL, was successfully configured by NASA Glenn Research Center to use IPSec and IPv6 technologies in space. The UK-DMC satellite is a member of the Disaster Monitoring Constellation (DMC), used for observing the Earth for major disasters and commercial land monitoring. The five DMC satellites in orbit rely on standard IP networking to send mission-critical imagery to ground stations and to interact with terrestrial networks. The DMC effectively extends the Internet to orbit, and its farsighted adoption of IP has made it possible to take the backbone of the Internet even further into space. Internet technologies new to the space environment, such as IPv6 and IPSec, can be tested using the Cisco 3251 Mobile Access Router provided by Cisco Systems as an experimental payload on the UK-DMC satellite. On Thursday morning NASA Glenn was able to reach across the Internet to the UK-DMC satellite from Cleveland, Ohio via SSTL's Guildford, England, Mission Control Centre, using mobile routing. The 3251 router in orbit was configured and tested during a 12-minute period while the UK-DMC satellite passed over the ground station. The Cisco Systems router and firewall used in SSTL's Mission Control Network were given simple software upgrades to add IPv6 capabilities to allow this end-to-end IPv6 testing to take place [SUR200701].

Juniper Networks has been supporting Northrop Grumman from 2005 to 2007 in developing the Next Generation Processor/Router (NGPR) for the Transformation Communications Satellite (TSAT) System which will use IPv6 routing on board the satellites to connect users with the Global Information Grid (GIG). The TSAT system will provide U.S. warfighters with high-bandwidth, networked connectivity by extending the reach of the U.S. Department of Defense's

GIG through advanced satellite communications. As video communications is integrated into robots, soldiers, and UAVs, and networkcentric warfare becomes the organizing principle of U.S. warfighting, front-line demands for bandwidth are rising sharply. The TSAT System is part of a larger effort by the U.S. military to address this need. The final price tag on the entire TSAT program has been quoted at anywhere from $14 to 25 billion through 2016, which includes the satellites, the ground operations system, the satellite operations center, and the cost of operations and maintenance [DEF200801].

9.4 IPv6 Implementation Scenarios and Issues

Figure 9.1 depicts some basic scenarios of satellite-based IPv6 support. Case (a) and (b) represent traditional environments where the satellite link supports either a clear channel that is used to connect, say, two IPv4 routers, or is IP-aware, such as the typical case of a VSAT. (In each case, the LHS "cloud" could also be the IPv4 Internet or the IPv6 Internet.)

In case (c) the satellite link is used to connect as a transparent link two IPv6 routers; the satellite link is not (does not need to be) aware that it is transferring IPv6 PDUs. In case (d) the satellite system is IPv4-aware, so the use of that environment to support IPv6 requires IPv6 to operate in a tunneled-mode over the non-IPv6 cloud, which is a capability of IPv6.

In case (e) the satellite infrastructure needs to provide a gateway function between the IPv4 and the IPv6 world (this could entail repacking the IP PDUs from the v4 format to the v6 format).

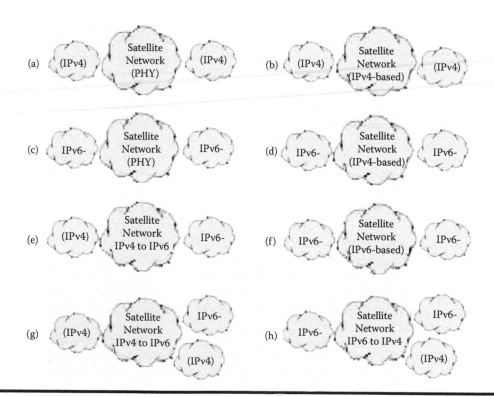

Figure 9.1   Support of IPv6 in satellite networks.

Case (f) is the ideal long-term scenario where the "world has converted to IPv6" and "so did the satellite network."

In case (g) the satellite IP-aware network provides a conversion function to support both IPv4 (as a baseline) and IPv6 (as a "new technology") handoffs. Possibly a dual-stack mechanism is utilized. In case (h) the satellite IPv6-aware network provides a support function for IPv6 (as a baseline) and also a conversion function to support legacy IPv4 islands.

To successfully deploy IPv6 in satellite networks, one needs to identify technical issues that may be a problem in that environment. For example the use of IPSec (mandatory in IPv6) precludes or complicates the use of TCP PEP. Satellite links may be unidirectional, which may have protocol implications. Also, there may be issue related to traditional (MPE) encapsulation in the IPv6 context.

One then needs to identify which is the best architecture for the construction of IPv6-aware satellite systems. Furthermore there may be some "missing" functionality in IPv6 which is needed to properly run over a satellite environment.

Strategies for optimal transition from IPv4 to IPv6 in a satellite environment need to be defined, along with specific transition mechanics.

As a minimum, in an IP-aware satellite environment (namely, an environment where router platforms are used, such as in the case of VSATs), the satellite operator will need to upgrade the router to support IPv6. Other IP-based components (for example, IP Encapsulators, IP-based DVB-S/DVB-S2 receivers, etc.) will also need to be upgraded. If Internet access is included, ISPs with IPv6 support will have to be added to the mix. Perhaps a migration to Unidirectional Lightweight Encapsulation (ULE) (IETF RFC 4326, June 2005) may be desirable. Network management systems may also need to be upgraded to IPv6.

Studies undertaken in Europe in 2004 on deploying IPv6 over satellite led to the following conclusions (also see Appendix A) [FRI200401]:

- MPE encapsulation lacks native IPv6 support (at this juncture)
- IPv6 stateless address autoconfiguration does not work on a number of satellite architectures
- Proprietary functionality, including PEP often does not work in IPv6 environments
- Mandatory IPSec support can make PEP use difficult or non-feasible (using encryption from hub-to-terminal instead of end-to-end can address some of the PEP issues)
- Introduction of IPv6 should be done at first using the dual-stack mode
- Non-IPv6 elements (e.g., DVB-S/DVB-S2) should be tunneled
- Satellite equipment vendors need to start adding IPv6 functionality (in modems, receivers, network management systems, VSAT platforms, IPv6 support in PEP, ULE support for DVB-S/DVB-S2).

References

[6WI200801] 6WIND, Multicast and Mobility over Satellite Using IPv6, *InfoWeek*, January 2008, http://www.informationweek.com

[AMO200801] J. J. Amoss, D. Minoli, *Handbook of IPv4 to IPv6 Transition Methodologies For Institutional & Corporate Networks,* Taylor & Francis, New York, 2008.

[CAM200801] D. Campbel, "IPv6 Over Satellite: Pie in the Sky?," February 27, 2008, *CircleID*, Online Magazine, http://www.circleid.com/posts/82277_ipv6_over_satellite/

[DEF200801] Defense Industry Daily, "Special Report: The USA's Transformational Communications Satellite System (TSAT)," January 27, 2008, http://www.defenseindustrydaily.com

[DIR200801] Directorate-Generals Information Society "IPv6: Enabling the Information Society," European Commission Information Society, Europe Information Society Portal, February 18, 2008, http://ec.europa.eu/information_society/policy/ipv6/index_en.htm

[ESA200701] European Space Agency, Project: Preparation for IPv6 in Satellite Communications, August 2007, Paris, France.

[FP5200701] FP5 Project Record: SATIP6: Satellite Broadband Multimedia System for IPv6, The European Commission, Community Research, Cordis (CORDIS operates a daily on-line News Service covering European research, innovation and related activities. All of these articles are stored in the CORDIS RTD News database.) http://cordis.europa.eu/data/PROJ_FP5/ACTIONeqDndSESSIONeq112422005919ndDOCeq1992ndTBLeqEN_PROJ.htm

[FRI200401] W. Fritsche, "Deploying IPv6 over Satellite," IPv6 Cluster Meeting, IABG, Manchester, September 29, 2004.

[LAD200601] L. Ladid, "European IPv6 Roadmap 2006 Recommendations, European IPv6 Task Force, IPv6 Task Force Steering Committee, IST-2-004572-CA IPv6 TF-SC Euro-v6-Roadmap, 8/10/2006.

[LOU200701] R. C. Lou, A. J. S. Esguevillas, B. de la C. Diego, B. Carro, L. Fan, Z. Sun, "IPv6 Networks over DVB-RCS Satellite Systems," International Journal of Satellite Communications and Networking, published online October 18, 2007 by Wiley InterScience (www.interscience.wiley.com).

[MIC200701] Changes to IPv6 in Windows Vista and Windows Server "Longhorn," http://www.microsoft.com/technet/community/columns/cableguy/cg1005.mspx, January 2, 2007.

[SUR200701] Surrey Satellite Technology Limited (SSTL) Press Release. "Cisco Router on UK-DMC First to Use IPv6 Onboard a Satellite in Orbit," 29.03.2007, Tycho House, Guildford, UK.

[YUN200701] A. Yun, D. Elkouss, E. Callejo, L. Liang, L. Fan and Z. Sun, "IP Networking over Next-Generation Satellite Systems," International Workshop, Budapest, July 2007, 10.1007/978-0-387-75428-4_16.

Appendix A*

Preparation for IPv6 in Satellite Communications

European Space Agency
Contract Report
Contract Number 17629/03/NL/ND

The work described in this report was done under ESA contract. Responsibility for the contents resides in the authors or organizations that prepared it.

Executive Summary

This investigation has been carried out under a contract awarded by the European Space Agency (ESA), contract number 17629/03/NL/ND. The ESA study manager is Frank Zeppenfeldt.

IABG mbH

Authors: Wolfgang Fritsche
       Karl Mayer

2004

IABG
Department for Communication Networks
Einsteinstrasse 20
D-85521 Ottobrunn
Germany

Telephone:
National: 089 6088 2897
International: +49 89 6088 2897

Contract No. 17629/03/NL/ND

---

* Included with permission.

	Facsimile:	
Issue 1	National:	089 6088 2845
06.07.2004	International:	+49 89 6088 2845

Document History

Version	Date	Section	Description
1.0	06.07.2004	General	Initial version

A.1 Introduction

This document is the executive summary report of ESTEC Contract Number 17629/03/NL/ND entitled "Preparation for IPv6 in Satellite Communications." The objective of the project is to support users, provider, and manufacturer in the introduction of IPv6 in satellite networks. Therefore, the project identified any IPv6 specific protocol issues with satellite communication and outlined appropriate transition scenarios for the various satellite network architectures. Within two major trials it demonstrated various aspects of IPv6 over satellite and described the lessons learns. Finally, it gives a roadmap and recommendations for the next steps required to allow for a smooth integration of IPv6 over satellite.

The project has been performed by a project team of IABG. IABG has many years of experience in the areas of Advanced IP Services such as IP over satellite and IPv6. The operation of an own teleport, the participation in many satellite related projects, the active contribution to the IETF standardization, as well as the membership of the Global IPv6 Forum, the European and German IPv6 Task Force, and the IPv6 Cluster are some of the key activities of IABG related to this project.

The authors gratefully acknowledge the support of Gerhard Gessler and the Teleport team of IABG, Frank Zeppenfeldt, Roberto Donadio, and Fausto Vieira of the European Space Agency, as well as Professor Peter Kirstein and the SILK project partners.

A.2 Project Objectives

The new Internet Protocol IPv6 has been in the process of standardization within the Internet Engineering Task Force for about 10 years now, and has reached a level of maturity that allows the start of the deployment phase. This deployment is clearly led by the Asian region, due to their severe shortness on global IPv4 addresses, their progressed deployment of the 3rd Generation cellular networks, and their strong commitment to go for IPv6. It will be followed by deployment in the U.S. and in Europe.

Because many regions with IPv4 address shortage also have a bad terrestrial network infrastructure, IPv6 over satellite can be an attractive solution for them. However, contrary to the investigation done for terrestrial and 3G cellular networks, the integration of IPv6 in satellite networks has not been analyzed in the same detail. Due to their specific characteristics, the frequent use of unidirectional links, or the deployment of proprietary components like Performance Enhancing Proxies, such an analysis is important.

Hence, this project conducted a solid analysis of the link layer, network layer, transport layer, and management protocols used in satellite networks, and identify their issues concerning the introduction of IPv6. Knowing these protocol issues, it has been possible to assess their impact on various satellite network architectures and to specify appropriate transition scenarios. Finally, it has been investigated in which way IPv6 affects current satellite services or allows for new ones.

Besides the theoretical analysis, a second key objective of the project has been to demonstrate IPv6 over satellite communication in two pilot scenarios, both of which were used the IPv6-capable DVB-S equipment developed and implemented in two parallel ESA projects.

Finally, the project produced a roadmap including recommendations for future research, development, and standardization activities, as well as for IPv6 over satellite-related trials required to support a smooth introduction of IPv6 in satellite networks.

The project embraced six major work packages, namely:

■ Identification of satellite specific protocol issues for IPv6
■ Impact of IPv6 on existing and future satellite network architectures and services
■ Definition and Preparation of IPv6 demonstration over satellite
■ Pilot demonstration of IPv6 over satellite
■ Identification of IPv6 roadmap and recommendation
■ Dissemination of project activities and results

The work undertaken for the project consisted partly of analysis and research, and partly on practical work related with the pilot demonstration. For the latter part, the experience of IABG's Teleport operation, as well as the operation of several testbeds for Advanced IP Services, provided a valuable input.

A.3 Satellite Specific Protocol Issues for IPv6

A.3.1 Link Characteristics

Compared to IPv4, IPv6 modifies the IP header format (e.g., by introducing 128-bit IP addresses) and adds new functionality to the IP layer, such as IPv6 Neighbor Discovery and IPv6 stateless address autoconfiguration.

This new functionality as well as many other protocols used for routing and IP multicast in the Internet expects certain characteristics from the underlying links. Most protocols require bidirectional links, which means that the IP layer sees the same interface for sending and receiving, some of them additionally require link multicast (each node attached to a link can send packets to each other). For example, IPv6 Neighbor Discovery mechanisms like address resolution, neighbor unreachability detection, Duplicate Address Detection (DAD), and router and prefix discovery require bidirectional links. Furthermore, DAD even expects full link multicast of the underlying architecture. Furthermore the stateless autoconfiguration is a new feature of IPv6, allowing the configuration of nodes without an additional DHCP server. As it makes use of prefix discovery and DAD, it also requires bidirectional links with full link multicast support.

However, in many satellite networks only unidirectional links are used, such as in hybrid satellite networks with a terrestrial return link or in DVB-S/RCS architectures. In this case, appropriate solutions have to be developed and implemented to cope with this issue.

One solution to address this issue is the UDLR mechanism described in IETF RFC 3077. In satellite network architectures with a unidirectional forward and a bidirectional return link, UDLR emulates a single bidirectional interface with link multicast support. In RFC 3077 it is explicitly stated that UDLR is not designed for architectures deploying unidirectional links in opposite directions. Hence, UDLR can be used, for example, in architectures based on a DVB-S forward link and a return link via Internet or PSTN, but cannot be used for architectures using a DVB-S forward link and a return link via DVB-S, SCPC, or DVB-RCS.

For the latter architectures, other solutions have to be taken into account. For example, two unidirectional, physical interfaces can be integrated in a single bidirectional, logical interface provided to the IP layer. The logical interface concept still cannot help to provide full link multicast in hub and spokes architectures. Therefore, additional functionality has to be implemented in hubs in order to redistribute multicast packets received from spokes to other spokes.

A.3.2 Satellite Specific Link Layers

A.3.2.1 DVB-S Link Layer

DVB-S links are unidirectional links and hence, in architectures deploying DVB-S links, the issues addressed in Section A.3.1 arise. Currently, mainly the Multiprotocol Encapsulation (MPE) mechanism is used to encapsulate IP packets in MPEG-2 TS frames. In principle, MPE can be used for IPv4 and IPv6, but in the standard it is not specified clearly how to signal the receiver the IP version encapsulated.

For this purpose, the IETF ipdvb WG specified a new encapsulation protocol, the Ultra Lightweight Encapsulation (ULE). ULE supports natively the encapsulation of IPv4 and IPv6 in MPEG-2 TSs, and consumes, compared to MPE less satellite bandwidth. Several prototype implementations already exist for ULE. Currently, a dynamic address resolution between IPv6 address, MAC address, and PID value is specified neither for MPE nor for ULE.

A.3.2.2 DVB-RCS Link Layer

On DVB-RCS links, either IP over MPE or IP over AAL5/ATM can be used. IPv6 runs smoothly over AAL5/ATM, however, using IP over MPE, the IPv6 issues discussed in the previous section apply. As the DVB-RCS standard does not specify the use of Ethernet bridging, this option cannot be deployed for transporting IPv6 traffic over IPv4-only DVB-RCS devices.

Principally, a DVB-S/RCS system is well suited for the provision of bidirectional satellite links. However, depending on the implementation of the DVB-S/RCS functionality, the IP layer will recognize DVB-S/RCS as a single bidirectional or two unidirectional interfaces. In the latter case UDLR or logical interfaces are measures to cope with this limitation.

Currently, the DVB-RCS connection control protocol, which could be used for the management of a DVB-S/RCS network, is only roughly specified. As the messages only contain 6-byte address fields, nodes cannot be addressed by IPv6 addresses.

A.3.2.2.1 Serial Line Protocols Used on SCPC Satellite Links

Several serial line protocols can be deployed on SCPC links (e.g., Cisco HDLC, PPP, and Frame Relay [FR]). All these serial line protocols are prepared for IPv6, and hence, IPv6 can be transported natively over SCPC satellite links.

A.3.3 Network Layer

A.3.3.1 Header Compression

The introduction of IPv6 in satellite-based networks using header compression could only cause problems if an early version as specified in RFC 1144 is used. This version is anyway not the best choice for satellite networks, as the compression of UDP traffic, RTP/UDP traffic and plain IP traffic are not supported. If header compression as specified in RFC 2507 or ROHC (RFC 3095) is deployed in the satellite network, IPv6 can be expected to work.

A.3.3.2 IPv6 Multicast

While in IPv4 networks with IGMPv3 a separate protocol is used for group membership management, IPv6 provides this by the Multicast Listener Discovery (MLD) integrated in ICMPv6. Contrary to IGMPv3 the current MLD version 1 does not support source-specific IP multicast; however, this limitation will be removed in version 2 of MLD.

Two variants of Protocol Independent Multicast (PIM) are available: PIM-DM (PIM dense mode) and PIM-SM (PIM sparse mode). For PIM-DM an Internet draft is available that is specified for IPv4 and IPv6. RFC 2117 specifying PIM-SM includes many IPv6-related open issues, which are addressed in a revised version. Other multicast protocols like Source Specific Multicast (SSM), DVMRP, MOSPF, BGF-4-, and BGMP support both, IPv4 and IPv6.

In case not all nodes in the satellite networks are multicast capable, multicast relay solutions like the one from OmniCast can be used. These are mostly proprietary and often don't support IPv6.

A.3.3.3 IPv6 Multihoming

Multihoming can allow satellite terminals to associate with different satellite hubs, or with a satellite hub and a terrestrial upstream service provider.

In principle, multihoming support already existed for IPv4. However, due to the large availability of globally unique addresses, and more advanced functionality like Mobile IPv6 or HIP, multihoming can be done more efficiently with IPv6.

A.3.3.4 IPv6 Mobility Support

The IP mobility area can be divided into host mobility, network mobility, and mobile ad hoc networks (MANETs). The currently used protocol to address host mobility is Mobile IP (MIP), and exists for IPv4 (MIPv4) and IPv6 (MIPv6). Contrary to MIPv4, MIPv6 integrates by default an already optimized routing possibility between a mobile host and its communication partner. In order to secure the required control information to establish the optimized routing, a security mechanism called Return Routability (RR) is deployed. For this security mechanism, after each movement of the mobile node to a new point of attachment, messages have to be exchanged between the mobile host, its communication partner, and the home agent. If part of these messages will be sent over satellite links, the handoff delay time of the mobile host will increase.

The IETF nemo WG is currently standardizing a protocol for supporting network mobility, which is basically an extension to MIPv6, but doesn't make use of optimized routing. Therefore, there are no major differences for supporting IPv4 or IPv6 mobile networks in satellite environments.

In MANETs the topology of the network itself is changing, and hence, satellite links do not seem to be good candidates for building MANETs. The MANET routing protocols OLSR and AODV address IPv4 and IPv6.

A.3.4 Enhanced Transport Layer Protocols

Due to high delay and high loss rate, TCP performs badly on satellite links. Two options exist to work around these issues. The first uses enhanced TCP protocols in end hosts, and the second deploys separate gateways—so-called Protocol Enhancing Proxies (PEPs) or TCP accelerators—on one side or both sides of the satellite link.

> **Enhanced TCP:** The enhanced TCP protocols TCP Tahoe, TCP Reno, TCP Vegas, TCP NewReno, and TCP Santa Cruz show no difference between operating above IPv4 and IPv6, and can therefore be used for enhancing TCP performance in satellite networks.
>
> **PEPs:** Some PEP technologies break the end-to-end transparency of the Internet, which could affect IPv6 more severe than IPv4, since IPv6 users expect to get back their end-to-end transparency and deploy services relying on exactly this expectation. Moreover, full IPv6 implementations have mandatory support of IPsec; that is, most probably IPsec will be used more widely with IPv6. However, IPsec works together with PEPs only in certain constellations, and hence, the simultaneous use of IPsec and PEPs could generate more problems in IPv6 networks.
>
> Furthermore, PEP devices mostly do not support IPv6, which means they do not implement an IPv6 stack, and their management is not IPv6-ready. Hence, TCP traffic based on IPv6 cannot be accelerated.

A.3.5 Network Management and AAA Issues

Management information bases (MIBs) used in satellite networks today are often not IPv6-ready, which means objects cannot contain IPv6 addresses. This includes standardized MIBs like the DVB-RCS MIB, as well as proprietary MIBs used by many satellite equipment manufacturers like SkyStream or Harmonic. Furthermore, many management applications do not support IPv6 and have to be prepared. This includes the enhancement of input and output fields, the preparation of command line parsers, and configuration file parsers, as well as allowing the exchange of IPv6 relevant management information between application and components. SNMP entities like SNMP agents and SNMP managers need to support IPv6. SNMPv1 does not support IPv6 and, hence, SNMPv2 or SNMPv3 has to be used in IPv6 satellite networks. All SNMP versions require bidirectional links. The Net-SNMP tool package contains several management applications, SNMP agents, and SNMP managers that support IPv4 and IPv6.

When used in IPv6 networks, all components of an AAA (authentication, authorization, and accounting) framework have to support IPv6, including the protocol entities of nodes, the messages transporting AAA information, and the databases storing IPv6 parameters. The protocols COPS, Diameter, and RADIUS are currently foreseen from the IETF for this purpose and can be regarded as IPv6 compliant.

A.4 Impact of IPv6 on Satellite Network Architectures and Services

Due to DVB-S encapsulation mechanisms and management functionality without IPv6 support, but often also due to lack of IPv6 capable protocol stacks in general, the majority of the currently available DVB-S and DVB-RCS equipment is not IPv6 ready. Using DVB-S equipment in

Ethernet bridging mode is at least an option to conceal IPv6 packets from the devices, however, even this alternative is not possible on the DVB-RCS return link. This lack of IPv6 functionality requires transition methods for introducing IPv6 in satellite networks until IPv6-ready equipment is available.

Three main classes of transition mechanisms are available: the dual stack mechanisms, which deploy IPv4 and IPv6 on a node simultaneously, the tunneling mechanisms, which encapsulate IPv6 within IPv4, and the translation mechanisms, which provide in specific gateways a translation functionality from one IP version to the other.

A.4.1 Investigation of Various Satellite Architectures

The project has investigated transition methods for various satellite architectures, such as using SCPC duplex links or deploying DVB-S on the forward link, and SCPC, DVB-S, DVB-RCS, the Internet, or the PSTN on the return link. Moreover, the ETSI BSM architecture has been considered concerning IPv6 transition. For all these architectures, trunking scenarios, star architectures, and meshed architectures have been taken into account.

For example, architectures with SCPC duplex links can be used straightforward to transport IPv6 natively because the serial line protocols used on SCPC links already support IPv6.

When interconnecting IPv6 networks by DVB-S links, an IPv6 over IPv4 tunnel could be configured from the DVB hub station to each branch station. On SCPC and PSTN return links native IPv6 transport is possible. On DVB-RCS and DVB-S return links from the branch stations to the hub station, an IPv6 over IPv4 tunnel has to be configured as well. Provided branch station and hub station are connected by an IPv6 Internet, the return link automatically will be native IPv6, otherwise, a tunnel is also required. Some tunneling mechanisms like 6over4 require the underlying support of IPv4 Multicast, which is not always given in satellite networks.

When connecting IPv6 and IPv4 networks via satellite links, translation devices (e.g., NAT-PT boxes or proxies) have to be inserted at the boundary of both networks.

The ETSI BSM architecture supports by definition the transport of IPv4 and IPv6 between its boundaries, however, it does not specify in detail what this support of IPv6 looks like. In case first BSM networks are based on IPv4, transition methods are applied inside the architecture to make the BSM network visible to the outside as IPv4 and IPv6 network.

A.4.2 Investigation of Modified Service Offerings

A.4.2.1 IPv6 Stateless Address Autoconfiguration (SAS)

With IPv6 SAS each IPv6 node automatically configures a link-local IPv6 address on each IPv6 interface after startup without the need for a configuration server. Moreover, when IPv6 hosts receive Router Advertisements, they get information about default router and prefixes on-link, and hence, they can assign site-local and global addresses to their interfaces. All these features allow a plug-and-play mechanism of IPv6 nodes (e.g., satellite terminals) without the need for manual configuration.

A.4.2.2 End-to-End IP Addressing

Due to the huge address space in IPv6, every IPv6 node can be addressed by one or more global IPv6 addresses and, hence, there is no need for deploying NAT devices and Application Level Gateways (ALGs). This brings back the end-to-end transparency to the Internet, which means that IP nodes can be connected end-to-end without intermediate gateways modifying parts of the IP packets. As applications and protocols like IPsec, SIP, and H.323 have difficulties in architectures with broken end-to-end transparency, introducing IPv6 in satellite networks supports a large scale deployment of IPsec and voice-over IP applications.

A.4.2.3 Mobile IPv6 Route Optimization

Contrary to MIPv4, MIPv6 already has an optimized routing functionality between the mobile host and the communication partner integrated in the basic protocol. Therefore, using MIPv6 the routing to and from mobile nodes will happen in an efficient way and consume less of the costly satellite bandwidth.

A.4.2.4 Mandatory IPsec

Full IPv6 implementations have to support IPsec—that is, with IPv6 it is more likely that a communication partner will support IPsec. Furthermore, due to the removal of NAT boxes in IPv6 networks IPsec can be easily deployed in a larger scale. This will make IPsec an attractive candidate to secure information sent over vulnerable wireless links, such as satellite networks.

A.4.2.5 Cryptographically Generated Addresses (CGAs)

The use of CGAs provides a mechanism to include information about the public key of a sender into its IP address. This is mainly done by hashing the public key and using 64 bits of this hash within the identifier part of the sender's IPv6 address; that is, the generation of CGAs is only possible for IPv6. By having a combination of IPv6 address and public keying information integrated in the CGA, it can serve as a kind of certificate, proving that a certain public key belongs to a certain IPv6 address. Consequently, a public key infrastructure is no longer required. Once the communication partner receives the public key of a sender in a certified way, the sender could authenticate its messages with its private key.

For example, satellite terminals communicating with the hub could authenticate their messages using CGAs.

A.4.3 Detailed Transition Plans

A.4.3.1 Transition Plan for a DVB-S/SCPC Teleport

A detailed transition plan for a DVB-S/SCPC scenario a teleport has been examined, which required first to investigate the current status of the network including the IPv6 deficiencies of devices, applications, and protocols. Understanding these deficiencies, in a first approach to integrate IPv6 quickly, the use of IPv6 over IPv4 tunnels have been proposed on DVB-S links interconnecting IPv6 networks. For the connection of IPv6 and IPv4 networks translation devices at the boundaries of the IPv6 networks are recommended.

For the long term vision of a native IPv6 scenario, devices and applications without IPv6 support need to be upgraded or replaced by IPv6-capable ones. In the example, this included upgrading the IOS of Cisco routers, replacing Harmonic DVB equipment by IPv6-capable DVB equipment, and upgrading operating systems and applications of management PCs.

Transition effort and costs are mainly due to the costs of new IPv6 capable DVB-S devices and the costs of the IPv6 training required for the teleport personnel.

A.4.3.2 Transition Plan for a DVB-S/RCS Teleport

The detailed transition plan for the DVB-S/RCS teleport is similar to the transition plan for the DVB-S/SCPC teleport. Also, here a first-step tunneling and translation are used as short term solutions, and upgrading or replacing devices and applications are used for a long-term solution. In the example, the DVB equipment of SkyStream and the Nera Satlink 1900 DVB-S/RCS terminals have to be upgraded or replaced by IPv6-capable ones. Furthermore, an upgrade of the IOS of Cisco routers is needed, and a replacement or upgrade of the HP disk cluster, the Oracle database system, and the operating system of management PCs is required.

Transition effort and costs are mainly due to the costs of new DVB-S and DVB-RCS devices, and the costs for IPv6 training of the network personnel.

A.5 IPv6 Demonstration over Satellite

A.5.1 Demonstration Scenario 1: IABG

The first pilot demonstration of IPv6 over satellite deployment has been established and performed at IABG's Teleport.

The architecture outlined in Figure A.5.1 includes a DVB-S-based forward link and a terrestrial return link. The DVB-S forward satellite link is realized by an IPv6-capable ULE-enabled 6WIND DVB gateway (6W) at the hub station and PCs at the branch stations (RX-PC1 and RX-PC2), which are equipped with Pent@Value DVB receiver cards supporting ULE. For the return links simple Ethernet links have been used.

This demonstration especially has been selected to illustrate proper functionality of advanced IPv6 features and applications over an DVB-S architecture, such as the usage of IPsec to encrypt one satellite link between hub station and branch station 1 or Mobile IPv6 to enable a mobile node (MN) to roam between a WLAN at branch station 1 and a WLAN at branch station 2. As applications, audio and video conferencing sessions between a correspondent node (CN) at the hub station and the MN have been demonstrated.

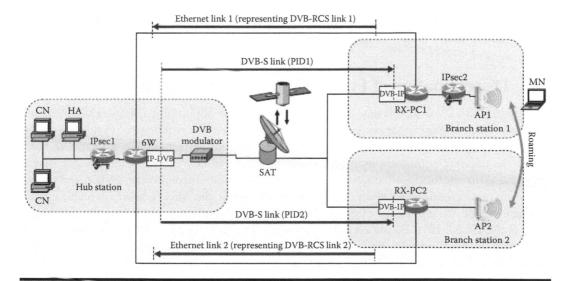

Figure A.5.1   Demonstration setup at IABG Teleport.

Prior to performing advanced demonstrations the performance of the IPv6 over DVB-S link is evaluated. Latency measurement between hub station and branch station 1 resulted in an average latency of about 380 ms. Measuring packet interarrival time (PIT) at branch station 1 resulted in 92% of packets with a PIT of 1ms, 5% with a PIT of 55 ms, and the rest with a PIT arbitrarily distributed between 1ms and 100 ms. These results are mainly related to the buffer management and the MPEG-2 TS packing functionality of the DVB-S devices. TCP throughput measurement between hub station and branch station 1 resulted in an average throughput rate of about 1.35 Mbps. The limiting factor was the bandwidth delay product due to a default TCP window size of 64 kByte.

Performing advanced demonstrations showed that IPsec, Mobile IPv6, and audio and video conferencing operated smoothly together in an architecture with IPv6 over DVB-S links. However, using the MIPv6 route optimization functionality over satellite links caused handoff delays in the order of 5s.

A.5.2 Demonstration Scenario 2: SILK

Within the second IPv6 over satellite pilot demonstration IPv6 has been integrated into the SILK network.

The SILK project has been founded in order to allow an Internet-based communication of academic and educational institutions residing in the Central Asian and Caucasian region with the rest of the world. For this purpose the national research networks (NRENs) in the 8 SILK countries—Armenia, Azerbaijan, Georgia, Kazakhstan, Kyrgyz Republic, Tajikistan, Turkmenistan, and Uzbekistan—have been connected via satellite to the European research network GEANT. Technically, the SILK satellite network architecture represents a DVB-S/SCPC hub and spoke architecture, with a hub located at the Deutsches Elektronen-Synchrotron (DESY) Institute in Hamburg.

The SILK user community has a high interest in getting also IPv6 connectivity to GEANT. In order to send IPv6 over satellite natively via the existing SILK architecture, a separate DVB-S carrier transporting IPv6 traffic has been set up and the hub station has been enhanced by an ULE-enabled, IPv6-capable open DVB-S gateway (ODG) from GCS. Each remote station has been enhanced with an additional Linux PC equipped with an ULE-enabled, IPv6-capable DVB-S receiver card from Pent@Value.

As this pilot demonstration has been set up for several months, and a large community of real users participated in the trial, the SILK demonstration has the character of a precommercial deployment of native IPv6 over satellite.

A.6 Dissemination of Project Activities and Results

In order to push IPv6 deployment in satellite networks, it is important to increase the awareness on this subject. Involving the critical mass and the key player is a requirement to help increasing and disseminating experience in this area, to rise funding for performing the outstanding tasks, to get the required equipment manufacturer integrating and helpful IPv6 functionality, and to initiate user to ask for the advantageous IPv6 services over satellite networks. Hence, the various dissemination activities performed during the project covers the whole area of IPv6 and satellite communication. To name a few of them, dissemination has been done within fora and task forces, such as the Global IPv6 Forum and the European IPv6 Task Force, within conferences, such as the 6NET Spring Workshop 2004, the German IPv6 Summit 2004, or the Asia-Pacific Advanced Network Conference 2004, within other projects, such as 6NET, SEINIT, SATIP6, or SILK, or within standardization bodies, such as the IETF ipdvb WG.

A.7 Key Results and Recommendations

In summary, one can say that many protocols used in satellite networks are already prepared for IPv6, however, some key deficiencies are the lack of IPv6 support in DVB-S devices using MPE, the lack of IPv6 support in management functionalities, the lack of IPv6 support in PEPs, as well as the lack of IPv6 neighbor discovery on unidirectional satellite links and links without support of link multicast. While some of these deficiencies can be solved short-term by appropriate transition mechanisms, others need new functionality to be specified and implemented.

In the following the main recommendations for the next steps in the various areas required for a smooth integration of IPv6 in satellite networks are listed.

A.7.1 Protocol Level Viewpoint

- The MPE standard has to be prepared to support IPv4 and IPv6, which requires, for example, a specification of how a DVB receiver distinguishes between both IP versions.
- Satellite-specific MIB definitions, management applications, and SNMP entities have to be enhanced to support IPv6. Moreover, SNMPv2 or SNMPv3 have to be used in networks since SNMPv1 is not IPv6 ready.
- DVB equipment manufacturers have to enhance their products with IPv6-ready MPE or/ and ULE functionality, with IPv6-ready network management, an IPv6 stack, and IPv6-capable interfaces.
- PEPs have to be enhanced for IPv6 support, which includes the implementation of an IPv6 stack, IPv6 capable management applications and MIBs, and IPv6-ready proprietary protocols used over the satellite link. Furthermore the use of PEP could be more problematic in IPv6 networks as one can expect a broader deployment of IPsec.
- Using MIPv6 with satellite networks could increase the handoff times of mobile nodes.

A.7.2 Address Resolution and Configuration

- In current satellite networks, address resolution is configured statically by the satellite network administrator at the DVB-S sender side for resolution between IP addresses and link layer identifiers and PID values. This functionality also has to be provided for IPv6.
- Furthermore, tables are used in order to dynamically advertise the mapping of IPv4 addresses to link layer addresses and PIDs. These table based mechanisms also need to be implemented for IPv6.
- Finally, it has to be investigated as to which satellite network architectures IPv6 stateless address autoconfiguration could be integrated.

A.7.3 System and Architectural Viewpoint

- In order to be able to support more broadly advanced IPv6 features such as the stateless address autoconfiguration, more detailed investigations about the applicability and usability of UDLR or the logical interface concept for the various satellite architectures need to be done.
- In this context it may be helpful to allow the use of UDLR in some architecture also for unidirectional return links.

A.7.4 Recommendable Standardization, Dissemination, and Deployment Activities

■ Dissemination activities on IPv6 over satellite need to continue in order to raise the awareness of users, provider and manufacturer on this subject, train them on the various aspects of IPv6, and collect their requirements.

■ The standardization on IPv6 over satellite needs to continue. For example, the IETF ipdvb WG needs to investigate address resolution mechanisms, ETSI needs to investigate in more detail the integration of IPv6 in ETSI BSM as well as to specify the IPv6 support in MPE.

■ New IPv6 functionality, such as IPv6-capable PEPs, DVB-S/RCS, or management systems need to be prototyped and tested.

■ Beyond deployment of IPv6 over satellite in the research community, real users have to test IPv6 over satellite in precommercial environments.

Index